# Progress in Littorinid
# and Muricid Biology

# Developments in Hydrobiology 56

*Series editor*

H. J. Dumont

# Progress in Littorinid and Muricid Biology

Proceedings of the Second European Meeting on Littorinid Biology,
Tjärnö Marine Biological Laboratory, Sweden, July 4—8, 1988

*Edited by*
K. Johannesson, D. G. Raffaelli and C. J. Hannaford Ellis

*Reprinted from Hydrobiologia, vol. 193 (1990)*

Kluwer Academic Publishers

Dordrecht / Boston / London

## Library of Congress Cataloging-in-Publication Data

European Meeting on Littorinid Biology (2nd : 1988 : Tjärnö Marine
  Biological Laboratory)
    Progress in littorinid and muricid biology : proceedings of the
  Second European Meeting on Littorinid Biology, Tjärnö Marine
  Biological Laboratory, July 4-8, 1988 / edited by K. Johannesson,
  D.G. Raffaelli, and C.J. Hannaford Ellis.
       p.    cm. -- (Developments in hydrobiology ; 56)
    "Reprinted from Hydrobiologia, vols. 193 (1990)."
    ISBN-13:978-94-010-6741-6       e-ISBN-13:978-94-009-0563-4
    DOI: 10.1007/978-94-009-0563-4


    1. Littorina--Congresses.  2. Littorinidae--Congresses.
  3. Muricidae--Congresses.   I. Johannesson, K. (Kerstin)
  II. Raffaelli, D. G. (Dave G.)  III. Ellis, C. J. Hannaford (Celia
  J. Hannaford)  IV. Title.  V. Series.
  QL430.5.L58E87   1988
  594'.32--dc20                                     90-4234
                                                       CIP

ISBN-13:978-94-010-6741-6

Published by Kluwer Academic Publishers,
P.O. Box 17, 3300 AA Dordrecht, The Netherlands.

Kluwer Academic Publishers incorporates
the publishing programmes of
D. Reidel, Martinus Nijhoff, Dr W. Junk and MTP Press.

Sold and distributed in the U.S.A. and Canada
by Kluwer Academic Publishers,
101 Philip Drive, Norwell, MA 02061, U.S.A.

In all other countries, sold and distributed
by Kluwer Academic Publishers Group,
P.O. Box 322, 3200 AH Dordrecht, The Netherlands


*Printed on acid-free paper.*

*This volume is dedicated to Professor Vera Fretter,*
*whose pioneering work on prosobranch biology*
*has laid the foundations for all of us.*

# Contents

# Preface

Littoral gastropods of the families Littorinidae and Muricidae are well studied compared to most marine taxa, yet there remain many basic problems concerning their taxonomy, ecology and evolutionary biology. In other words, we know these snails well enough to realize just how little we really know about them. This awareness prompted the First European Meeting on Littorinid Biology held at the British Museum in London on 26th November 1986, and the discussion continued through the Second Meeting on Littorinid Biology, held at the Tjärnö Marine Biological Laboratory, Sweden, from 4th to 8th July 1988. During the Tjärnö meeting, it was agreed to have a third meeting at Dale, Pembrokeshire, U.K. in 1990.

Twenty-two people attended the Tjärnö meeting, and a further ten contributed as co-authors to the papers that were presented. These covered research in progress in a broad range of topics, and geographical areas. Unfortunately, Cesare Sacchi and Domenico Voltolina, as well as Elisabeth Boulding were not able to attend the meeting in person, but their contributions were ably presented by David Reid and Richard Palmer, respectively. We also regret that one of us, C.E., and several of our Russian colleagues, did not have the opportunity to come.

Besides the presentation of papers and the discussion sessions that followed, there were two field excursions; a tramp around the island of Saltö and a boat trip to an exposed site at Ursholmen, to see Swedish snail populations in the field. Both trips turned into something of a 'treasure hunt', through the unpredicted and almost total collapse of *Littorina* and *Nucella* populations in the area just a few weeks before the meeting due to a bloom of toxic algae. Only a few live snails were found by the whole party of experienced snail pickers! This catastrophe was a sad reminder of the devastating effects of man's impact on the marine environment and on the urgent need for applied research on eutrophication of coastal waters coupled with political decisions on restrictions of environmental pollution. Otherwise we may soon have nothing left to study.

All who attended felt that the meeting was very successful and this was in no small part due to the staff of the Tjärnö Laboratory, especially the Director Lars Afzelius who provided superb laboratory facilities, and to Margareta Davidsson who served traditional Swedish dishes which made even the English breakfast seem pale in comparison. We are also most grateful to Marianne Saur and Bo Johannesson for the invaluable help with all kinds of practical arrangements during the meeting. Lastly we wish to thank all the participants for their excellently prepared talks and for creating a stimulating atmosphere for long and informal discussions during the evening sessions.

*Tjärnö July 1988*

KERSTIN JOHANNESSON
DAVE RAFFAELLI
CELIA HANNAFORD ELLIS

# List of participants

BACKELJAU Thierry, RUCA, Laboratorium voor Algemene Dierkunde, Groenborgerlaan 171, B-2020 Antwerpen, Belgium

COOK Lawrence, Department of Environmental Biology, University of Manchester, Manchester M13 9PL, England, United Kingdom

DYTHAM Calvin, Department of Pure and Applied Zoology, University of Leeds, Leeds LS2 9JT, England, United Kingdom

GRAHAME John, Department of Pure and Applied Zoology, University of Leeds, Leeds LS2 9JT, England, United Kingdom

HAVENHAND Jonathan, Kristineberg Marine Biological Station, Pl. 2130, S-450 34 Fiskebäckskil, Sweden

IMRIE David, Department of Environmental Biology, Williamson Building, University of Manchester, Oxford Road, Manchester M13 9PL, England, United Kingdom

JOHANNESSON Bo, Tjärnö Marine Biological Laboratory, University of Gothenburg, Pl. 2781, S-452 00 Strömstad, Sweden

JOHANNESSON Kerstin, Tjärnö Marine Biological Laboratory, University of Göteborg, Pl. 2781, S-452 00 Strömstad, Sweden

KARLSSON Willy, Tjärnö Marine Biological Laboratory, University of Göteborg, P. 2781, S-452 00 Strömstad, Sweden

KNIGHT Andrew, Department of Human Sciences, Loughborough University of Technology, Loughborough, Leicestershire LE11 3TU, England, United Kingdom

MCMAHON Robert, Section of Comparative Physiology, Department of Biology, The University of Texas at Arlington, Box 19498, Arlington, TX 76019, USA

NORTON Trevor, Department of Marine Biology, University of Liverpool, Port Erin, Isle of Man

PALMER Richard, Department of Zoology, University of Alberta, Edmonton, AB T6G 2E9, Canada

RAFFAELLI Dave, Culterty Field Station, University of Aberdeen, Newburgh, Ellon, Aberdeenshire AB4 0AA, Scotland, United Kingdom

REID David, Department of Zoology, British Museum (Natural History), Cromwell Road, London SW7 5BD, England, United Kingdom

ROOS Annika, Department of Zoology, University of Uppsala, P.O. Box 561, S-751 22 Uppsala, Sweden

SAUR Marianne, Tjärnö Marine Biological Laboratory, University of Gothenburg, Pl. 2781, S-452 00 Strömstad, Sweden

SUNDBERG Per, Department of Zoology, University of Göteborg, Box 25059, S-400 31 Göteborg, Sweden

TAYLOR John, Department of Zoology, British Museum (Natural History), Cromwell Road, London SW7 5BD, England, United Kingdom

WARD Robert, Department of Human Sciences, Loughborough University of Technology, Loughborough, Leicestershire LE11 3TU, England, United Kingdom

WARÉN Anders, Swedish Natural History Museum, Roslagsväg 120, S-104 05 Stockholm, Sweden

WARMOES Thierry, RUCA, Laboratorium voor Algemene Dierkunde, Groenborgerlaan 171, B-2020 Antwerpen, Belgium

WILLIAMS Gray, Department of Marine Biology, University of Liverpool, Port Erin, Isle of Man

*Back row:* A. Knight, I. Dumon, T. Warmoes, R. Ward, J. Havenhand, G. Williams, D. Imrie.

*Middle row:* R. McMahon, T. Norton, W. Karlsson, T. Backeljau, J. Taylor, B. Johannesson, J. Knight, J. Grahame.

*Front row:* D. Reid, C. Dytham, A. Roos, M. Saur, K. Johannesson, D. Raffaelli, L. Cook, A. Warén, R. Palmer.

*Absent:* P. Sundberg.

*Hydrobiologia* **193**: 1–19, 1990.
*K. Johannesson, D. G. Raffaelli and C. J. Hannaford Ellis (eds), Progress in Littorinid and Muricid Biology.*
© 1990 *Kluwer Academic Publishers.*

# A cladistic phylogeny of the genus *Littorina* (Gastropoda): implications for evolution of reproductive strategies and for classification

David G. Reid

*Department of Zoology, British Museum (Natural History), London SW7 5BD, England, UK*

*Key words:* biogeography, Atlantic, Pacific, reproductive strategy, larval development

## Abstract

Gross anatomical characters of all 18 species of *Littorina* are used to construct a phylogenetic hypothesis for the genus, by the method of cladistic analysis. The resulting cladogram suggests that of the four subgenera of *Littorina*, one (*Littorina*) is paraphyletic. It is uncertain whether the genus *Mainwaringia* should be included in *Littorina*. It is shown that the non-planktotrophic *Littorina* species in the northern Atlantic comprise a monophyletic group, with the sister-species *L. kurila* and/or *L. subrotundata* in the northern Pacific. Invasion of the Atlantic by a minimum of two Pacific species, across the Arctic migration route established during the late Cenozoic, is sufficient to account for the modern distribution of the subgenera *Littorina* and *Neritrema*. The importance of the cladogram as a basis for hypotheses of adaptation is illustrated by a discussion of spawn and development in *Littorina*.

## Introduction

Cladistic analysis is the method of inferring the phylogenetic history of a group from the evidence of shared derived characters (synapomorphies). The phylogeny is reconstructed using the principle of maximum parsimony (e.g. Hennig, 1966; Wiley, 1981; Felsenstein, 1982; Ax, 1987). The resulting cladogram is an hypothesis of phylogenetic relationships among taxa (although a species-level cladogram is not necessarily equivalent to a phylogenetic tree, because species can be ancestral to others, Wiley, 1981), which can be tested as more characters become available. It is thus a basis for hypotheses about the evolution of the group, its historical biogeography (e.g. Humphries & Parenti, 1986) and the evolution and adaptive significance of individual characters (Coddington, 1988). The cladogram can also be expressed as an hierarchical, phylogenetic classification (e.g. Wiley, 1981).

In a recent study, anatomical characters were used in a cladistic analysis of the relationships among species representing the 32 subgenera of the family Littorinidae (Reid, 1989). The most well known genus, *Littorina*, was represented by five taxa (*L. striata* King & Broderip, *L. keenae* Rosewater, *L. plena* Gould, *L. littorea* (*L.*), *L. obtusata* (*L.*)). These were shown to comprise a monophyletic group, defined by the synapomorphies of paraspermatic nurse cells without rods, and of two or more spiral loops in the egg groove of the pallial oviduct. This definition restricted the genus to 18 or 20 species, all occurring in the Northern Hemisphere. It excluded many other species previously classified as *Littorina* on the basis of superficially similar shells, and is in broad agreement with other recent classifications of the family (Bandel & Kadolsky, 1982; Reid, 1986a). However, these synapomorphies of *Littorina* were not unique, and there was some uncertainty as to whether two species of *Mainwaringia* (Reid, 1986b) should be included; this is discussed further below.

The present study aims to produce a species-level cladogram for the genus *Littorina*. The analysis is based largely on gross anatomical characters, with special emphasis on the details of the reproductive system which distinguish species in this group. The resulting cladogram will be used in an analysis of the historical biogeography of *Littorina*, to be described elsewhere (Reid, in press). The importance of a phylogenetic hypothesis in discussion of the evolution and adaptive significance of individual characters will be illustrated using the examples of spawn and developmental types. Littorinids show a wide diversity of reproductive strategies, and most types occur within the genus *Littorina*. In the absence of a phylogenetic hypothesis, previous discussions have been based on the concept of an optimal strategy in a particular habitat (e.g. Mileikowsky, 1975). A new, evolutionary interpretation is now possible. In addition, the species-level cladogram will permit an appraisal of the subgeneric classification of *Littorina* proposed by Reid (1989), by testing the monophyly of the two subgenera (*Littorina* and *Neritrema*) which are represented in the analysis by more than a single species.

Previously, the only information on the phylogeny of *Littorina* was derived from estimates of genetic similarity based on analysis of isoenzymes, and from palaeontology and the comparison of Recent shells. Genetic analysis has so far been limited to some of the Atlantic species (Heller, 1975; Ward & Warwick, 1980; Moyse *et al.*, 1982; Ward & Janson, 1985; Warmoes, 1986) and two species in the northeastern Pacific (Mastro *et al.*, 1982; see review by Ward, this volume). The fossil record of *Littorina* is rather poor, the earliest undoubted species occurring in the Lower Miocene of northwestern America (*L. sookensis* Clark & Arnold, 1923; Reid, 1989). In Europe, *Littorina* first appears in the late Pliocene (*L. littorea* in the Red Crag of

eastern England (Harmer, 1920–25), earlier records should be referred to other genera; Reid, 1989). Similar patterns, of older occurrence in the northern Pacific and sudden appearance in the northern Atlantic in the late Cenozoic, are known in many other molluscs, providing evidence for a mainly unidirectional exchange of marine fauna (Durham & MacNeil, 1967; Einarsson *et al.*, 1967). These migrations were permitted by the opening of the Bering Strait, which has occurred intermittently since the late Pliocene, and of seaways through the Canadian Archipelago (Hopkins, 1967; Herman & Hopkins, 1980). It has therefore been suggested that the northern Atlantic *Littorina* species are of Pacific origin (Golikov & Tzvetkova, 1972; Berger, 1978).

Climatic deterioration during the Pleistocene prevented further exchange of temperate fauna across the Arctic, thus isolating the *Littorina* species in the two oceans. Some authors have identified pairs of sibling species, believed to have been formed by allopatric speciation following this vicariant event. These Pacific-Atlantic pairs are: *L. squalida* Broderip & Sowerby and *littorea*, *L. kurila* Middendorff and '*obtusata*', and *L. sitkana* Philippi and '*saxatilis*' (Olivi) (Rosewater, 1963; Golikov & Tzvetkova, 1972; Berger, 1978). However, these pairs were established on the basis of similarities in shells, which are notoriously variable in the genus (e.g. Raffaelli, 1979; Janson, 1982; Backeljau *et al.*, 1984; Seeley, 1986) and therefore a poor guide to relationships. Furthermore, the evidence from shell similarities is itself equivocal, for both *L. aleutica* Dall and *L. kurila* in the Pacific resemble *L. 'obtusata'* in the Atlantic. The cladistic analysis will test the suggested relationships.

To add to the uncertainty surrounding the phylogeny of *Littorina*, the species of the *obtusata* and *saxatilis* complexes in the Atlantic have since been re-evaluated. Although still the subject of debate, the *L. obtusata* complex is believed to comprise two species (*L. obtusata* and *L. mariae*; Sacchi & Rastelli, 1966; Goodwin & Fish, 1977; Moyse *et al.*, 1982) and the *L. saxatilis* group four species (*L. saxatilis*, *L. neglecta* Bean, *L. arcana*

Hannaford Ellis and *L. nigrolineata* Gray; Heller, 1975; Raffaelli, 1979; Hannaford Ellis, 1979; Fretter, 1980; Fretter & Graham, 1980; Fish & Sharp, 1985; Janson & Ward, 1985; Ward & Janson, 1985; but see K. & B. Johannesson, this volume and B. & K. Johannesson, this volume). These species have been defined largely by features of their anatomy, of which characters of the penis, oviduct, method of development, and radula are most important. In contrast, their Pacific congeners have been neglected, and hitherto their anatomy has been virtually unknown.

**Material and methods**

Three stages are involved in the process of cladistic analysis: selection of taxa belonging to the ingroup and outgroup, the choice of characters and their coding into ordered character states, and finally the analysis of the data to construct the most parsimonious cladogram.

*Material*

Preserved material of all 18 of the recognized species of *Littorina* was obtained for dissection. These were: *L. (Liralittorina) striata, L. (Planilittorina) keenae* ( = *L. planaxis* Philippi; Rosewater, 1978), *L. (Littorina) brevicula* Philippi, *L. (L.) littorea, L. (L.) mandshurica* Schrenck, *L. (L.) plena, L. (L.) scutulata* Gould, *L. (L.) squalida, L. (Neritrema) aleutica, L. (N.) arcana, L. (N.) kurila, L. (N.) mariae* Sacchi & Rastelli, *L. (N.) neglecta, L. (N.) nigrolineata, L. (N.) obtusata, L. (N.) saxatilis, L. (N.) sitkana, L. (N.) subrotundata* (Carpenter) ( = *Algamorda newcombiana* (Hemphill)) (subgeneric assignments after Reid, 1989). In addition, *Mainwaringia rhizophila* Reid was included in the ingroup. (The other species of this genus, *M. leithii* (Smith) is similar in shell and radular characters (Reid, 1986b), but no preserved material was available). *Mainwaringia* is a genus of uncertain affinity which shares synapomorphies with *Littorina*, and could be a member of the same clade (Reid, 1989).

The phylogenetic analysis of the subgenera of the Littorinidae (Reid, 1989) showed that the genus *Nodilittorina* is the most probable sister-group of *Littorina*. Accordingly, representatives of the three subgenera of *Nodilittorina* were included in the outgroup. These were *N. (Fossarilittorina) meleagris* (Potiez & Michaud), *N. (Nodilittorina) pyramidalis* (Quoy & Gaimard) and *N. (Echinolittorina) dilatata* (d'Orbigny); the first two are the type species of their respective subgenera, the third closely related to its type species (*N. (E.) tuberculata* (Menke)).

Material of all species examined is deposited in the British Museum (Natural History), and additional specimens of northeastern Pacific species were borrowed from the National Museum of Natural History, Smithsonian Institution (Washington, D.C.). A total of 273 specimens of *Littorina* were dissected for the examination of reproductive anatomy and the alimentary tract. Paraspermatic nurse cells were examined by light microscopy in 32 living and 60 preserved specimens. Radulae were examined, mainly by light microscopy, in 55 specimens. Further details of techniques are given by Reid (1986a, 1989). Shell microstructure was investigated using techniques of acetate peels of ground sections, scanning electron microscopy of etched sections, and X-ray analysis for the identification of aragonite and calcite (Taylor & Reid, this volume).

*Characters used in cladistic analysis*

The characters employed in the analysis (Table 1) are briefly described below. More detailed descriptions and a full account of the comparative morphology of the Littorinidae can be found in Reid (1989).

*1. Shell shape*
The majority of *Littorina* species have turbinate shells, but two extreme shapes are defined: flat-spired (shell height/height of aperture, both measured parallel to coiling axis, < 1.20, Fig. 1k, o, p), and tall-spired (> 1.70, Fig. 1c–f). Extreme variation in shell shape occurs in at least

4

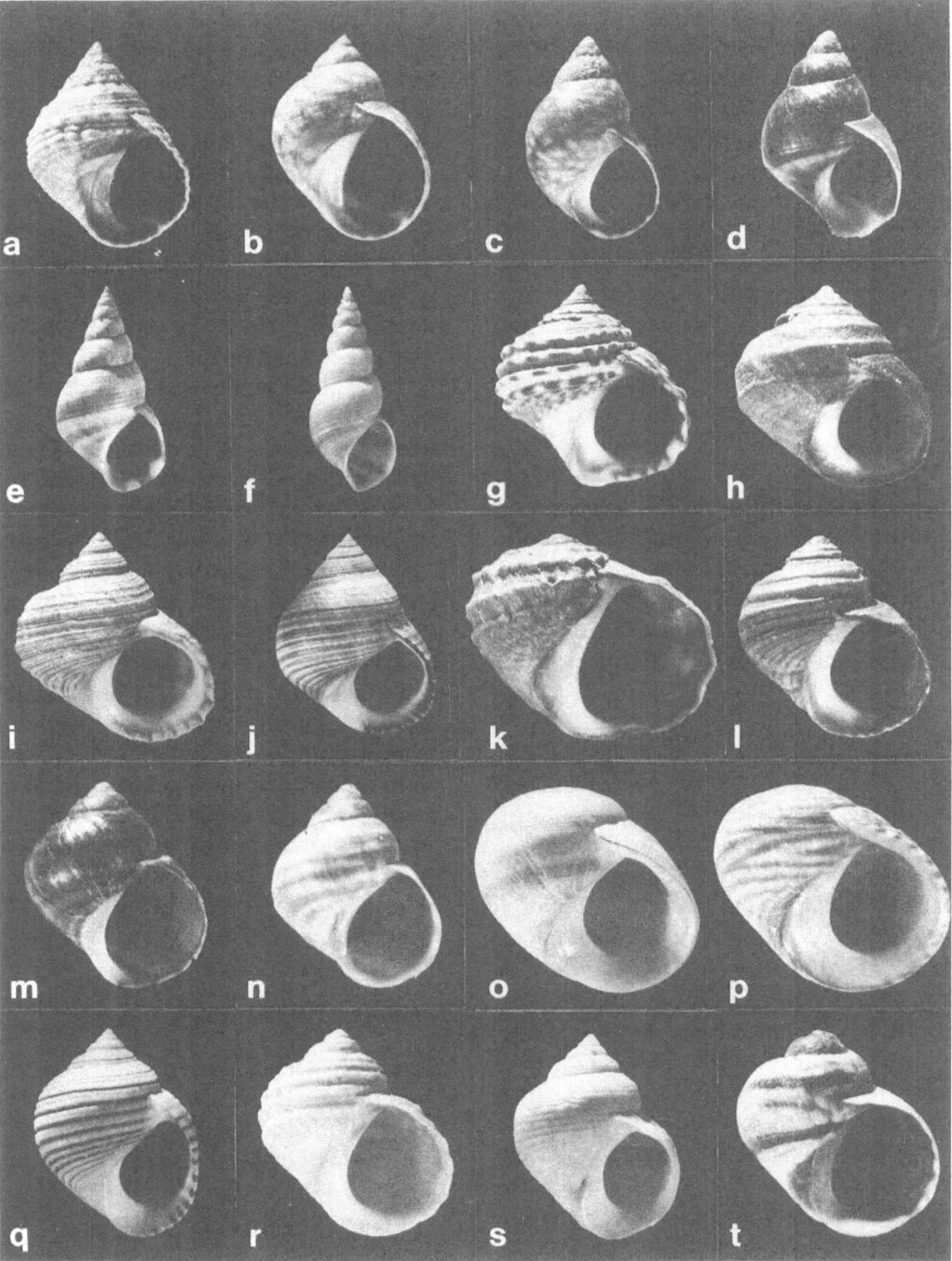

*Fig. 1.* Shells of all recognized Recent species of *Littorina* and *Mainwaringia* (shell heights in parentheses): (a) *Littorina striata*, Madeira (9.6 mm); (b) *L. keenae*, San Diego, California (12.3 mm); (c) *L. scutulata*, San Luis Obispo Bay, California (11.4 mm); (d) *L. plena*, San Luis Obispo Bay, California (5.5 mm); (e) *Mainwaringia rhizophila*, Santubong, Sarawak (13.7 mm); (f) *M. leithii*, Bombay, India (7.3 mm); (g) *L. brevicula*, Yokohama, Honshu (13.7 mm); (h) *L. mandshurica*, Ohmu, Hokkaido (12.0 mm); (i) *L. squalida*, Japan (28.0 mm); (j) *L. littorea*, Lyme Regis, Dorset (27.6 mm); (k) *L. aleutica*, Nizki I., Alaska (10.5 mm); (l) *L. sitkana*, Sooke, British Columbia (14.1 mm); (m) *L. kurila*, Adak I., Alaska (8.9 mm); (n) *L. subrotundata*, Nahcotta,

two species, *L. obtusata* (Backeljau *et al.*, 1984; Seeley, 1986) and *L. saxatilis* (Raffaelli, 1979, as *L. rudis*; Smith, 1981; Janson, 1982), but the most frequent shapes are flat-spired and turbinate respectively.

## 2. Shell sculpture
Nodulose sculpture occurs in only two *Littorina* species (Fig. 1a, k), but is frequent in the outgroup.

## 3. Mineralogy
Three unordered states of this character are defined. In *Nodilittorina* species, in *L. striata* and *L. keenae* the shell consists of three aragonitic layers of crossed-lamellar structure, with the outermost part of the outer layer showing a very fine crossed-lamellar structure with prominent growth lines. In *M. rhizophila* the shell consists of three aragonitic layers of normal crossed-lamellar structure. In the remaining *Littorina* species there are two aragonitic layers of crossed-lamellar structure, and in addition a thick outer calcitic layer of irregular prismatic structure. Shell mineralogy in the family is discussed in detail by Taylor & Reid (this volume).

## 4. Colour pattern of head
The pigmentation of the head in species of *Nodilittorina* is characteristic, with two or three longitudinal black lines on each tentacle (Reid, 1989). This pattern also occurs in four *Littorina* species.

## 5. Longitudinal division of foot
In most members of the ingroup and outgroup the mesopodial sole of the foot is divided longitudinally by a deep groove, but this division is absent in *M. rhizophila*.

## 6. Hermaphroditism
Uniquely in the ingroup, *M. rhizophila* is a protandrous hermaphrodite (Reid, 1986b).

## 7. Penial vas deferens
In many littorinids the penial vas deferens is an open groove, but this is superficially closed as a duct (an epithelial connection to the surface remains) in *M. rhizophila* and one member of the outgroup.

## 8. Bifurcation of penial base
The base is not bifurcate in any members of the ingroup, but this character serves to define the topology of the outgroup.

## 9. Simple penial glands
Simple, unicellular, subepithelial glands are present in the penis of all *Littorina* species and *M. rhizophila*, but are grouped to form a glandular disc in two members of the outgroup.

## 10. Mamilliform penial glands
These complex subepithelial glands (Linke, 1933: Fig. 34) are present in most *Littorina* species, but there appears to have been a trend for multiplication of their number within the genus. There is intraspecific variation in the number of penial glands: 2 or rarely 1 in *L. keenae*, 1–6 in *L. brevicula*, 1–5 in *L. mandshurica*, 26–44 in *L. squalida*, 16–36 in *L. littorea* (Fretter & Graham, 1980), 12–15 in *L. aleutica*, 5–19 in *L. sitkana*, 9–17 in *L. kurila*, 7–10 in *L. subrotundata*, 13–54 in *L. obtusata*, 5–15 in *L. mariae* (Goodwin & Fish, 1977), 3–12 in *L. nigrolineata* (Heller, 1975; Sacchi, 1975), 16–34 in *L. arcana*, 8–23 in *L. saxatilis* (both Hannaford Ellis, 1979), 1–9 in *L. neglecta* (Fish & Sharp, 1985). Five states of this character are defined: absent, 1, 2, not more than 6, and up to 7 or more glands. (The presence of numerous minute mamilliform glands on the penial filament of *Nodilittorina pyramidalis* is an autapomorphy (Reid, 1989), and only the single large gland on the base has been counted for the purpose of this analysis.)

Washington (4.5 mm); (o) *L. obtusata*, Hoy Sound, Orkney Is (14.3 mm); (p) *L. mariae*, South Uist, Outer Hebrides (11.6 mm); (q) *L. nigrolineata*, St Marys, Scilly Is (17.9 mm); (r) *L. arcana*, Anglesey, Wales (11.5 mm); (s) *L. saxatilis*, Barmouth, Wales (17.3 mm); (t) *L. neglecta*, Narragansett, Rhode Island, U.S.A. (4.9 mm). All specimens in British Museum (Natural History).

## 11. Paraspermatic nurse cells

In all *Littorina* species and in *M. rhizophila* the nurse cells in the seminal vesicle contain large granules only, whereas in *Nodilittorina* species there are in addition pointed, refractile rod-pieces.

## 12. Oviducal sperm groove

Partial closure of the sperm groove in the pallial oviduct is an autapomorphy of *M. rhizophila* (Reid, 1986b).

## 13. Position of bursa copulatrix

Two positions of the bursa copulatrix are defined: at the anterior end of the pallial oviduct (close to the opening of the latter into the mantle cavity) or alternatively halfway or further back along the straight section (the jelly gland or brood pouch) of the pallial oviduct. The bursa is in a posterior position in the oviducts illustrated in Fig. 2a–d, f. The state of this character in *L. squalida* is uncertain; in available specimens the bursa is situated

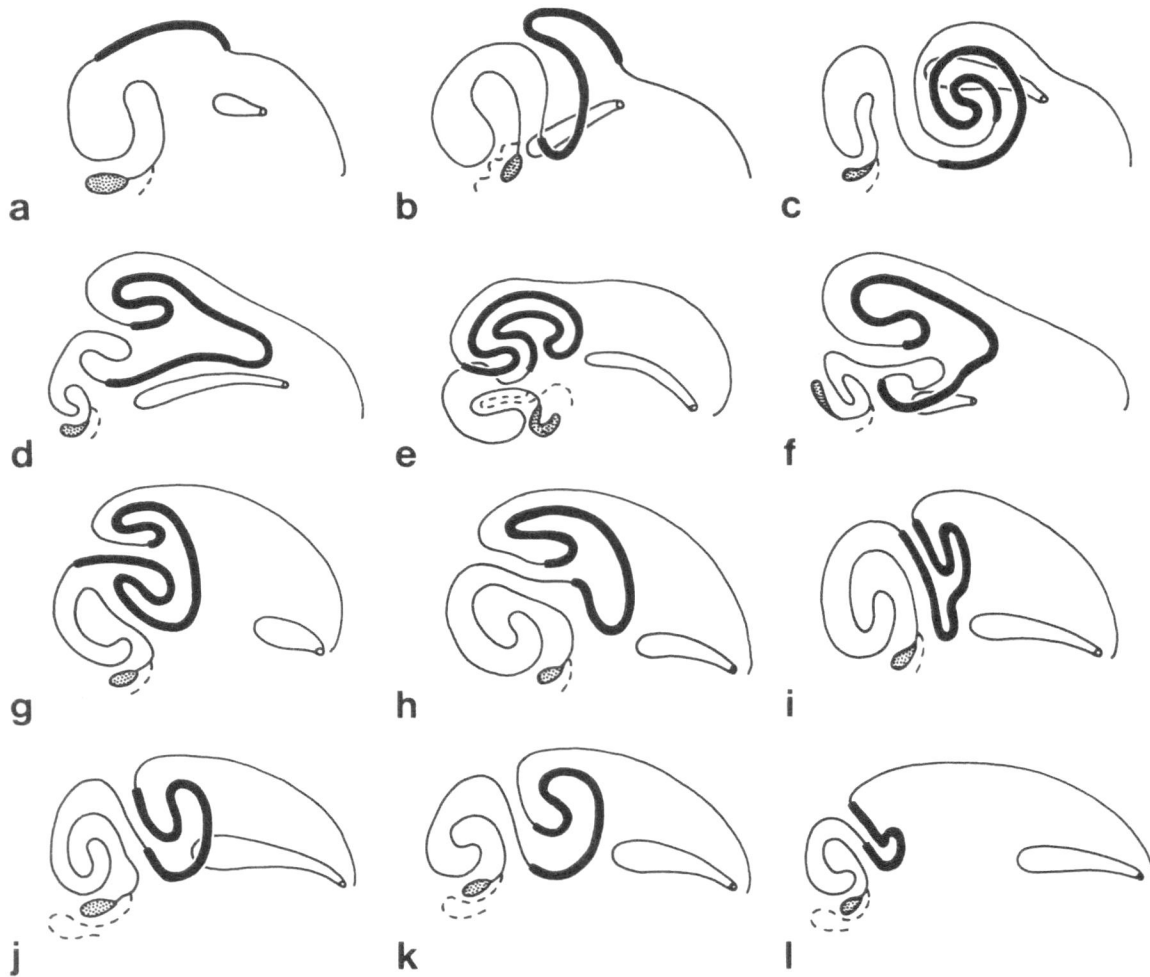

*Fig. 2.* Diagrams of pallial oviducts of *Nodilittorina*, *Littorina* and *Mainwaringia* species. Key: continuous line, spiral route of egg groove from renal oviduct (at left) to opening into mantle cavity (at right); thickened part of line, opaque capsule gland; dashed line, renal oviduct; stippled sac, seminal receptacle; unshaded sac, bursa copulatrix. State of character 14 given in parentheses after species name. (a) *Nodilittorina meleagris* (0); (b) *Littorina striata* (1); (c) *L. keenae* (2); (d) *L. scutulata* (3); (e) *Mainwaringia rhizophila* (3); (f) *L. littorea* (3) (*brevicula*, *squalida*, *mandshurica* all similar); (g) *L. aleutica* (3); (h) *L. sitkana* (4); (i) *L. kurila* (5) (*subrotundata* similar); (j) *L. obtusata* (5) (*mariae* similar); (k) *L. nigrolineata* (5) (*arcana* similar); (l) *L. saxatilis* (6) (*neglecta* similar).

about halfway along the straight section, but is very small, perhaps as a consequence of immaturity.

## 14. Coiling of egg grove

Spiral coiling of the egg groove within the lumen of the pallial oviduct is a unique synapomorphy of the Littorinidae, and the form of the spiral is a valuable phylogenetic character at the generic level (Reid, 1986a, 1989). The genus *Littorina* is unusual in showing a rather wide range of form. Seven character states are defined (Fig. 2) and arranged in a hypothetical evolutionary sequence. States 0 to 3 represent a sequence of increasing complexity, from the simple loop of the outgroup (state 0, Fig. 2a), to two consecutive loops, of albumen followed by capsule gland (state 1, Fig. 2b), increased coiling of the capsule gland (state 2, Fig. 2c), and then the introduction of a third loop (of either albumen or capsule gland) between the other two, which projects into the coil of the capsule gland (state 3, Fig. 2d–g). In the remaining three states the coiling pattern is simplified and the capsule gland reduced in size, associated with the loss of pelagic egg capsules. The loss of the third loop of state 3 defines state 4 (Fig. 2h). In state 5 further reduction in size of the capsule gland has occurred (Fig. 2i–k). The trend culminates in the simple pattern of the ovoviviparous species, with a still more reduced capsule gland (state 6, Fig. 2l). The evidence from other characters does not contradict this proposed sequence of character states, as discussed below.

## 15. Jelly gland

The jelly gland, occupying the distal straight section of the pallial oviduct, is small in those species releasing individual pelagic egg capsules. In species releasing capsules embedded in a pelagic gelatinous mass (*L. keenae*, Schmitt, 1979, as *L. planaxis*), in a benthic mass (*L. sitkana*, Kojima, 1958a, as *L. atkana*; Buckland-Nicks *et al.*, 1973; *L. kurila*, E.G. Boulding, pers. commun.) or laying non-encapsulated eggs in a benthic mass (*L. obtusata*, *L. mariae*, *L. nigrolineata*, *L. arcana*; e.g.

Goodwin & Fish, 1977; Hannaford Ellis, 1979) the jelly gland is greatly enlarged and packed with fleshy septa (Linke, 1933). In the ovoviviparous species (*L. saxatilis*, *L. neglecta*) the brood pouch is large and septate, but thin-walled; it is homologous with the jelly gland (Hannaford Ellis, 1979) and these species are therefore included in the group with an enlarged jelly gland.

## 16. Egg capsules

Four states of this character have been defined. Members of the outgroup share with *Littorina striata* cupola-shaped pelagic capsules which are sculptured by concentric rings (Fig. 3a). In *L. scutulata* and *L. plena* the capsules are flattened, with a double rim (Buckland-Nicks *et al.*, 1973; Murray, 1979, 1982; Mastro *et al.*, 1982; Fig. 3g, i). In other *Littorina* species with pelagic capsules (Linke, 1933; Kojima, 1957, 1958b, c, d; Fretter & Graham, 1962; Yamaguchi, 1967; Murray, 1979; Schmitt, 1979), in *L. sitkana* (Kojima, 1958a; Buckland-Nicks *et al.*, 1973) and in *L. kurila* (E.G. Boulding, pers. commun.) the shape is biconvex (Fig. 3b–f, h). Biconvex capsules are also found in *M. rhizophila* (Fig. 3j). In *L. obtusata* and *L. mariae*, which produce a benthic, gelatinous spawn (Goodwin & Fish, 1977; Goodwin. 1979), and in the ovoviviparous species *L. saxatilis* and *L. neglecta*, the membrane covering each egg is spherical, and it appears that capsules are much reduced or absent. Further study of the composition of the egg coverings is necessary. The states of this character were specified as unordered in the analysis, because of the uncertain derivation of the type with a double rim. The spawn is unknown in *L. aleutica*, but because both jelly and capsule glands are large, it can be predicted that (as in *L. sitkana*) a benthic gelatinous spawn containing embedded capsules may be produced.

## 17. Eggs per capsule

Eggs are usually encapsulated singly in littorinids, but in five *Littorina* species the pelagic capsules typically contain several eggs. The ranges of the numbers of eggs per capsule have been quoted as: 1–14 in *L. scutulata*, 4–41 in *L. plena* (both

8

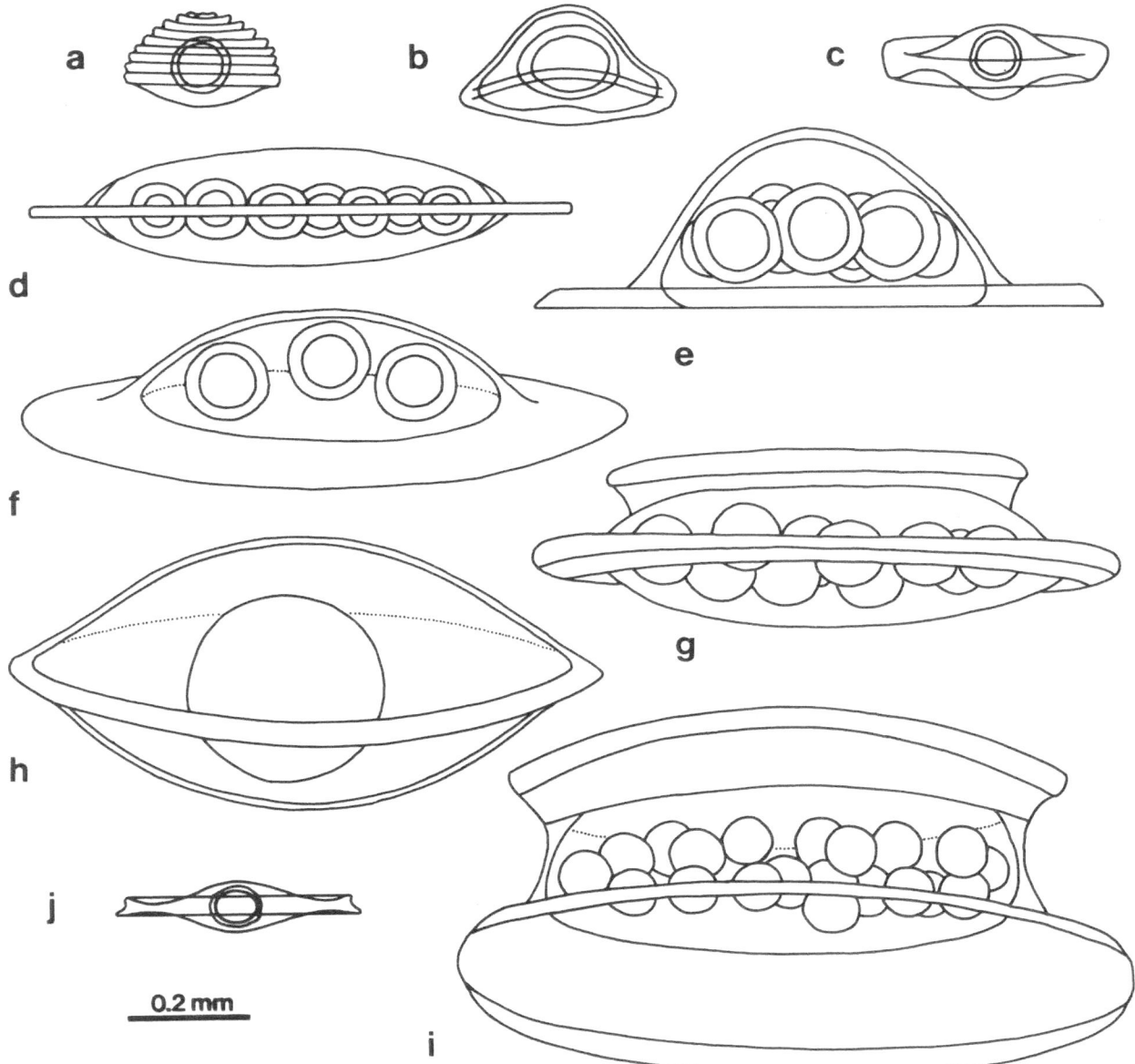

*Fig. 3.* Egg capsules of *Littorina* and *Mainwaringia* species. (a) *Littorina striata* (original); (b) *L. brevicula* (after Kojima 1957); (c) *L. keenae* (after Schmitt, 1979); (d) *L. squalida* (after Kojima, 1958c); (e) *L. mandshurica* (after Kojima, 1958d); (f) *L. littorea* (after Fretter & Graham, 1962); (g) *L. scutulata* (after Murray, 1979, 1982); (h) *L. sitkana* (after Buckland-Nicks *et al.*, 1973); (i) *L. plena* (after Murray, 1979, 1982); (j) *Mainwaringia rhizophila* (original).

Murray 1979, 1982), 1, but sometimes 2–5, in *L. brevicula* (Kojima, 1957; Yamaguchi, 1967), 9–12 in *L. mandshurica* (Kojima, 1958d), 14–15 in *L. squalida* (Kojima, 1958b, c) and 1–9 in *L. littorea* (Linke, 1933). This character is coded as unknown in those species lacking egg capsules.

### 18. Development

For some species there are direct observations of the type of development, whether planktotrophic or non-planktotrophic (e.g. *L. scutulata*, *L. plena*, *L. keenae*, all Murray, 1979; *L. brevicula*, Kojima, 1957; *L. subrotundata*, Matthews, 1978; *L. sit-*

*kana*, Buckland-Nicks *et al.*, 1973; *L. kurila*, E.G. Boulding, pers. commun.; European species reviewed by Fretter & Graham, 1980). In other cases the type of development can be predicted from the protoconch. In planktotrophic species this is usually less than 0.44 mm in diameter, sculptured with spiral ridges and tubercles, consists of 2 to 4 whorls, and is terminated by a sinusigera rib. In non-planktotrophic species the protoconch is larger, unsculptured, comprised of 1–2 whorls, and has a simple lip (review by Reid, 1989). By this means the development of *L. striata* (Rosewater, 1981: Plate 6E) and *M. rhizophila* (Reid, 1986b) can be predicted to be planktotrophic. Another indicator of type of development is egg diameter, which is 0.074–0.130 mm (excluding the albuminous covering) in planktotrophic species (Linke, 1933; Kojima, 1957; Murray, 1979; pers. obs. of *L. striata*), but 0.175–0.255 mm in non-planktotrophic species (Thorson, 1946; Buckland-Nicks *et al.*, 1973; Goodwin & Fish, 1977). On this

basis *L. squalida* (0.095 mm, Kojima, 1958b, c) and *L. mandshurica* (0.120 mm, Kojima, 1958d) are likely to be planktotrophic. Neither development, protoconch nor egg size are known for *L. aleutica*.

### 19. Ovoviviparity
Retention of embryos in the brood pouch of the pallial oviduct (the modified jelly gland) occurs only in *L. saxatilis* and *L. neglecta*.

### 20. Rachidian tooth shape
An index of the shape of the rachidian tooth is provided by the ratio of its total length (when mounted flat and viewed from above) to its width at the mid-point. Two shapes are defined: square (index < 1.15, see Fig. 4c–f) and normal (> 1.15).

### 21. Rachidian cusps
The number of cusps on the rachidian tooth is either 3 or 5, and in the latter case the outermost cusps are usually small (Fig. 4c–i).

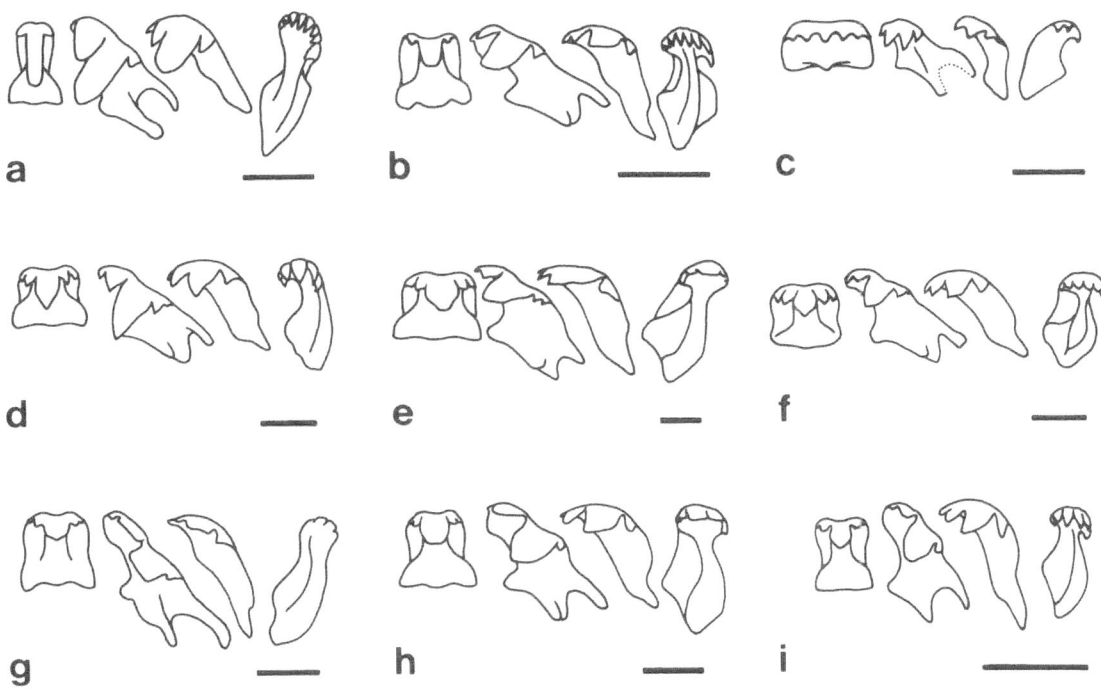

*Fig. 4.* Radulae of *Littorina* and *Mainwaringia* species. Radulae mounted flat, with outer marginal folded outwards, viewed from above. From left, teeth are rachidian, lateral, inner marginal, outer marginal. Scale bars = 0.050 mm. (a) *Littorina striata*; (b) *L. scutulata*; (c) *Mainwaringia rhizophila*; (d) *L. brevicula*; (e) *L. squalida*; (f) *L. sitkana*; (g) *L. obtusata*; (h) *L. nigrolineata*; (i) *L. neglecta*.

10

*Table 1.* List of characters and character states for cladistic analysis of species of *Littorina*. The plesimorphic states are those for the clade *Littorina*. The consistency indices are those for each character on the cladogram (Fig. 5). Abbreviations: U, character states unordered); *, autapomorphic character, excluded from analysis.

| No. | Character | Character states | Plesio-morphic state | Consistency index |
|---|---|---|---|---|
| 1 | Shell shape | 0: flat-spired<br>1: turbinate<br>2: tall-spired | 1 | 0.500 |
| 2 | Shell sculpture | 0: nodulose<br>1: not nodulose | 1 | 0.333 |
| 3 | Mineralogy (U) | 0: aragonite only; outer fine crossed-lamellar layer<br>1: outer calcitic layer<br>2: aragonite only; 3 normal crossed-lamellar layers | 0 | 1.000 |
| 4 | Colour pattern of head | 0: not so<br>1: *Nodilittorina* pattern | 0 | 0.500 |
| 5 | Longitudinal division of foot (*) | 0: divided<br>1: not divided | 0 | – |
| 6 | Hermaphroditism (*) | 0: gonochoristic<br>1: protandrous | 0 | – |
| 7 | Penial vas deferens | 0: open<br>1: superficial closure | 0 | 0.500 |
| 8 | Bifurcation of penial base | 0: absent<br>1: present | 0 | 1.000 |
| 9 | Simple penial glands | 0: absent<br>1: scattered<br>2: glandular disc | 1 | 1.000 |
| 10 | Mamilliform penial glands | 0: absent<br>1: 1 gland<br>2: 2 glands<br>3: 1–6 glands<br>4: up to 7 or more glands | 1 | 0.571 |
| 11 | Paraspermatic nurse | 0: with rod-pieces<br>1: rod-pieces absent | 1 | 1.000 |
| 12 | Oviducal sperm groove (*) | 0: open<br>1: partially closed | 0 | – |
| 13 | Position of bursa | 0: more posterior<br>1: anterior | 0 | 0.333 |
| 14 | Coiling of egg groove | states 0–6, see Fig. 2 | 1 | 1.000 |
| 15 | Jelly gland | 0: small<br>1: large | 0 | 0.500 |
| 16 | Egg capsules (U) | 0: cupola<br>1: biconvec disc<br>2: absent<br>3: 2 peripheral rims | 0 | 1.000 |
| 17 | Eggs per capsule | 0: 1 egg<br>1: typically more than 1 | 0 | 0.333 |
| 18 | Development | 0: planktotrophic<br>1: non-planktotrophic | 0 | 1.000 |
| 19 | Ovoviviparity | 0: absent<br>1: brooding in oviduct | 0 | 1.000 |
| 20 | Rachidian tooth shape | 0: normal<br>1: square | 0 | 0.500 |

*Table 1.* (continued)

| No. | Character | Character states | Plesio-morphic state | Consistency index |
|-----|-----------|-----------------|---------------------|-------------------|
| 21 | Rachidian cusps | 0: 3 cusps<br>1: 5 cusps | 0 | 1.000 |
| 22 | Lateral and inner marginal cusp shape | 0: pointed<br>1: blunt | 0 | 1.000 |
| 23 | Outer marginal tooth shape | 0: neck and basal projection<br>1: elongate rectangular | 0 | 1.000 |
| 24 | Outer marginal cusp shape | 0: pointed<br>1: blunt | 0 | 0.500 |
| 25 | Position of salivary glands (*) | 0: posterior to nerve ring<br>1: constricted by nerve ring | 0 | – |
| 26 | Size of salivary glands | 0: small<br>1: large | 0 | 0.500 |

## 22. *Lateral and inner marginal cusp shape*

The cusps of these teeth may be either blunt (Fig. 4g) or pointed. Cusp shape may be difficult to determine when teeth are orientated as in the standard views of Fig. 4.

## 23. *Outer marginal tooth shape*

In the genus *Nodilittorina* and in four *Littorina* species the outer marginal tooth shows a narrow neck beneath the cusp-bearing head, and a strong basal projection on the outer side (Fig. 4a, b). In contrast, other *Littorina* species show a shape of more uniform width, termed elongate rectangular.

## 24. *Outer marginal cusp shape*

The cusps of the outer marginal teeth may be blunt (Fig. 4e–h) or pointed (Fig. 4a–d, i). This is the only character in the analysis which distinguishes *L. neglecta* from *L. saxatilis*. However, because juvenile *L. saxatilis* show pointed cusps like *L. neglecta* (Raffaelli, 1979), this may not be a reliable character by which to separate the two. It is possible that *L. neglecta* is only a small ecotype of *L. saxatilis* (B. & K. Johannesson this volume, K. & B. Johannesson this volume).

## 25. *Position of salivary glands*

*Mainwaringia rhizophila* is unique among species of *Nodilittorina* and *Littorina* in possessing constricted salivary glands, with glandular material on both sides of the circumoesophageal nerve ring (Reid, 1986b). In other species the glandular material is entirely posterior to the nerve ring.

## 26. *Size of salivary glands*

Salivary glands are small in *Nodilittorina* and some *Littorina* species, consisting of coiled strands running beneath the mid-oesophagus. In other *Littorina* species these glands are much enlarged, together equivalent in width to the mid-oesophagus, and spreading around its sides, or even covering the radular sac on the dorsal side.

## *Cladistic analysis*

The 26 characters and 65 character states chosen for use in the analysis are listed in Table 1. The complete matrix of character states for the 19 ingroup and 3 outgroup taxa is given in Table 2. Both tables include autapomorphic characters (numbers 5, 6, 12, 25) for the sake of completeness.

The data were analysed using version 2.4.1 of the PAUP (phylogenetic analysis using parsimony) program (Swofford, 1985). All characters were unweighted, and numbers 3 and 16 specified as unordered, as discussed above. Before running

*Table 2.* Character states for cladistic analysis of species of *Littorina* and *Mainwaringia*. For list of characters see Table 1. The three subgenera of *Nodilittorina* comprise the outgroup.

| Taxon | Character number | | | | | | | | | | | | | | | | | | | | | | | | | |
|---|---|---|---|---|---|---|---|---|---|---|---|---|---|---|---|---|---|---|---|---|---|---|---|---|---|---|
| | 1 | 2 | 3 | 4 | 5 | 6 | 7 | 8 | 9 | 10 | 11 | 12 | 13 | 14 | 15 | 16 | 17 | 18 | 19 | 20 | 21 | 22 | 23 | 24 | 25 | 26 |
| *N. (Fossarilittorina)* | 1 | 1 | 0 | 1 | 0 | 0 | 1 | 0 | 0 | 0 | 0 | 0 | 0 | 0 | 0 | 0 | 0 | 0 | 0 | 0 | 0 | 0 | 0 | 0 | 0 | 0 |
| *N. (Echinolittorina)* | 1 | 0 | 0 | 1 | 0 | 0 | 0 | 1 | 2 | 1 | 0 | 0 | 0 | 0 | 0 | 0 | 0 | 0 | 0 | 0 | 0 | 0 | 0 | 0 | 0 | 0 |
| *N. (Nodilittorina)* | 1 | 0 | 0 | 1 | 0 | 0 | 0 | 1 | 2 | 1 | 0 | 0 | 1 | 0 | 0 | 0 | 0 | 0 | 0 | 0 | 0 | 0 | 0 | 0 | 0 | 0 |
| *L. striata* | 1 | 0 | 0 | 0 | 0 | 0 | 0 | 0 | 1 | 0 | 1 | 0 | 0 | 1 | 0 | 0 | 0 | 0 | 0 | 0 | 0 | 0 | 0 | 0 | 0 | 0 |
| *L. keenae* | 1 | 1 | 0 | 0 | 0 | 0 | 0 | 0 | 1 | 2 | 1 | 0 | 0 | 2 | 1 | 1 | 0 | 0 | 0 | 0 | 0 | 0 | 0 | 0 | 0 | 0 |
| *L. scutulata* | 2 | 1 | 1 | 0 | 0 | 0 | 0 | 0 | 1 | 0 | 1 | 0 | 0 | 3 | 0 | 3 | 1 | 0 | 0 | 0 | 0 | 0 | 0 | 0 | 0 | 1 |
| *L. plena* | 2 | 1 | 1 | 0 | 0 | 0 | 0 | 0 | 1 | 1 | 1 | 0 | 0 | 3 | 0 | 3 | 1 | 0 | 0 | 0 | 0 | 0 | 0 | 0 | 0 | 1 |
| *L. rhizophila* | 2 | 1 | 2 | 0 | 1 | 1 | 1 | 0 | 1 | 1 | 1 | 1 | 1 | 3 | 0 | 1 | ? | 0 | 0 | 1 | 1 | 0 | 1 | 0 | 1 | 0 |
| *L. brevicula* | 1 | 1 | 1 | 0 | 0 | 0 | 0 | 0 | 1 | 3 | 1 | 0 | 0 | 3 | 0 | 1 | 0 | 0 | 0 | 1 | 1 | 0 | 1 | 0 | 0 | 1 |
| *L. mandshurica* | 1 | 1 | 1 | 0 | 0 | 0 | 0 | 0 | 1 | 3 | 1 | 0 | 0 | 3 | 0 | 1 | 1 | 0 | 0 | 1 | 1 | 0 | 1 | 0 | 0 | 1 |
| *L. squalida* | 1 | 1 | 1 | 0 | 0 | 0 | 0 | 0 | 1 | 4 | 1 | 0 | ? | 3 | 0 | 1 | 1 | 0 | 0 | 1 | 1 | 0 | 1 | 1 | 0 | 1 |
| *L. littorea* | 1 | 1 | 1 | 0 | 0 | 0 | 0 | 0 | 1 | 4 | 1 | 0 | 0 | 3 | 0 | 1 | 1 | 0 | 0 | 1 | 1 | 0 | 1 | 1 | 0 | 1 |
| *L. aleutica* | 0 | 0 | 1 | 0 | 0 | 0 | 0 | 0 | 1 | 4 | ? | 0 | 1 | 3 | 1 | ? | ? | ? | 0 | 1 | 1 | 0 | 1 | 1 | 0 | 1 |
| *L. sitkana* | 1 | 1 | 1 | 0 | 0 | 0 | 0 | 0 | 1 | 4 | 1 | 0 | 1 | 4 | 1 | 1 | 0 | 1 | 0 | 1 | 1 | 0 | 1 | 1 | 0 | 1 |
| *L. kurila* | 1 | 1 | 1 | 0 | 0 | 0 | 0 | 0 | 1 | 4 | 1 | 0 | 1 | 5 | 1 | 1 | ? | 1 | 0 | 1 | 1 | 0 | 1 | 1 | 0 | 1 |
| *L. subrotundata* | 1 | 1 | 1 | 0 | 0 | 0 | 0 | 0 | 1 | 4 | 1 | 0 | 1 | 5 | 1 | ? | ? | 1 | 0 | 1 | 1 | 0 | 1 | 1 | 0 | 1 |
| *L. obtusata* | 0 | 1 | 1 | 0 | 0 | 0 | 0 | 0 | 1 | 4 | 1 | 0 | 1 | 5 | 1 | 2 | ? | 1 | 0 | 0 | 1 | 1 | 1 | 1 | 0 | 1 |
| *L. mariae* | 0 | 1 | 1 | 0 | 0 | 0 | 0 | 0 | 1 | 4 | 1 | 0 | 1 | 5 | 1 | 2 | ? | 1 | 0 | 0 | 1 | 1 | 1 | 1 | 0 | 1 |
| *L. nigrolineata* | 1 | 1 | 1 | 1 | 0 | 0 | 0 | 0 | 1 | 4 | 1 | 0 | 1 | 5 | 1 | ? | ? | 1 | 0 | 0 | 1 | 0 | 1 | 1 | 0 | 1 |
| *L. arcana* | 1 | 1 | 1 | 1 | 0 | 0 | 0 | 0 | 1 | 4 | 1 | 0 | 1 | 5 | 1 | ? | ? | 1 | 0 | 0 | 1 | 0 | 1 | 1 | 0 | 1 |
| *L. saxatilis* | 1 | 1 | 1 | 1 | 0 | 0 | 0 | 0 | 1 | 4 | 1 | 0 | 1 | 6 | 1 | 2 | ? | 1 | 1 | 0 | 1 | 0 | 1 | 1 | 0 | 1 |
| *L. neglecta* | 1 | 1 | 1 | 1 | 0 | 0 | 0 | 0 | 1 | 4 | 1 | 0 | 1 | 6 | 1 | 2 | ? | 1 | 1 | 0 | 1 | 0 | 1 | 0 | 0 | 1 |

the analysis the autapomorphic characters were removed, because they cannot contribute information about relationships, and they would artificially increase the consistency index of the cladogram. In addition, two taxa (*L. mariae* and *L. arcana*) were removed, since they showed character states identical to other taxa (*L. obtusata* and *L. nigrolineata* respectively). To find the shortest (most parsimonious) tree or trees, the heuristic algorithm (SWAP = GLOBAL MULPARS option) was employed. For rooting the ingroup tree, the ROOT = OUTGROUP option was used, enabling the program to make decisions about character state polarity by outgroup comparison, which ensured global parsimony (Maddison *et al.*, 1984; Swofford, 1985). As recommended by Platnick (1987) a range of HOLD and ADDSEQ values were used to check that no shorter trees could be found. The MINF option

was employed for the reconstruction of character state changes on the branches of the cladogram, and alternative reconstructions were examined by specifying the CSPOSS and BLRANGE commands (Swofford, 1985; Platnick, 1987).

The analysis was repeated without *Mainwaringia rhizophila*, a doubtful member of the ingroup.

## Results

A single most parsimonious tree topology was found, of length 52 steps, which included four unresolved trichotomies (Fig. 5). The consistency index (the sum of the number of derived character states divided by the total number of character state changes on the tree) was relatively high, 0.673, indicating that 32.7% of the character state changes could be ascribed to homoplasy. The plesiomorphic character states within the clade *Littorina*, as inferred by the PAUP program, are listed in Table 1, together with the consistency indices of the individual characters. For charac-

ters 2, 4, 11 and 14 the plesiomorphic states have been judged by reference to more distant outgroups (Reid, 1989). Alternative character state reconstructions were found for characters 1,

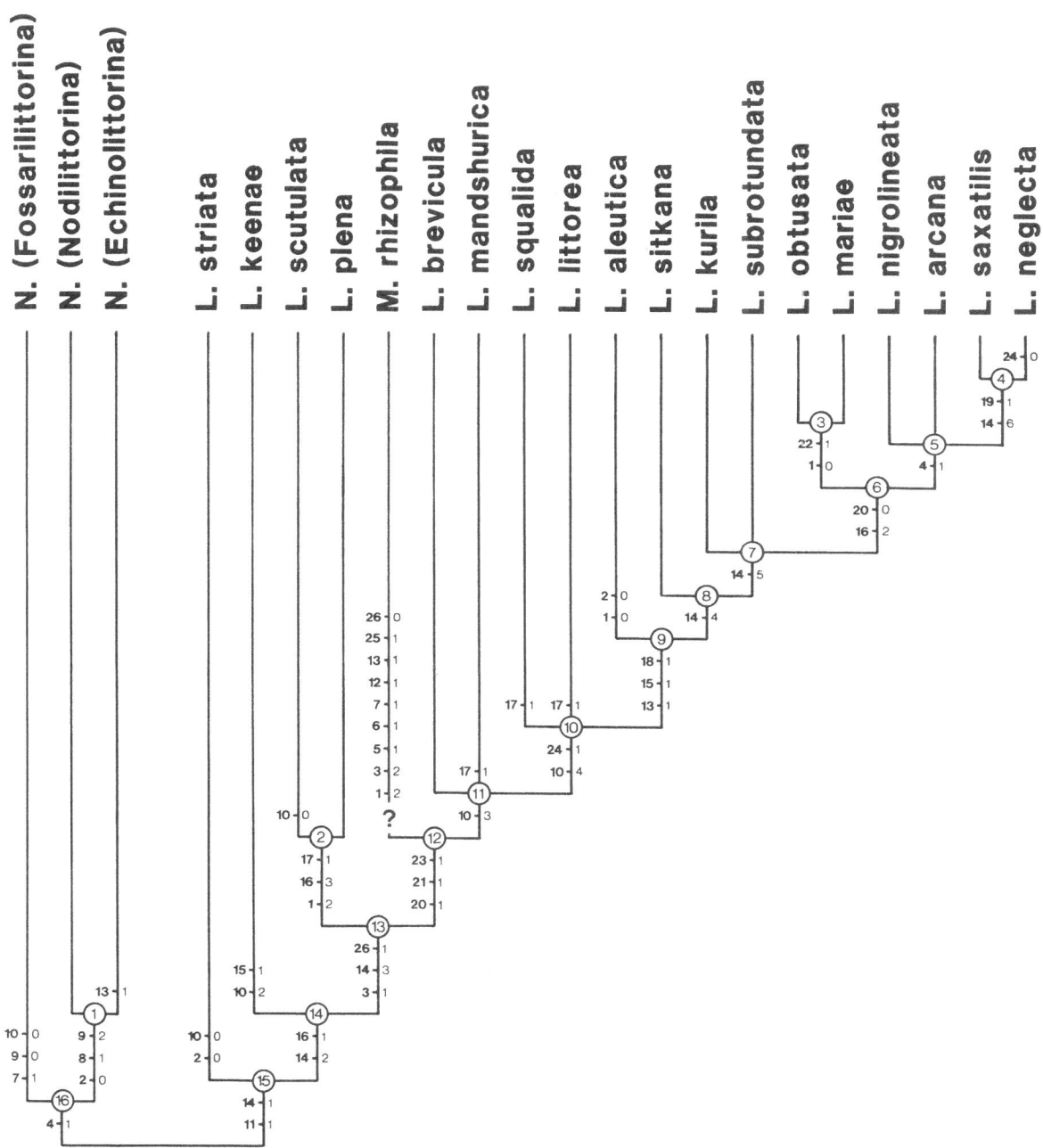

*Fig. 5.* Cladogram of 18 species of *Littorina* and one of *Mainwaringia*. See text for discussion of the uncertain position of *Mainwaringia rhizophila*. Synapomorphies are indicated by character number (bold type) and character state (see Table 1). Nodes are numbered arbitrarily for reference. Consistency index (excluding autapomorphic characters) = 0.673.

14

2, 10, 13, 17, 18 and 26. For all except number 18 there were no reasons for rejecting the reconstructions generated by the program (MINF optimization method), and these are shown in Fig. 5. For character 18, however, non-planktotrophic development was reconstructed for *L. aleutica*, because it has a large jelly gland and therefore probably a benthic gelatinous egg mass, which elsewhere in the genus is correlated with non-planktotrophic development. The two taxa removed before the analysis have been restored in the cladogram.

The number of mamilliform penial glands (character 10) shows a general increase from primitive to more derived species of *Littorina*. Although numbers of glands are positively correlated with shell size within some species (Heller, 1975; Raffaelli, 1979; Janson, 1982), this is not clearly the case when comparisons are made between species.

The hypothesized sequence of 7 character states for character 14 (coiling of egg groove) is maintained in the cladogram, so that this character has a consistency index of 1. If this character was specified as unordered, the PAUP program produced at least 100 trees (the default maximum), all of length 52 steps. The topologies were similar to that of Fig. 5, but the resolution was poorer. Since no shorter trees were found by this means, the weight of evidence from the other characters does not conflict with the proposed sequence of character 14.

When the analysis was repeated without *M. rhizophila*, the topology of the remaining taxa was unchanged; the tree length was reduced to 47 steps, while the consistency index increased to 0.727.

**Discussion**

*Topology of cladogram and classification of Littorina*

In a cladistic analysis of the subgenera of the Littorinidae, Reid (1989) recognized four subgenera of *Littorina*. These, with the species by which they were represented in the analysis, were:

*L. (Liralittorina) striata*, *L. (Planilittorina) keenae*, *L. (Littorina) littorea* and *L. (Neritrema) obtusata*. The present analysis, based on a similar set of characters, produced the same hypothesis of relationships among these four taxa. The first two of the subgenera are monotypic. *Neritrema*, as defined by Reid, corresponds to a monophyletic group on the cladogram (clade 9, Fig. 5). However, the six species included by Reid in the subgenus *Littorina* (*scutulata, plena, brevicula, mandshurica, littorea, squalida*) form a paraphyletic group or, if *M. rhizophila* is truly a member of *Littorina*, a polyphyletic one.

The phylogenetic relationships of *Mainwaringia* are still uncertain, as discussed in detail by Reid (1989). In the phylogenetic analysis of the subgenera of the Littorinidae, the most parsimonious cladogram placed *Mainwaringia* in a relatively basal position, distant from *Littorina*. However, subjective weighting of certain characters (paraspermatic nurse cells without rods, coiling of pallial egg groove, shape of egg capsules, square rachidian tooth) suggested the possibility of its inclusion within *Littorina*. Among *Littorina* species, *Mainwaringia* is unique in its exclusively tropical distribution, and unusual in its small size and brackish, muddy habitat. These facts could account for some of the characters which appear to exclude *Mainwaringia* from *Littorina*. The undivided foot, for example, could be an adaptation for crawling on mud. The absence of an outer calcitic layer in the shell could be a consequence of the tropical distribution (Taylor & Reid, this volume). Until more evidence is available, *Mainwaringia* has been retained as a genus, separate from *Littorina*. If its position within *Littorina* is confirmed, it should become a subgenus of the latter. It may then be desirable to remove *L. scutulata* and *L. plena* to a new subgenus, to avoid polyphyly of the subgenus *Littorina*.

The biogeography of *Littorina* will be discussed in detail in another paper (Reid, in press). Here, it may be pointed out that although node 10 is an unresolved trichotomy, this is consistent with *L. littorea* in the Atlantic being the sister-species of *L. squalida* in the Pacific, as proposed by previous authors (Rosewater, 1963; Golikov & Tzvetkova, 1972; Berger, 1978). The other supposed pairs of sibling species (*sitkana* and '*saxatilis*', *kurila* and '*obtusata*') are not supported. Instead, the six Atlantic species of *L. (Neritrema)* form a monophyletic group (clade 6), with *L. kurila* and/or *L. subrotundata* as sister-taxa. A minimum of two species may therefore have migrated from the Pacific to the northern Atlantic. The main episode of trans-Arctic migration took place during the late Pliocene (3 to 4 Ma ago; Durham & MacNeil, 1967; Einarsson *et al.*, 1967; Hopkins, 1967). If it is assumed that both stocks of *Littorina* reached the northern Atlantic at this time, it is interesting that *L. littorea*, with planktotrophic development, has not speciated further, while clade 6, with non-planktotrophic, non-pelagic development, has undergone rapid speciation to produce six species. This is in agreement with the recognized potential of species with non-planktotrophic development for geographical isolation and allopatric speciation (Berger, 1973; Jablonski & Lutz, 1983; Janson & Ward, 1984).

The close relationship between members of the *L. obtusata* and *L. saxatilis* groups (clade 6), and more distant relationship with *L. littorea*, is supported by electrophoretic evidence (Warmoes, 1986; Ward, this volume). The topology of the *L. saxatilis* group (clade 5) is incompletely resolved on the cladogram. Genetic analysis is clearly a more appropriate method for the determination of relationships between such closely related species. Such analyses suggest that *L. neglecta* and *L. saxatilis* are the most similar of the group (and perhaps conspecific; K. & B. Johannesson, this volume), that *L. arcana* is very close to these, and *L. nigrolineata* more distant (Heller, 1975; Ward & Warwick, 1980; Ward & Janson, 1985; Warmoes, 1986; Ward, this volume). In terms of its morphology and development *L. arcana* does not differ from the hypothetical ancestor of *L. saxatilis* and *L. neglecta*.

It must be emphasized that this phylogenetic hypothesis for the genus *Littorina* is based on a relatively limited number of easily observed morphological characters, and relies heavily on features of the reproductive system. The hypothesis will be tested as more characters are discovered. In particular the use of genetic characters may provide new information about relationships at the infrageneric level.

*Evolution of spawn and developmental types*
The cladogram is not only an hypothesis of phylogenetic relationships, but is also a source of hypotheses about the evolution of individual characters. The concept of adaptation is central to evolutionary theory, and has been defined in cladistic terms as 'apomorphic function promoted by natural selection' (Coddington, 1988). A character for which an adaptive function has been proposed should therefore be shown to be apomorphic (by reference to a cladogram derived from other characters, to avoid circularity), and its function compared with that of the plesiomorphic state in the appropriate outgroup. A weaker form of argument has commonly been used in the past, involving correlations between the character and the ecological conditions under which it is presumed to be adaptive. This argument is implicit in the concept of 'optimal strategies', in which alternative strategies may be overlooked and in which it is generally assumed that the trait is free to evolve without phylogenetic 'constraints' (see Stearns, 1977, 1984, for discussion in relation to reproductive strategies).

The diversity of spawn and developmental types in the Littorinidae is well known (reviews by Bandel, 1974; Mileikowsky, 1975; Reid, 1986a). Of all littorinid genera, this diversity is greatest in *Littorina*, which also occupies the widest range of habitats and climatic zones (Reid, 1989). In seeking adaptive explanations for the range of life history parameters in littorinids, authors have sought correlations with habitat (e.g. Underwood, 1974; Mileikowsky, 1975; Hughes & Roberts, 1981). These are not always clear, and indeed sympatric species can show, for example, different breeding seasons (Palant & Fishelson, 1968) and spawn types (Buckland-Nicks *et al.*, 1973). The cladogram of *Littorina* permits a more rigorous, evolutionary approach.

Several prosobranch families are represented by species with non-planktotrophic development at high latitudes, and planktotrophic development nearer the tropics, and among prosobranchs in general there is an increasing proportion of planktotrophic forms from the poles towards the tropics, at least to a latitude of 40° N or S (Thorson, 1950, 1965; Spight, 1981). Thorson (1950) explained these observations in terms of short periods of algal productivity and slow larval growth rates at low temperatures, selecting for suppression of planktotrophy at high latitudes. An alternative view is that predation on planktonic larvae, by pelagic predators and benthic filter feeders, is more intense at higher latitudes, while severe benthic predation in the tropics favours planktotrophic development (Highsmith, 1985).

Using the cladogram, it is possible to propose adaptive functions for some of the reproductive and developmental characters of *Littorina*. As an example, consider the number of eggs per capsule (and, correlated with this, because eggs are of similar size, the diameter of the discoidal capsule) in planktotrophic species. Removal of character 17 from the data matrix does not change the topology of the cladogram. From this topology, a parsimony argument suggests that the small capsules with single eggs are probably plesiomorphic and the large capsules with many eggs are apomorphic. An adaptive explanation must account for the function of the latter. The species with single eggs in small capsules (*L. striata*, *L. keenae*, *L. brevicula*, *M. rhizophila*) are, with the exception of *L. keenae*, the only members of the clade with all or a significant part of their distribution in tropical to subtropical biogeographical regions (Reid, 1989). The outgroup, *Nodilittorina*, also inhabits warm water and has capsules of the plesiomorphic type. In contrast, the species showing the apomorphic state (*L. scutulata*, *L. plena*, *L. mandshurica*, *L. squalida*, *L. littorea*) inhabit temperate and cold regions. One possible hypothesis is that a larger capsule is less susceptible to capture by pelagic predators and benthic filter feeders. The apomorphic state may therefore be adaptive in relation to the greater severity of

such predation at high latitudes, in agreement with the argument of Highsmith (1985) discussed above. This hypothesis could be tested experimentally, by comparing predation on the two capsule types. The exception, *L. keenae*, inhabits the temperate coast of California. Although the individual capsules are small and contain single eggs, the size of the propagule is increased in another way, many capsules being enclosed together in an ephemeral, pelagic, gelatinous mass (Schmitt, 1979). This may be an alternative strategy of adaptation to the problem of predation on pelagic capsules, and again could be tested experimentally.

A similar argument can be used for the character of non-planktotrophic development. Characters 14, 15 and 18 (which are directly or indirectly related to type of development) are first excluded from the data matrix. The resulting cladogram is less well resolved; clade 9 (subgenus *Neritrema*) remains intact only if character 13 is taken as state 0 in *L. squalida* (this choice in effect assumes a closer relationship of this species with *L. littorea* than with clade 9, which seems likely because of close resemblances between the shells), otherwise *L. squalida* is added to clade 9. In either case non-planktotrophic development is therefore an apomorphic character within the genus and an adaptive explanation for it can be sought. All members of *Neritrema* are found in temperate and cold regions, and this developmental strategy could be an alternative response to predation on pelagic stages (Highsmith, 1985) or an adaptation to the problems of limited planktonic food supply (Thorson, 1950). Assuming that *L. squalida* is not a member of clade 9, this character is a probable example of a phylogenetic 'constraint' on developmental strategy, because, once lost, planktotrophic larval stages are not easily regained (Strathmann, 1978, but see Reid, 1989).

Within *Neritrema* an evolutionary sequence can be proposed, in which primitively pelagic capsules were first embedded in a benthic, protective, gelatinous mass (e.g. *L. sitkana*); capsules therefore became redundant and were reduced or lost (e.g. *L. obtusata*), and finally the egg mass was retained in the oviduct, resulting in ovoviviparity (e.g. *L. saxatilis*). Without a phylogenetic analysis, there is the danger that such a sequence could be read in the wrong direction, as in the earlier suggestion that a benthic spawn is primitive in *Littorina* (Fretter, 1980). Hypotheses of adaptation based upon such a misconception would be misleading.

Another example of the use of cladograms in formulating hypotheses about adaptation is given by Taylor & Reid (this volume). It is hoped that the cladogram presented here will permit a similar approach to the problem of adaptation in other investigations of the comparative biology of *Littorina* species.

## Acknowledgements

This work was carried out during the tenure of a Senior Research Fellowship at the British Museum (Natural History). I should like to thank J.D. Taylor for his encouragement and help. I gratefully acknowledge financial support from the Percy Sladen Fund of the Linnean Society for field work in Alaska and Japan. For help and hospitality on collecting trips I thank A.J. Kohn (University of Washington, Seattle), E.G. Boulding (Friday Harbor Laboratories), A. Matsukuma (National Science Museum, Tokyo), K. Aketa (Akkeshi Marine Biological Station), K. Takada (Abashiri) and B.S. Morton (University of Hong Kong). I am grateful to T. Habe, A. Matsukuma and S. Uozumi for kindly sending specimens of *Littorina* from Japan, to G.J. Vermeij for specimens from the Aleutian Islands, to J. Dyson for specimens from the Canary Islands and to D.M. Bohmhauer for the loan of material from the National Museum of Natural History (Washington, D.C.). My thanks to G.N.G. Summons (BMNH) for photography of the shells. I should also like to acknowledge the use of a manuscript on *Littorina* species of the northeastern Pacific by the late J. Rosewater.

## References

Ax, P., 1987. The Phylogenetic System: the Systematization of Organisms on the Basis of their Phylogenies. J. Wiley & Sons, Chichester, 340 pp.

Backeljau, T., M. De Meyer. L. Janssens & R. Proesmans, 1984. Prosobranch and shelled opisthobranch molluscs from Store Ekkerøya, Varangerfjorden (northern Norway). Fauna norvegica Ser. A 5: 1–5.

Bandel, K., 1974. Studies on Littorinidae from the Atlantic. Veliger 17: 92–114.

Bandel, K. & D. Kadolsky, 1982. Western Atlantic species of *Nodilittorina* (Gastropoda: Prosobranchia): comparative morphology and its functional, ecological, phylogenetic and taxonomic implications. Veliger 25: 1–42.

Berger, E. M., 1973. Gene-enzyme variation in three sympatric species of *Littorina*. Biol. Bull. 145: 83–90.

Berger, V. Ya., 1978. Euryhalinity and the evolution of *Littorina*. Morphol. sistematika i evolyutsiya zhivotnikh, Leningrad: 46–47. (For abstract of Russian original see M. K. Jacobson & K. J. Boss, 1985. Veliger 27: 341).

Buckland-Nicks, J., F.-S. Chia & S. Behrens, 1973. Oviposition and development of two intertidal snails *Littorina sitkana* and *Littorina scutulata*. Can. J. Zool. 51: 359–365.

Clark, B. L. & R. Arnold, 1923. Fauna of the Sooke formation, Vancouver Island. Univ. Calif. Publs Bull. Dep. Geol. 14: 123–234.

Coddington, J. A., 1988. Cladistic tests of adaptational hypotheses. Cladistics 4: 3–22.

Durham, J. W. & F. S. MacNeil, 1967. Cenozoic migrations of marine invertebrates through the Bering Strait region. In D. M. Hopkins (ed.), The Bering Land Bridge. Stanford University Press, Stanford (Calif.): 326–349.

Einarsson, T., D. M. Hopkins & R. R. Doell, 1967. The stratigraphy of Tjornes, northern Iceland and the history of the Bering land bridge. In D. M. Hopkins (ed.), The Bering Land Bridge. Stanford University Press, Stanford (Calif.): 312–325.

Felsenstein, J., 1982. Numerical methods for inferring evolutionary trees. Q. Rev. Biol 57: 379–404.

Fish, J. D. & L. Sharp, 1985. The ecology of the periwinkle, *Littorina neglecta* Bean. In P. G. Moore & R. Seed (eds), The Ecology of Rocky Coasts. Hodder & Stoughton, London: 143–156.

Fretter, V., 1980. Observations on the female genital duct of British *Littorina* spp. J. moll. Stud. 46: 148–153.

Fretter, V. & A. Graham, 1962. British Prosobranch Molluscs: their functional anatomy and ecology. Ray Soc., London, 755 pp.

Fretter, V. & A. Graham, 1980. The prosobranch molluscs of Britain and Denmark. Part 5. Marine Littorinacea. J. moll. Stud. Suppl. 7: 243–284.

Golikov, A. N. & N. L. Tzvetkova, 1972. The ecological principle of evolutionary reconstruction as illustrated by marine animals. Mar. Biol. 14: 1–9.

Goodwin, B. J., 1979. The egg mass of *Littorina obtusata* and

18

*Lacuna pallidula* (Gastropoda: Prosobranchia). J. moll. Stud. 45: 1–11.

Goodwin, B. J. & J. D. Fish, 1977. Inter- and intraspecific variation in *Littorina obtusata* and *L. mariae* (Gastropoda: Prosobranchia). J. moll. Stud. 43: 241–254.

Hannaford Ellis, C. J., 1979. Morphology of the oviparous rough winkle *Littorina arcana* Hannaford Ellis, 1978, with notes on the taxonomy of the *L. saxatilis* species-complex (Prosobranchia: Littorinidae). J. Conch., Lond. 30: 43–56.

Harmer, F. W., 1920–25. The Pliocene Mollusca of Great Britain, being supplementary to S. V. Wood's Monograph of the Crag Mollusca, 2. Palaeontographical Society, London: 485–900.

Heller, J., 1975. The taxonomy of some British *Littorina* species with notes on their reproduction (Mollusca: Prosobranchia). J. linn. Soc., Zool. 56: 131–151.

Herman, Y. & D. M. Hopkins, 1980. Arctic oceanic climate in late Cenozoic time. Science 209: 557–562.

Hennig, W., 1966. Phylogenetic Systematics. University of Illinois Press, Urbana (Ill.), 208 pp.

Highsmith, R. C., 1985. Floating and algal rafting as potential dispersal mechanisms in brooding invertebrates. Mar. Ecol. prog. Ser. 25: 169–179.

Hopkins, D. M., 1967. The Cenozoic history of Beringia – a synthesis. In D. M. Hopkins (ed.), the Bering Land Bridge. Stanford University Press, Stanford (Calif.): 451–484.

Hughes, R. N. & D. J. Roberts, 1981. Comparative demography of *Littorina rudis*, *Littorina nigrolineata* and *Littorina neritoides* on three contrasted shores in north Wales. J. anim. Ecol. 50: 251–268.

Humphries, C. J. & L. R. Parenti, 1986. Cladistic Biogeography. Clarendon Press, Oxford, 98 pp.

Jablonski, D. & R. A. Lutz, 1983. Larval ecology of marine benthic invertebrates: paleobiological implications. Biol. Rev. 58: 21–89.

Janson, K., 1982. Phenotypic differentiation in *Littorina saxatilis* Olivi (Mollusca, Prosobranchia) in a small area on the Swedish West coast. J. moll. Stud. 48: 167–173.

Janson, K. & R. D. Ward, 1984. Microgeographical variation in allozyme and shell characters in *Littorina saxatilis* Olivi (Prosobranchia: Littorinidae). Biol. J. linn. Soc. 22: 289–307.

Janson, K. & R. D. Ward, 1985. The taxonomic status of *Littorina tenebrosa* Montagu as assessed by morphological and genetic analyses. J. Conch., Lond. 32: 9–15.

Kojima, Y., 1957. On the breeding of a periwinkle, *Littorivaga brevicula* (Philippi). Bull. biol. Stn Asamushi 8: 59–62.

Kojima, Y., 1958a. On the breeding of periwinkles *Littorivaga brevicula* (Philippi) and *Littorivaga atkana* (Dall). Venus 19: 224–229.

Kojima, Y., 1958b. On the floating egg capsules of periwinkles, *Littorina squalida* Broderip et Sowerby and *Nodilittorina pyramidalis* (Quoy et Gaimard). Venus 19: 233–237.

Kojima, Y., 1958c. A new type of the egg capsule of a periwinkle, *Littorina squalida* Broderip et Sowerby. Bull. biol. Stn Asamushi 9: 39–41.

Kojima, Y., 1958d. On the planktonic egg capsules of *Littorivaga mandschurica* (Schrenk) and *Littoraria strigata* (Lischke). Venus 20: 81–86.

Linke, O., 1933. Morphologie und Physiologie des Genitalapparates der Nordseelittorinen. Helgoländer wiss. Meeresunters. 19(5): 1–60.

Maddison, W. P., M. J. Donoghue & D. R. Maddison, 1984. Outgroup analysis and parsimony. Syst. Zool. 33: 83–103.

Mastro, E., V. Chow & D. Hedgecock, 1982. *Littorina scutulata* and *Littorina plena*, sibling species status of two prosobranch gastropod species confirmed by electrophoresis. Veliger 24: 239–246.

Matthews, R. S., 1978. Natural history and biology of the salt-marsh periwinkle *Littorina newcombiana*. Unpublished student report, Oregon Institute of Marine Biology, University of Oregon.

Mileikowsky, S. A., 1975. Types of larval development in Littorinidae (Gastropoda: Prosobranchia) of world ocean, and ecological patterns of their distribution. Mar. Biol. 30: 129–136.

Moyse, J., J. P. Thorpe & E. Al-Hamadani, 1982. The status of *Littorina aestuarii* Jeffreys: an approach using morphology and biochemical genetics. J. Conch., Lond. 31: 7–15.

Murray, T. E., 1979. Evidence for an additional *Littorina* species and a summary of the reproductive biology of *Littorina* from California. Veliger 21: 469–474.

Murray, T. E., 1982. Morphological characterization of the *Littorina scutulata* species complex. Veliger 24: 233–238.

Palant, B. & L. Fishelson, 1968. *Littorina punctata* (Gmelin) and *Littorina neritoides* (L.), (Mollusca, Gastropoda) from Israel: ecology and annual cycle of genital system. Israel J. Zool. 17: 145–160.

Platnick, N. I., 1987. An empirical comparison of microcomputer parsimony programs. Cladistics 3: 121–144.

Raffaelli, D. G., 1979. The taxonomy of the *Littorina saxatilis* species-complex, with particular regard to the systematic position of *Littorina patula* Jeffreys. J. linn. Soc., Zool. 65: 219–232.

Reid, D. G., 1986a. The Littorinid Molluscs of Mangrove Forests in the Indo-Pacific Region: the Genus *Littoraria*. British Museum (Natural History), London, 228 pp.

Reid, D. G., 1986b. *Mainwaringia* Nevill, 1885, a littorinid genus from Asiatic mangrove forests, and a case of protandrous hermaphroditism. J. moll. Stud. 52: 225–242.

Reid, D. G., 1989. The comparative morphology, phylogeny and evolution of the gastropod family Littorinidae. Phil. Trans. r. Soc., Lond. B 324: 1–110.

Reid, D. G., in press. Trans-Arctic migration and speciation induced by climatic change: the biogeography of *Littorina* (Mollusca:Gastropoda). Bull. mar. Sci.

Rosewater, J., 1963. Problems of species analogues in world Littorinidae. Rep. am. malac. Un. Pacif. Div. 30: 5–6.

Rosewater, J., 1978. A case of double primary homonymy in Eastern Pacific Littorinidae. Nautilus 92: 123–125.

Rosewater, J., 1981. The family Littorinidae in tropical West Africa. Atlantide Rep. 13: 7–48.

Sacchi, C. F., 1975. *Littorina nigrolineata* Gray (Gastropoda, Prosobranchia). Cah. Biol. mar. 16: 111–120.

Sacchi, C. F. & M. Rastelli, 1966. *Littorina mariae*, nov. sp.: les différences morphologiques et écologiques, entre 'nains' et 'normaux' chez l'"espèce' *L. obtusata* (L.) (Gastr. Prosobr.) et leur signification adaptive et évolutive. Atti Soc. ital. Sci. nat. 105: 351–370.

Schmitt, R. J., 1979. Mechanics and timing of egg capsule release by the littoral fringe periwinkle *Littorina planaxis* (Gastropoda: Prosobranchia). Mar. Biol. 50: 359–366.

Seeley, R. H., 1986. Intense natural selection caused a rapid morphological transition in a living marine snail. Proc. natn Acad. Sci. USA 83: 6897–6901.

Smith, J. E., 1981. The natural history and taxonomy of shell variation in the periwinkles *Littorina saxatilis* and *Littorina rudis*. J. mar. biol. Ass. UK 61: 215–241.

Spight, T. M., 1981. Latitude and prosobranch larvae: whose veligers are found in tropical waters? Ecosynthesis 1: 29–52.

Stearns, S. C., 1977. The evolution of life history traits: a critique of the theory and a review of the data. Annu. Rev. Ecol. Syst. 8: 145–171.

Stearns, S. C., 1984. The tension between adaptation and constraint in the evolution of reproductive patterns. In W. Engels & A. Fischer (eds), Advances in Invertebrate Reproduction. Elsevier Science Publishers, Amsterdam, vol. 3: 387–398.

Strathmann, R. A., 1978. The evolution and loss of feeding larval stages of marine invertebrates. Evolution 32: 894–906.

Swofford, D. L., 1985. PAUP: Phylogenetic Analysis Using Parsimony. Version 2.4. Champaign (Ill.). Mimeo.

Thorson, G., 1946. Reproduction and larval development of Danish marine bottom invertebrates, with special reference to the planktonic larvae in the Sound (Øresund). Meddr. Kommn. Havunders., Serie Plankton 4(1): 1–523.

Thorson, G., 1950. Reproduction and larval ecology of marine bottom invertebrates. Biol. Rev. 25: 1–45.

Thorson, G., 1965. The distribution of benthic marine Mollusca along the N.E. Atlantic shelf from Gibraltar to Murmansk. Proc. First Europ. malac. Congr. (1962): 5–25.

Underwood, A. J., 1974. The reproductive cycles and geographical distribution of some common eastern Australian prosobranchs (Mollusca: Gastropoda). Aust. J. mar. Freshwat. Res. 25: 63–88.

Ward, R. D. & K. Janson, 1985. A genetic analysis of sympatric subpopulations of the sibling species *Littorina saxatilis* (Olivi) and *Littorina arcana* Hannaford Ellis. J. moll. Stud. 51: 86–94.

Ward, R. D. & T. Warwick, 1980. Genetic differentiation in the molluscan species *Littorina rudis* and *Littorina arcana* (Prosobranchia: Littorinidae). Biol. J. linn. Soc. 14: 417–428.

Warmoes, T., 1986. Een inleidende systematische en taxonomische studie van het genus *Littorina* (Gastropoda, Prosobranchia). Licentiaatsthesis, Universitaire Instelling Antwerpen, 128 pp.

Wiley, E. O., 1981. Phylogenetics: the Theory and Practice of Phylogenetic Systematics. J. Wiley, N.Y., 439 pp.

Yamaguchi, M., 1967. Egg capsules of a periwinkle, *Littorina brevicula*, in plankton samples. Venus 25: 73–76.

## Notes added in proof

1. Examination of the syntypes of *Littorina kurila* Middendorff has shown that the oviducts are of the form illustrated in Fig. 2h. It is therefore believed to be a junior synonym of *L. sitkana* Philippi. A new name may be necessary for the species here referred to as *L. kurila*.

2. E. G. Boulding has now reported (pers. comm.) that the egg capsules of *L. 'kurila'* are thin and spherical. Character 16 in this species may therefore be more appropriately coded as state 2. This would not alter the topology of the cladogram.

*Hydrobiologia* **193**: 21–27, 1990.
*K. Johannesson, D. G. Raffaelli and C. J. Hannaford Ellis (eds), Progress in Littorinid and Muricid Biology.*
© 1990 *Kluwer Academic Publishers.*

# Distribution of the species of rough periwinkle (*Littorina*) in Great Britain

P.J. Mill & J. Grahame
*Department of Pure and Applied Biology, University of Leeds, Leeds LS2 9JT, England, UK*

*Key words:* Biogeography *Littorina saxatilis*, *Littorina arcana*, *Littorina nigrolineata*, *Littorina neglecta*

## Abstract

The status of the four currently recognised species in the *Littorina saxatilis* species-complex, i.e. *Littorina nigrolineata* Gray, *L. arcana* Hannaford Ellis, *L. saxatilis* (Olivi) and *L. neglecta* Bean is reviewed briefly, with notes on their characteristic features and location on the shore. Since the taxonomy of these rough periwinkles has only become stable relatively recently much of the previously published information on distribution is of little or no use. In this paper their distribution around the coastline of England and Wales (with some notes on Scotland) is described and discussed. *L. saxatilis* is found in a wide range of habitats from exposed peninsulas to estuaries on all shores where there is a suitable rocky or stony substrate; also in salt marsh pools. *L. arcana* has a more restricted distribution and is notably absent along much of the south English coast, central Cardigan Bay and possibly northern Scotland; it is not found in estuaries. *L. nigrolineata* has an even more restricted distribution, although it occurs both on exposed coasts and in estuaries; it has only been found by us in one locality on the east coast. *L. neglecta* is probably fairly widely distributed but we have few details so far. The implications of the different patterns of distribution are discussed.

## Introduction

Four species of rough periwinkle are currently recognised on the shores of north western Europe. Two of these, *Littorina nigrolineata* Gray and *L. arcana* Hannaford Ellis, are oviparous, laying egg masses; the other two, *L. saxatilis* (Olivi) and *L. neglecta* Bean, are ovoviviparous, giving birth to young 'crawlaways'.

Since the taxonomy of the rough periwinkles has only become stable comparatively recently much of the published information on distribution is of little or no use. This applies in particular to that regarding *L. saxatilis*, many records of which

will have included *L. arcana* and probably the other two species as well. The records for *L. nigrolineata* should all be acceptable and probably those for *L. neglecta* also but there appears to be little specific information on the latter. *L. arcana* has only recently been described (Hannaford Ellis, 1978, 1979). The species all present problems of identification, often only resolved by careful examination of fresh soft parts. Consequently an understanding of their distribution requires an examination of newly collected material, and to this end extensive field investigations have been carried out around the coastline of England, Wales and southeastern Scotland.

We have also examined material which has been sent to us from other regions in Scotland as well as from other localities such as the Scilly Isles and the Channel Islands. We have also looked at material from some sites in France. Mostly we have been concerned with the three largest species.

Until recently the above species were all included within *Littorina 'saxatilis'*. Within this complex *L. nigrolineata* was initially distinguished as a species by Gray (1839) but was later relegated to a subspecies (Dautzenberg & Fischer, 1912) and then a variety (James, 1968); it was reassigned species status by Heller (1975) and Sacchi (1975). Similarly *L. neglecta* was originally accorded species status by Bean (1844), reduced to a subspecies by James (1968) and given species status again by Heller (1975). Recently, however, Johannesson & Johannesson (this volume) have queried its status as a species. *L. saxatilis* was described from the lagoons at Venice by Olivi (1792) (originally as *Turbo saxatilis*), whereas the rough periwinkle in north-western Europe was attributed to *L. rudis* (Maton, 1797) (originally as *Turbo rudis*), with the type locality at Cargreen on the River Tamar in Devon, England. It is only comparatively recently that *L. arcana* has been separated from *L. rudis* by Hannaford Ellis (1978), and more recent still that *L. rudis* has been generally accepted as a junior synonym of *L. saxatilis* (e.g. Janson, 1985; Mill & Grahame, 1988; Grahame & Mill, 1989).

The existence of two other species has been suggested in recent years. *L. tenebrosa* was originally described by Montagu (1803) as a separate species (*Turbo tenebrosa*) 'but not without some doubt'. He described it as 'found on the mud, and on rocks near high-water mark, and even in ditches subject to the daily flux of the tide'. It was considered a subspecies by Dautzenberg & Fischer (1912) and by James (1968) but Fretter & Graham (1980) reinstated it as a species although they noted that its 'standing.... has yet to be unequivocally demonstrated'. They described it as 'found on weed, often permanently submerged, in sheltered situations with reduced salinity', which is much more restrictive than Montagu's description. At the present time it does not appear to be accepted as separate from *L. saxatilis* (Janson & Ward, 1985). The other 'species', *L. patula*, was described by Heller (1975). It is now realized that Heller included morphs of both *L. saxatilis* and *L. arcana* in his '*L. patula*' and thus *L. patula* is not now considered to be a valid species (Hannaford Ellis, 1978, 1979; see also Raffaelli, 1979).

## Species recognition

*Littorina nigrolineata* is the easiest of the four species to recognise in the field. Its shell bears ridges which are rather flattened and are wider than the grooves which separate them. The grooves usually (but not in all localities) contain a dark line – hence the specific epithet. Occasionally a shallow, secondary groove, containing a dark line, can be seen on some of the wider ridges.

*Littorina arcana* and *L. saxatilis* both have a number of different morphs and their shells may be ridged or smooth. With few exceptions they are easily confused in the field and can only be distinguished reliably on the basis of the presence of a jelly gland in mature females of *L. arcana*, in contrast to the brood pouch of *L. saxatilis* (Hannaford Ellis, 1979). The jelly gland of *L. arcana* occupies about half of the overall length of the pallial oviduct, whereas the brood pouch of *L. saxatilis* contributes about two thirds of the overall length. In immature animals the relative proportions are similar, with the brood pouch of *L. saxatilis* being relatively longer and narrower than the jelly gland of *L. arcana*. In contrast to the findings of Hannaford Ellis (1979) we have found that, on many shores, the extent of the ciliary field is not a good character on which to separate the two species. Although we have been rigorous in only accepting a definite identification at a site on the basis of the presence of mature females, there has been no occasion where identified immature *L. arcana* have occurred in our samples without mature individuals also being present. Furthermore, most of our collections have been made in May and August to October, when mature females of *L. arcana* should be in evidence (Hannaford Ellis, 1983).

*L. neglecta* is smaller than the other species and is rather globular, and it is generally found in empty barnacle shells. Its shell often has a wide, dark band running into the aperture. It can also be distinguished from the other ovoviviparous species, *L. saxatilis*, in that it matures (i.e. females have a full brood pouch) at a smaller size.

## Observations

### Location on the shore

*L. saxatilis* and *L. arcana* occur highest up the shore and, where sympatric, appear to be inter-mingled except on very exposed shores, where the former species may be absent. In extremely dry habitats *L. saxatilis* may extend further up the shore, as at St Ann's Head in Pembrokeshire. They both overlap with *L. nigrolineata*, although the latter does not extend as far up the shore as the upper limit of *L. saxatilis* and *L. arcana* and extends further down than do either of these two species. This is similar to the description in Heller (1975). At Westdale Bay, Dyfed, the juveniles of *L. nigrolineata* and some juveniles of *L. saxatilis* can be identified by shell sculpturing which, from observation of adults, we know to be characteris-tic of the species. At this site juveniles of these two species may be found in the zone occupied by *L. neglecta*, the lowest of the four species. *L. neglecta* normally occupies empty barnacle shells.

### Distribution

*Littorina saxatilis.* This species occurs in a very wide range of habitats. It is found on all rocky shores where a suitably stable habitat exists and occurs on either boulders or cliffs (or both) at all sites investigated, except at one site on North Foreland in Kent (Botany Bay), where the chalk cliffs are extremely soft, and at one site in Dorset (Burton Bradstock), where the cliffs are very

friable and are probably too far back from the sea (and with no boulders at their base). It is present on very exposed shores, such as at Portland Bill (Dorset) and St Catherine's Point (Isle of Wight) and also on more sheltered shores such as at Cawsand in Cornwall. It also occurs in sheltered estuarine regions, where it may be found under fairly small stones (e.g. the Gann in Dyfed and Poole Harbour in Dorset), and in pools in salt marshes (e.g. Sandy Haven in Dyfed). At the east end of the Fleet (on the landward side of Chesil beach) in Dorset small specimens are found in fairly small gravel. It clearly shows tolerance to a wide range of exposure and salinity and also exhibits a very wide size range. On shores where it is sympatric with *L. arcana* it may not be present at the most exposed sites; this was noted at Great Castle Head (Dyfed) (Grahame & Mill, 1986) (Fig. 1).

*Littorina arcana* has a more restricted distribu-tion than the above species. We have found it on the northeast coast from Fife as far south as Humberside. Moving southwards and clockwise around the coast we have not found it at any of the sites that we have visited on the southeast and south coasts until Lyme Regis in Dorset. This, in spite of the fact that we have made collections from the most exposed sites on the south coast, i.e. regions where *L. arcana* would be most likely to occur. It is present westwards from Lyme Regis around the Devon and Cornish coasts and into Somerset. It is also present on Lundy Island and in southwest Wales, including both St Ann's and St David's peninsulas in Dyfed, as well as the islands off the west Dyfed coast. However, it is absent from much of Cardigan Bay, reappearing on the Lleyn peninsula in North Wales and in Anglesey. Further north we have just two records on the west coast, at Seil Bay near Oban (Strathclyde) and at Polbain in Highland. It is found on exposed coasts and, as mentioned above, is present to the exclusion of *L. saxatilis* on the most exposed region of Great Castle Head in Dyfed. We have never found it in estuarine regions (Fig. 2).

*Littorina nigrolineata* has an even more re-stricted range than does *L. arcana*. We have

24

found it in west Cornwall and at sites on the St Ann's and St David's peninsulas in Dyfed. We have also received specimens from Lundy Island, Guernsey, the Isle of Man (1 specimen, but K. Johannesson (pers. commun.) has also found it there) and three sites in western Scotland (Seil Bay in Strathclyde and Loch Sunart and Polbain in Highland). Our only record from the east coast is from North Berwick (Lothian). It occurs on exposed shores where, as noted above, it is found generally rather lower down the shore than *L. saxatilis* and *L. arcana* and, as with the former of these, it does not reach the most exposed regions of Great Castle Head. We have also collected large specimens of this species from sheltered and even estuarine sites in Milford Haven (Dyfed) (Fig. 3).

*L. neglecta.* We have not yet made a detailed appraisal of the distribution of this species but have found it at sites as far apart as Robin Hood's Bay (North Yorkshire) and Great Castle Head (Dyfed); at the latter site it was present on both the most exposed region of the headland and more sheltered rocks. Further details of its distribution are given in the discussion.

The papers of James (1968), Heller (1975), Raffaelli (1979), Atkinson & Warwick (1983), Fish & Sharp (1985) and Knight & Ward (1986) contain useful information on distribution and this is included, where it helps to complete the picture, in Figs 1–3. These data corroborate our findings. We have also studied material from the British Museum (Natural History) which indicates that *L. nigrolineata* is also present at various localities on both the west and east coasts of Scotland and at Sharpness Bay/Point in Tyne and Wear. We have not yet investigated the Scottish sites but were unable to locate any specimens at the Tyne and Wear site.

## Discussion

Where *L. arcana* is not present, such as at Portland Bill in Dorset, *L. saxatilis* occupies the most exposed sites; whereas on Great Castle Head in Dyfed, where *L. saxatilis* is sympatric

*Fig. 1.* Distribution of *Littorina saxatilis* ●—— Present – site visited, ○—— Present – specimens received, ◇—— Present – in literature (records from Atkinson & Warwick, 1983; Knight & Ward, 1986), -—— Absent – site visited

with *L. arcana*, the latter species exclusively occupies the most exposed aspects. Furthermore, Grahame & Mill (1989), in a comparison of the morphometrics of south English coast allopatric *L. saxatilis* with south-west English and Welsh sympatric *L. saxatilis* and *L. arcana*, have found differences not only between the two species where they are sympatric (see also Grahame & Mill, 1986) but also between allopatric and sympatric *L. saxatilis*. We have attributed this latter to character displacement.

It is difficult at this stage to reach an understanding of why *L. arcana* does not extend eastwards along the south English coast beyond about Lyme Regis. The egg mass of *L. arcana*

*Fig. 2.* Distribution of *Littorina arcana* ●—— Present – site visited, ○—— Present – specimens received, ◇—— Present – in literature (records from Atkinson & Warwick, 1983; Knight & Ward, 1986) Absent from other sites visited (see Fig. 1)

*Fig. 3.* Distribution of *Littorina nigrolineata* ●—— Present – site visited, ○—— Present – specimens received, ◇—— Present – in literature (records from James, 1968; Heller, 1975; Raffaelli, 1979; Knight & Ward, 1986) Absent from other sites visited (see Fig. 1)

appears poorly formed and fragile (own observations) and we speculate that it requires a stable substrate with plenty of shelter for survival. If this is so the extent of Chesil Beach (an approximately 23 km stretch of pebbles without backing cliffs or boulders) to the west of Portland Bill possibly imposes a barrier to such an oviparous species. Also, the very extensive sandy stretches on the southern half of the English east coast could present an even more formidable barrier. These two barriers might be the explanation for its absence in this region. However, it is also absent from central Cardigan Bay. Here long sandy stretches at the northeastern end of the bay could

provide a barrier but there is no obvious barrier at the southeastern end of the bay. On the French coast an extensive search at Omaha Beach, Normandy which is about opposite Sussex, failed to show up either *L. arcana* or *L. nigrolineata* although *L. saxatilis* was present on pebbles, boulders, cliffs and pier supports. However, both of the former species are found further west. We have received a specimen of *L. arcana* from Finistère and Fretter & Graham (1980) report that *L. nigrolineata* is present on west Channel and Atlantic shores in France. Hence it may be that northern France shows the same phenomenon as is found on the south coast of England.

A number of 'southern' marine invertebrates and algae reach an eastward limit at various points along the south coast of England and the north coast of France and Crisp & Southward (1958) have suggested temperature as a possible limiting factor. However, *L. arcana* does not appear to fit into this category as its range extends into Scotland and it is also found on the northeast coast of England, where temperatures are generally lower than along the south coast.

There is some indication that *L. arcana* does not reach as far as extreme northern Scotland and it may be that this is beyond the northern limit of the species (on the mainland of Europe it is not found for example in Sweden (K. Johannesson, pers. commun.)). However, if this proves to be the case we have two populations (east and west coasts) which may well be genetically separated.

The very restricted range of *L. nigrolineata* may indicate that this species is towards the northern limit of its range in Britain, but if so the presence of the east coast population(s) is puzzling. It is certainly not a case of a preference for the extreme exposure of western shores since, not only does it not occur on the most exposed regions of Great Castle Head, but it tends to occur lower down the shore than the above two species and we have also found it in very sheltered habitats.

A possible implication of the distributions of *L. saxatilis*, *L. arcana* and *L. nigrolineata* is that the last two species either have not yet reached the full extent of their potential range or, perhaps, that their egg masses require the shelter afforded by crevices in rocks and cannot survive on a mobile substrate such as is provided by the pebbles at Chesil Beach. However, other oviparous gastropods, such as *Nucella lapillus* (L.) (Bantock & Cockayne, 1975), occur all along the south coast. *N. lapillus* has a tough leathery egg case which is attached to rocks by a short stalk fairly low on the shore. This is in marked contrast to the jelly-like egg masses of the oviparous littorinids and is presumably much more resistant to adverse conditions. The oviparous littorinids may thus be limited by their inability to recruit and thereby establish a permanent breeding population.

As mentioned above, our knowledge of the dis-tribution of *L. neglecta* is very sparse but the early indications are that it is widespread. Apart from our records it has been recorded from a number of sites by James (1968) including the Solway Firth and the Shetland Islands. Raffaelli (1979) and Hannaford Ellis (1983) have recorded it from Anglesey and Fish & Sharp (1985) examined samples from 'a wide range of localities.... from the Isles of Scilly to the Shetland Islands.' Raffaelli (1979) has suggested that this species may have evolved from '*L. saxatilis*' by a process of neotony.

It is intended to explore Scotland in greater detail, especially to check on the range of *L. arcana* and *L. nigrolineata*; also to look at the Irish and French coasts; we should also like to look at the distribution of *L. neglecta* in more detail.

## Acknowledgements

We should like to thank the many individuals who have provided us with specimens, including Dr J. Adams, Mr A. Brown, Miss R. Bowman, Mr J.H. Crothers, Mr C. Dytham, Dr H. Hassall, Mr M. Kendall, Mr J. King, Dr R.G. Loxton, Miss J. Nuttall, Miss S. Paviour, Dr J. Rosewell, Mrs G. Rowe, Dr S.L. Sutton, Dr C. Todd, Mr A. Williams and Dr G. Williams. We also wish to thank the University of Leeds for a grant from the Research Fund.

## References

Atkinson, W. D. & T. Warwick, 1983. The role of selection in the colour polymorphism of *Littorina rudis* Maton and *Littorina arcana* Hannaford Ellis (Prosobranchia: Littorinidae). Biol. J. linn. Soc. 20: 1327–151.

Bean, W., 1844. A supplement of new species. In C. Thorpe, British Marine Conchology. Lumley, London.

Crisp, D. J. & A. J. Southward, 1958. The distribution of intertidal organisms along the coasts of the English Channel. J. mar. biol. Ass. UK 37: 157–208.

Dautzenberg, P. H. & H. Fischer, 1912. Mollusques provenant des campagnes de 'L'Hirondelle' et de 'la Princesse-Alice' dans les Mers du Nord. Resultats des campagnes 'Prince Albert de Monaco', 37: 187–201.

Fish, J. D. & L. Sharp, 1985. The ecology of the periwinkle *Littorina neglecta* Bean. In 'The Ecology of Rocky Coasts' P. G. Moore & R. Seed (eds). Hodder & Stoughton, London.

Fretter, V. & A. Graham, 1980. The prosobranch molluscs of Britain and Denmark. Part 5 – Marine Littorinacea. J. moll. Stud. Suppl. 7: 241–284.

Grahame, J. & P. J. Mill, 1986. Relative size of the foot of two species of *Littorina* on a rocky shore in Wales. J. Zool., Lond. 208: 229–236.

Grahame, J. & P. J. Mill, 1989. Shell shape variation in *Littorina saxatilis* (Olivi) and *L. arcana* Hannaford Ellis; a case of character displacement? J. mar. biol. Ass. UK, in press.

Gray, J. E., 1839. The zoology of Captain Beechey's Voyage. Molluscous animals and their shells. London.

Hannaford Ellis, C. J., 1978. *Littorina arcana* sp. nov.: a new species of winkle (Gastropoda: Prosobranchia: Littorinidae). J. Conch. 29: 304.

Hannaford Ellis, C., 1979. Morphology of the oviparous rough winkle, *Littorina arcana* Hannaford Ellis, 1978, with notes on the taxonomy of the *L. saxatilis* species-complex (Prosobranchia: Littorinidae). J. Conch. 30: 43–56.

Hannaford Ellis, C., 1983. Patterns of reproduction in four *Littorina* species. J. moll. Stud. 49: 98–106.

Heller, J., 1975. The taxonomy of some British *Littorina* species, with notes on their reproduction (Mollusca: Prosobranchia). J. linn. Soc., Zool. 56: 131–151.

James, B. L., 1968. The characters and distribution of the subspecies and varieties of *Littorina saxatilis* (Olivi, 1792) in Britain. Cah. Biol. mar. 9: 143–165.

Janson, K., 1985. A morphologic and genetic analysis of *Littorina saxatilis* (Prosobranchia) from Venice, and on the problem of *saxatilis-rudis* nomenlature. Biol. J. Linn. Soc. 24: 51–59.

Janson, K. & R. D. Ward, 1985. The taxonomic status of *Littorina tenebrosa* Montagu as assessed by morphological and genetic analyses. J. Conch. 32: 9–15.

Knight, A. & R. D. Ward, 1986. Purine nucleoside phosphorylase polymorphism in the genus *Littorina* (Prosobranchia: Mollusca). Biochem. Genet. 24: 405–413.

Maton, W. G., 1797. Observations relative chiefly to the natural history, picturesque scenery, and antiquities of the western counties of England, made in the years 1794 and 1796. Easton, Salisbury.

Mill, P. J. & J. Grahame, 1988. Esterase variability in the gastropods *Littorina saxatilis* (Olivi) and *L. arcana* Ellis. J. moll. Stud. 54: 347–353.

Montagu, G., 1803. Testacea Britannica; or, natural history of British shells, marine, land and fresh-water, including the most minute: systematically arranged. White, London.

Olivi, A. G., 1792. Zoologia Adriatica. Bassano.

Raffaelli, D., 1979. The taxonomy of the *Littorina saxatilis* species-complex, with particular reference to the systematic status of *Littorina patula* Jeffrys. J. linn. Soc., Zool. 65: 219–232.

Sacchi, C. F., 1975. *Littorina nigrolineata* (Gray), (Gastropoda, Prosobranchia). Cah. Biol. mar. 16: 111–120.

*Hydrobiologia* **193**: 29–40, 1990.
*K. Johannesson, D. G. Raffaelli and C. J. Hannaford Ellis (eds), Progress in Littorinid and Muricid Biology.*
© 1990 *Kluwer Academic Publishers.*

# Estimating the phylogeny in mollusc *Littorina saxatilis* (Olivi) from enzyme data: methodological considerations

Per Sundberg [1], Andrew J. Knight [2], Robert D. Ward [2] & Kerstin Johannesson [3]
[1] *University of Göteborg, Department of Zoology, P.O. Box 250 59, S-400 31 Göteborg, Sweden;*
[2] *Loughborough University of Technology, Department of Human Sciences, Loughborough, Leicestershire LE11 3TU, England, UK;* [3] *Tjärnö Marine Biological Laboratory, Pl. 2781, S-452 00 Strömstad, Sweden*

*Key words:* phylogeny, electrophoresis, genetic distance, *Littorina*

## Abstract

The evolutionary history of 19 populations of *Littorina saxatilis* (Olivi) was estimated by four different approaches. Three of these operate upon a population by population matrix of genetic distances: average linkage clustering, and two versions of the Fitch-Margoliash method. The fourth method was a maximum likelihood estimate based on differences in allele frequencies between populations. The study aims to assess how well each method estimates the phylogeny by including seven populations of the closely related species *L. arcana* Hannaford Ellis. The rationale behind this is that a good estimation technique should be able to separate these two monophyletic taxa.

The results show that, by our criteria, the maximum likelihood method yields the best estimate and the unconstrained Fitch-Margoliash technique gives reasonable estimates. Both average-linkage clustering and the Fitch-Margoliash method with evolutionary clock perform less well. We argue that this is expected since both these techniques are based on probably unrealistic assumptions such as the overall rate of evolutionary divergence being homogeneous over phyletic lines.

## Introduction

An increasing number of studies use molecular data to construct evolutionary trees and to solve various other biological questions. So far, such information is most readily provided by determination of gene frequencies, and enzyme electrophoresis has become an important and widely applied tool in biology. The technique has proven invaluable in taxonomy for problems such as distinguishing between sibling species and assessing the conspecific status of sympatric populations.

The genetic differentiation between populations or species, often estimated and summarized in the form of a genetic distance, can also form the basis for a reconstruction of the evolutionary history of populations and species.

It is clear that in a general and rather imprecise way genetic 'distance' does throw light on population history. If a set of populations cluster together because they have similar genetic constitutions, it is plausible to infer that they at least have some part of their ancestry in common. A model, however, is needed for the history and for the evolutionary behaviour of the underlying data for a vague statement like this to be made more precise. One such assumption is that the genetic differences between populations are due to some stochastic factor such as genetic drift. Given this supposition, a genetic distance can be constructed that will reflect the length of time since two populations diverged and can therefore work as the basis for phylogeny estimation.

Another approach is to consider the allozymes as discrete characters, either each locus as one character, or each allele as a character (Buth, 1984). A qualitative approach in the construction of phylogenetic trees has several advantages (e.g. Patton & Avise, 1983), but there are obvious problems with it when it comes to intra-specific comparisons where allelic differences between populations are quantitative rather than qualitative. That is, when the differences are in gene frequencies and not in the presence of different loci and alleles. Other objections to this approach are discussed in Swofford & Berlocher (1987).

From the literature, the analysis of electrophoretic data seems to proceed in four, rather uniform, steps: (1) interpretation of the electromorphs as genotypes, (2) computation of allele frequencies at various loci, (3) conversion of allele frequency data into a measure of genetic distance between populations, and (4) a cluster analysis of these populations on basis of the distances. The fourth step is employed to summarize the distances in a two-dimensional plane, but often also as a way of estimating the phylogeny. This paper relates to this fourth step, and sets out to assess four ways of estimating phylogeny on basis of allozyme data, including the commonly applied method of phenetic clustering. Fourteen European populations, three North American, and two South African populations of the rough periwinkle *Littorina saxatilis* (Olivi) were electrophoretically screened. In addition, eight populations of the closely related (Ward & Warwick, 1980; Ward & Janson, 1985) *L. arcana* Hannaford-Ellis were screened for the same enzymes. These populations were included as a way of testing the performance of the estimation techniques, as will be described below. *Littorina saxa-*

*tilis* is an intertidal gastropod essentially restricted to hard substrates, and saltmarshes, of the North Atlantic. There are, however, also several populations in the Mediterranean (Reid, pers. comm.; Janson, 1985), two reported from South Africa (Hughes, 1979), and some on the north west Atlantic coast of Africa (Johannesson, 1988). Several of the populations outside the main geographical distribution of this species are probably introduced by man, and founder effects in, for example, the Venice and South African populations support this view (Janson, 1985; Knight *et al.*, 1987).

## Material and methods

### Populations and enzymes

Nineteen populations of *Littorina saxatilis* were sampled: two from South Africa, 12 from the United Kingdom (including the Channel Islands and Northern Ireland), one each from Eire and Italy, two from the USA, and one from Canada. Eight populations of *L. arcana*, all from the UK, were assayed. The sampling localities are listed in Table 1. The gene frequencies for fifteen of the *L. saxatilis* and six of the *L. arcana* populations were obtained from Knight *et al.* (1987).

Sixteen enzyme loci were screened (Table 2), of which three (Mdh-1, Mdh-2, and Idh-1) were fixed identically in all populations.

### Genetic distance

Several measures of genetic distance between populations have been proposed. Two coefficients have for various reasons come to predominate in the systematic literature: Nei's distance (Nei, 1972; and others), and Rogers's (1972). Rogers's similarity coefficient has no biological interpretation, it is merely the mean geometric distance between allele frequencies. Nei's coefficient, on the other hand, aims to measure the biological characteristic of average number of (electrophoretically detectable) substi-

*Table 1.* Localities of the sampling sites.

| Site no. | Country | Location |
|---|---|---|
| *Littorina saxatilis* | | |
| 1 | South Africa | Langebaan lagoon, W. coast, Cape Province |
| 2 | South Africa | Knysna Lagoon, S. coast, Cape Province |
| 3 | Scotland | North Berwick harbour wall, E. Lothian |
| 4 | Scotland | Logan road, Isle of May, Fife |
| 5 | England | Rock shelf, Swanage, Dorset |
| 6 | England | Rock spit, Bude, Cornwall |
| 7 | England | Rock shelf, Watchet, Somerset |
| 8 | Channel Isles | Cliff face, Nez des Pas, Jersey |
| 9 | Wales | Under pebbles, Bangor, Menai Straits, Gwynedd |
| 10 | N. Ireland | Grand Causeway, Giant's Causeway, Derry |
| 11 | Eire | Limestone pavement, Doolin, Clare |
| 12 | Italy | M'dell Orte Canal wall, Venice |
| 13 | United States | Unknown location in state of Maine |
| 14 | United States | Breakwater, Stonningen, New Hampshire |
| 15 | Canada | Rocks, Churchill, Hudson's Bay, Manitoba |
| 16 | England | Harbour wall, Boscastle, Cornwall |
| 17 | England | Boulders, Robin Hood's Bay, N. Yorkshire |
| 18 | England | Pebbles, N. of Weston, Avon |
| 19 | England | Seawall, Portsmouth, Hampshire |
| *Littorina arcana* | | |
| I | Scotland | Roxborough Hotel seawall, Dunbar, E. Lothian |
| II | Scotland | Cliff face, Milsey Bay, E. Lothian |
| III | Scotland | North Berwick harbour wall, E. Lothian |
| IV | England | Rock face, Hartland Quay, Devon |
| V | England | Bay Hotel wall, Robin Hood's Bay, Yorkshire |
| VI | England | Rockface, Whitby, N. Yorkshire |
| VII | Wales | Rock face, Port Maria, Anglesey |
| VIII | Channel Isles | Slipway, Le petit Etacquerel, Jersey |

tutions per locus that have accumulated since the two populations diverged from their common ancestor, if 'the rate of gene substitution per locus is the same for all loci' (Nei, 1972: 283).

Nei's genetic distance based on allelic differences is expected to underestimate the genetic distance when closely related organisms are compared (Nei, 1978). It is, however, still assumed to be proportional to the real number of substitutions, unless the distance is large (*op. cit.*). One problem, though, with Nei's distance is that it is not metric and negative branch lengths may thus appear leading to difficulties in interpreting the tree. Farris (1981) has argued that a distance that is non-metric cannot be strictly clock-like and the straight correlation with antiquity of common

ancestry must be invalidated in many cases. He concludes that genetic distances based on electrophoretic data cannot be used to estimate phylogenies, but this view has been questioned, most recently by Swofford & Berlocher (1987). Felsenstein (1984) justified the use of distances by viewing it as a statistical problem. The calculated distance is an estimate, in the statistical sense, of the true distance and an estimate may well be wrong. Hence we may obtain negative branch lengths due to sampling problems and it does not jeopardize the approach as such. The reader is referred to Farris (1981, 1985) and Felsenstein (1984, 1986) and references therein for a discussion of this topic.

*Phylogeny estimates*

There are basically two kinds of approaches for estimating the evolutionary history of a group on the basis of protein-electrophoretic information, quantitative and qualitative. In the latter case, each locus, or allele, is treated as a separate character (Buth, 1984 and references in that paper) and the analysis moves forward like an ordinary cladistic analysis (e.g. Wiley, 1981; Eldredge & Cracraft, 1980). This approach has several advantages; the analysis is more heuristic in a sense and it provides the researcher with a better apprehension of the data. Another benefit is that additional characters can readily be added to the tree without recalculating an entire distance matrix. A qualitative Hennigian analysis deals with changes in character states, and will therefore explicitly reveal how these states have changed along branches. Furthermore, monophyletic groups, and hence genealogical relationships, are found by patterns in synapomorphies which are not equally self-evident in estimates based on distance matrices.

The advantages are counterbalanced by some weaknesses, primarily of an empirical and practical nature (discussed in Patton & Avise, 1983), and also of a theoretical nature (Swofford & Berlocher, 1987). For intra-specific population differences, as in this study, the qualitative

*Table 2.* Allele frequencies of 19 populations of *Littorina saxatilis* (1–19) and 8 populations of *L. arcana* (I–VIII).

| Locus | | 1 | 2 | 3 | 4 | 5 | 6 | 7 | 8 | 9 | 10 | 11 | 12 | 13 | 14 |
|---|---|---|---|---|---|---|---|---|---|---|---|---|---|---|---|
| *Pgi* | 140 | – | – | – | – | – | – | – | 0.02 | – | – | – | – | – | – |
| | 110 | 0.06 | – | – | – | – | – | – | – | – | – | 0.03 | – | – | – |
| | 100 | 0.56 | 0.29 | 0.83 | 0.80 | 0.64 | 0.68 | 0.67 | 0.70 | 0.65 | 0.71 | 0.74 | 1.00 | 0.85 | 0.80 |
| | 90 | 0.39 | 0.71 | 0.17 | 0.20 | 0.36 | 0.32 | 0.33 | 0.28 | 0.35 | 0.29 | 0.24 | – | 0.15 | 0.20 |
| *Mpi* | 120 | 1.00 | 1.00 | 0.77 | 0.65 | 0.58 | 0.69 | 0.42 | 0.57 | 0.62 | 0.80 | 0.61 | 0.61 | 0.61 | 0.89 |
| | 100 | – | – | 0.23 | 0.35 | 0.42 | 0.31 | 0.52 | 0.38 | 0.36 | 0.13 | 0.39 | 0.39 | 0.39 | 0.11 |
| | 75 | – | – | – | – | – | | 0.06 | 0.05 | 0.02 | 0.07 | – | – | – | – |
| *Aat-1* | 120 | 1.00 | 0.03 | 0.16 | 0.47 | 0.54 | 0.17 | – | 0.36 | – | 1.00 | 0.50 | – | 0.65 | 0.11 |
| | 100 | – | 0.97 | 0.84 | 0.53 | 0.46 | 0.83 | 1.00 | 0.64 | 1.00 | – | 0.50 | 1.00 | 0.35 | 0.89 |
| | 60 | – | – | – | – | – | – | – | – | – | – | – | – | – | – |
| *Pgm-1* | 105 | 0.99 | 0.78 | 0.16 | 0.08 | 0.02 | 0.19 | – | 0.10 | 0.04 | 0.07 | 0.18 | 0.22 | – | – |
| | 100 | – | 0.22 | 0.50 | 0.68 | 0.67 | 0.60 | 0.88 | 0.62 | 0.78 | 0.57 | 0.63 | 0.03 | 0.77 | 0.76 |
| | 85 | – | – | 0.34 | 0.24 | 0.31 | 0.21 | 0.12 | 0.29 | 0.18 | 0.36 | 0.19 | 0.75 | 0.23 | 0.24 |
| | 75 | 0.01 | – | – | – | – | – | – | – | – | – | – | – | – | – |
| *Pgm-2* | 100 | 1.00 | 1.00 | 0.33 | 0.50 | 0.56 | 0.41 | 0.52 | 0.38 | 0.26 | 0.47 | 0.74 | 0.78 | 0.73 | – |
| | 85 | – | – | 0.37 | 0.48 | 0.44 | 0.42 | 0.39 | 0.59 | 0.20 | 0.47 | 0.24 | – | 0.08 | 0.11 |
| | 70 | – | – | 0.30 | 0.02 | – | 0.17 | 0.10 | 0.03 | 0.54 | 0.06 | 0.02 | 0.22 | 0.19 | 0.77 |
| | 20 | – | – | – | | – | – | – | | | – | – | – | – | 0.12 |
| *Odh* | 110 | – | – | – | – | – | – | 0.02 | – | – | – | – | – | – | 0.02 |
| | 105 | – | – | 0.98 | 0.90 | 0.94 | 0.94 | 0.94 | 0.80 | 0.98 | 0.93 | 1.00 | 0.72 | 1.00 | 0.98 |
| | 100 | 1.00 | 1.00 | 0.02 | 0.10 | 0.06 | 0.05 | 0.04 | 0.20 | 0.02 | 0.05 | – | 0.28 | – | – |
| | 90 | – | – | – | – | – | 0.01 | – | – | – | – | – | – | – | – |
| | 80 | – | – | – | – | – | – | – | – | – | 0.02 | – | – | – | – |
| *Hdh* | 180 | – | – | – | – | – | – | – | – | – | – | – | – | – | – |
| | 145 | – | – | – | – | – | – | – | – | – | 0.04 | 0.03 | – | – | – |
| | 100 | 1.00 | 1.00 | 1.00 | 1.00 | 1.00 | 0.99 | 0.98 | 1.00 | 1.00 | 0.96 | 0.97 | 1.00 | 1.00 | 1.00 |
| | 60 | – | – | – | – | – | 0.01 | 0.02 | – | – | – | – | – | – | – |
| *Np* | 140 | – | – | – | – | – | 0.03 | 0.22 | – | 0.02 | 0.05 | 0.03 | – | – | – |
| | 100 | – | – | 0.41 | 0.40 | 0.35 | 0.40 | 0.33 | 0.15 | 0.39 | 0.30 | 0.32 | 0.31 | 0.75 | 0.94 |
| | 65 | 1.00 | 1.00 | 0.53 | 0.60 | 0.52 | 0.49 | 0.45 | 0.74 | 0.20 | 0.55 | 0.58 | 0.69 | 0.15 | –' |
| | 35 | – | – | 0.06 | – | 0.14 | 0.08 | – | 0.11 | 0.39 | 0.10 | 0.08 | – | 0.10 | 0.06 |
| *Aco-1* | 110 | 0.10 | – | 0.19 | 0.10 | – | 0.01 | – | 0.09 | – | 0.02 | – | – | – | – |
| | 100 | 0.90 | 1.00 | 0.81 | 0.90 | 0.96 | 0.90 | 0.92 | 0.84 | 1.00 | 0.93 | 1.00 | 1.00 | 0.94 | 0.96 |
| | 95 | – | – | – | – | 0.04 | 0.06 | 0.08 | 0.06 | – | 0.05 | – | – | – | 0.04 |
| | 75 | – | – | – | – | – | 0.03 | – | – | – | – | – | – | 0.06 | – |
| *Idh-2* | 100 | 1.00 | 1.00 | 0.89 | 0.90 | 1.00 | 0.98 | 1.00 | 1.00 | 1.00 | 1.00 | 1.00 | 1.00 | 1.00 | 1.00 |
| | 85 | – | – | – | – | – | 0.02 | – | – | – | – | – | – | – | – |
| | 75 | – | – | 0.11 | 0.10 | – | – | – | – | – | – | – | – | – | – |
| *Aat-2* | 150 | – | – | – | 0.03 | 0.02 | – | – | – | – | – | – | – | – | – |
| | 100 | 1.00 | 1.00 | 1.00 | 0.97 | 0.98 | 1.00 | 1.00 | 1.00 | 1.00 | 1.00 | 1.00 | 1.00 | 1.00 | 1.00 |
| *Aco-2* | 160 | – | 0.03 | – | – | – | – | – | – | – | – | – | – | – | – |
| | 100 | 1.00 | 0.97 | 1.00 | 1.00 | 1.00 | 1.00 | 1.00 | 1.00 | 1.00 | 1.00 | 1.00 | 1.00 | 1.00 | 1.00 |
| *Est-1* | 105 | 0.02 | – | – | – | – | – | – | – | – | – | – | – | – | – |
| | 100 | 0.98 | 1.00 | 1.00 | 1.00 | 1.00 | 1.00 | 1.00 | 1.00 | 1.00 | 1.00 | 1.00 | 1.00 | 1.00 | 1.00 |

*Table 2.* (continued)

| Locus | | 15 | 16 | 17 | 18 | 19 | I | II | III | IV | V | VI | VII | VIII |
|---|---|---|---|---|---|---|---|---|---|---|---|---|---|---|
| *Pgi* | 140 | – | – | – | – | – | – | – | – | – | – | – | – | – |
| | 110 | – | – | – | – | – | – | – | – | – | – | – | – | – |
| | 100 | 0.74 | 0.83 | 0.79 | 0.65 | 0.57 | 0.87 | 0.83 | 0.75 | 0.69 | 0.95 | 0.93 | 0.87 | 0.83 |
| | 90 | 0.26 | 0.17 | 0.21 | 0.35 | 0.43 | 0.13 | 0.17 | 0.25 | 0.31 | 0.05 | 0.07 | 0.13 | 0.17 |
| *Mpi* | 120 | 0.94 | 0.73 | 0.71 | 0.27 | 0.60 | 0.37 | 0.28 | 0.25 | 0.84 | 0.31 | 0.38 | 0.40 | 0.54 |
| | 100 | 0.06 | 0.21 | 0.28 | 0.42 | 0.40 | 0.63 | 0.70 | 0.75 | 0.16 | 0.69 | 0.62 | 0.60 | 0.46 |
| | 75 | – | 0.06 | 0.01 | 0.31 | – | – | 0.02 | – | – | – | | – | – |
| *Aat-1* | 120 | 0.19 | 0.12 | – | – | 0.33 | 0.23 | 0.21 | – | 0.29 | 0.31 | 0.70 | 0.82 | 0.36 |
| | 100 | 0.79 | 0.88 | 0.96 | 1.00 | 0.67 | 0.77 | 0.79 | 1.00 | 0.71 | 0.69 | 0.30 | 0.18 | 0.64 |
| | 60 | 0.02 | – | 0.04 | – | – | – | – | – | – | – | – | – | – |
| *Pgm-1* | 105 | 0.01 | 0.03 | 0.02 | – | 0.02 | 0.14 | 0.22 | 0.36 | 0.07 | 0.20 | 0.39 | 0.08 | 0.46 |
| | 100 | 0.70 | 0.90 | 0.64 | 0.69 | 0.77 | 0.64 | 0.73 | 0.47 | 0.86 | 0.39 | 0.39 | 0.70 | 0.54 |
| | 85 | 0.29 | 0.07 | 0.34 | 0.31 | 0.21 | 0.21 | 0.05 | 0.17 | 0.07 | 0.41 | 0.22 | 0.22 | – |
| | 75 | – | – | – | – | – | – | – | – | – | – | | – | – |
| *Pgm-2* | 100 | 0.52 | 0.52 | 0.47 | 0.77 | 0.52 | 0.74 | 0.71 | 0.70 | 0.69 | 0.99 | 1.00 | 0.47 | 0.73 |
| | 85 | 0.28 | 0.38 | 0.27 | 0.17 | 0.35 | 0.21 | 0.26 | 0.20 | 0.31 | – | – | 0.53 | 0.27 |
| | 70 | 0.19 | 0.10 | 0.26 | 0.06 | 0.13 | 0.05 | 0.03 | 0.10 | – | 0.01 | – | – | – |
| | 20 | – | – | – | | – | | | – | – | – | – | – | – |
| *Odh* | 110 | 0.04 | – | – | – | – | – | – | – | – | – | – | – | – |
| | 105 | 0.96 | 0.84 | 0.97 | 0.98 | 0.96 | 0.25 | 0.26 | 0.27 | 0.91 | 0.78 | 0.84 | 0.98 | 0.93 |
| | 100 | – | 0.16 | 0.03 | 0.02 | 0.04 | 0.75 | 0.74 | 0.73 | 0.09 | 0.22 | 0.16 | 0.02 | 0.07 |
| | 90 | – | – | – | – | – | – | – | – | – | – | – | – | – |
| | 80 | – | – | – | – | – | – | – | – | – | – | – | – | – |
| *Hdh* | 180 | – | – | – | – | – | 0.10 | 0.12 | 0.30 | – | 0.01 | – | – | – |
| | 145 | – | 0.05 | – | – | – | 0.15 | 0.17 | 0.20 | 0.17 | 0.73 | 0.25 | 0.15 | – |
| | 100 | 1.00 | 0.95 | 1.00 | 1.00 | 1.00 | 0.75 | 0.71 | 0.50 | 0.83 | 0.25 | 0.65 | 0.85 | 0.92 |
| | 60 | – | – | – | – | – | – | – | – | – | 0.01 | 0.10 | – | 0.08 |
| *Np* | 140 | 0.01 | 0.05 | – | – | – | 0.06 | 0.03 | 0.06 | – | – | – | 0.07 | 0.09 |
| | 100 | 0.65 | 0.30 | 0.77 | 0.19 | 0.27 | 0.94 | 0.96 | 0.94 | 1.00 | 0.96 | 1.00 | 0.90 | 0.82 |
| | 65 | 0.14 | 0.63 | 0.07 | 0.79 | 0.71 | – | – | – | – | – | – | 0.03 | 0.09 |
| | 35 | 0.20 | 0.02 | 0.16 | 0.02 | 0.02 | – | 0.01 | – | – | 0.04 | – | – | – |
| *Aco-1* | 110 | 0.01 | – | 0.09 | – | – | – | – | – | – | 0.11 | 0.02 | – | – |
| | 100 | 0.93 | 1.00 | 0.88 | 1.00 | 0.97 | 1.00 | 1.00 | 1.00 | 0.93 | 0.89 | 0.86 | 1.00 | 1.00 |
| | 95 | 0.06 | – | 0.03 | – | 0.03 | – | – | – | 0.07 | – | 0.12 | – | – |
| | 75 | – | – | – | – | – | – | – | – | – | – | – | – | – |
| *Idh-2* | 100 | 1.00 | 0.98 | 0.93 | 1.00 | 1.00 | 0.98 | 0.89 | 0.98 | 1.00 | 1.00 | 1.00 | 1.00 | 1.00 |
| | 85 | – | 0.02 | – | – | – | – | – | – | – | – | – | – | – |
| | 75 | – | – | 0.07 | – | – | 0.02 | 0.11 | 0.02 | – | – | – | – | – |
| *Aat-2* | 150 | – | – | – | – | – | – | – | 0.06 | 0.05 | – | – | – | – |
| | 100 | 1.00 | 1.00 | 1.00 | 1.00 | 1.00 | 1.00 | 1.00 | 0.94 | 0.95 | 1.00 | 1.00 | 1.00 | 1.00 |
| *Aco-2* | 160 | – | – | – | – | – | – | – | – | – | – | – | – | – |
| | 100 | 1.00 | 1.00 | 1.00 | 1.00 | 1.00 | 1.00 | 1.00 | 1.00 | 1.00 | 1.00 | 1.00 | 1.00 | 1.00 |
| *Est-1* | 105 | 0.09 | – | – | – | – | – | – | – | – | – | – | – | – |
| | 100 | 0.91 | 1.00 | 1.00 | 1.00 | 1.00 | 1.00 | 1.00 | 1.00 | 1.00 | 1.00 | 1.00 | 1.00 | 1.00 |

Three, *Mdh-1*, *Mdh-2*, and *Idh-1* are fixed identically in all populations.

approach is not likely to estimate the phylogeny since the populations differ essentially in allele frequencies and not by the presences and absences of particular alleles. The recorded allele frequencies thus become so ambiguous that they provide no useful evidence on phylogenetic relationships. It is furthermore questionable if one can interpret, say, more than 5% of a particular allele as the apomorphic state and less than 5% as the plesiomorphic. We carried out two analyses, first where the loci were treated as characters, and secondly, where the presence and absence pattern of alleles was treated as a character. Both analyses (using PAUP, D. Swofford, USA) led to large numbers of equally parsimonious trees and it was apparent that there were no synapomorphy patterns in the data matrix.

The phylogeny estimates in this study are therefore based on quantitative approaches that utilize differences and variation in allele frequencies. We have compared four different techniques, discussed below. Three are based on the interpopulation distances: the average-linkage clustering method (UPGMA), and the two versions of the Fitch & Margoliash (1967) method of which one version (FITCH in the PHYLIP-package, see below) has the branch lengths unconstrained and the other (KITSCH) assumes that the true branch lengths from the root of the tree to each tip are the same, i.e. the expected amount of evolution in any lineage is proportional to elapsed time. The fourth method is a maximum-likelihood estimation based on the direct count of allele frequencies (program CONTML in the PHYLIP-package).

There are two types of errors involved in a reconstructed tree: (1) topological errors and (2) errors in the branch length estimates. These two types of errors are intricately related, since the topology and branch lengths are estimated simultaneously. In the preferred tree, these two errors should be minimized, i.e. the constructed tree should be as close to the 'true' tree as possible. Since we do not know the phylogeny in this case, we need to use some other criterion for determining the validity of the estimate. We have included the closely related species *L. arcana* for

this purpose, with the rationale that a 'good' method should be able to separate the two species. Likewise, including the two South African populations together with the Venice population will act as a test of the examined approaches as these populations exhibit founder effects and deviate from the rest (Janson, 1985; Knight *et al.*, 1987).

### Unweighted Pair Group Arithmetic Average Clustering (UPGMA)

This method by Sokal & Michener (1958) is probably the most common approach for summarizing the distance matrix in two dimensions, and at the same time reaching a phylogeny estimate. The procedure has mainly been used for constructing phenograms in numerical taxonomy, but it can also be used for estimating phylogenies under certain assumptions (Nei, 1987). As such, however, it has been debated and criticized with one group (e.g. Farris, 1981) advocating that it is almost useless, and another (e.g. Nei, 1987 and references therein) arguing in favour of it. The technique will estimate the true phylogeny if the amount of protein differentiation reflects the length of time since their evolutionary divergence. The expected value of the distance measure should thus be proportional to evolutionary time. The algorithm for this clustering technique is described in detail in Sneath & Sokal (1973).

### Fitch and Margoliash's method: with and without evolutionary clock

Contrary to UPGMA, the method of Fitch and Margoliash allows for different evolutionary rates along lineages (Fitch & Margoliash, 1967). The method fits the distances in the matrix to the tree in such a way that the following sum of squares (SSQ) is minimized:

$$SSQ = \sum_i \sum_j n(D - d)^2/D^2$$

where $D$ is the input distance between populations $i$ and $j$, $d$ the resulting patristic distance as calculated from the tree, and $n$ the number of comparisons for populations pair $i$ and $j$.

There are two possible modes for the approach where one (program FITCH) fits a tree where the branch lengths are unconstrained in the sense that they may be of different lengths, and another (program KITSCH) which by contrast assumes that an 'evolutionary clock' is valid. That is, the true branch lengths from the root of the tree to each tip are the same since the expected amount of evolution in any lineage is proportional to elapsed time.

### Maximum likelihood method of Felsenstein

Felsenstein (1973, 1981) has devised a restricted maximum likelihood algorithm for estimating phylogenies on basis of allele frequencies. It is thus not based on some measure of genetic distance between species or populations, but instead assumes that population divergence in enzyme alleles is due to stochastic factors that are comparable to Brownian motions. The method is described in detail in Felsenstein (1981), see also Nei (1987).

### Computations

Nei's genetic distance was calculated by a program written by one of us (K.J.). The phylogeny estimates were obtained by using the three programs KITSCH, FITCH, and CONTML (for the maximum likelihood estimate) in J. Felsenstein's program package PHYLIP version 3.0 (J. Felsenstein, Department of Genetics, University of Washington, Seattle, USA). The UPGMA cluster analysis was carried out using a program described in Nei et al., 1985).

### Results

The application of the four methods to the distance matrix computed from the allele frequencies in Table 2 led to the evolutionary trees in Figs. 1A–1D. Two of the four trees (from the cluster analysis and Fitch and Margoliash's method with evolutionary clock) are rooted. This is because of the assumption of homogeneous rates of evolutionary divergence along the branches in the latter case, and part of the algorithm in the first. The maximum likelihood approach gives the best estimate, followed by the unconstrained Fitch and Margoliash's method, as judged from how well they manage to separate the two species. Phenetic clustering performs poorly in this respect, blending the populations from both species rather freely.

It is especially illuminating to examine how the different approaches place the two South African populations. Phenetic clustering (Fig. 1A) and the constrained Fitch-Margoliash method (Fig. 1B) both fail to recognize them as deviant populations, instead they place them in the root and thus suggest that the two South African populations are sister-populations to those remaining. Both the unconstrained Fitch-Margoliash method (Fig. 1C) and Felsenstein's approach (Fig. 1D), on the other hand, manage to recognize them correctly as highly differentiated populations, although still belonging to the same *L. saxatilis* gene-pool. Felsenstein's approach is the only one that clusters all the *L. arcana* populations separated from the *L. saxatilis* populations. Another interesting feature of these two approaches is that they produce trees similar in many respects, despite originating from completely different algorithms and with different underlying bases; e.g. in placing populations (Table 1) 9, 14, 15, and 17 together, with another cluster in common being 7, 16, and 18.

### Discussion

Distance data are widely used in evolutionary biology and systematics to infer degrees of relatedness among taxa. Subject to certain reservations (e.g. sampling errors and possible effects of natural selection), individuals from a single sympatric interbreeding population may be ex-

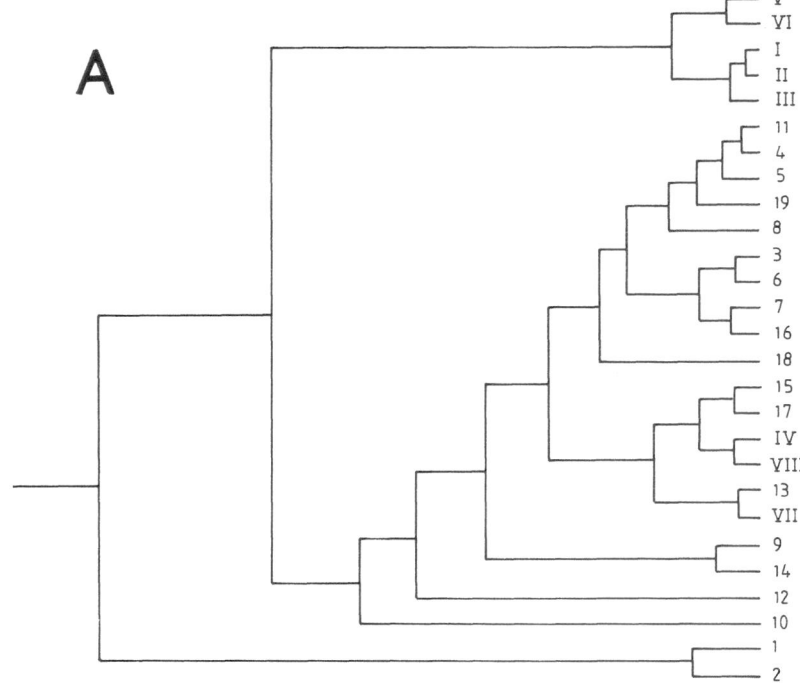

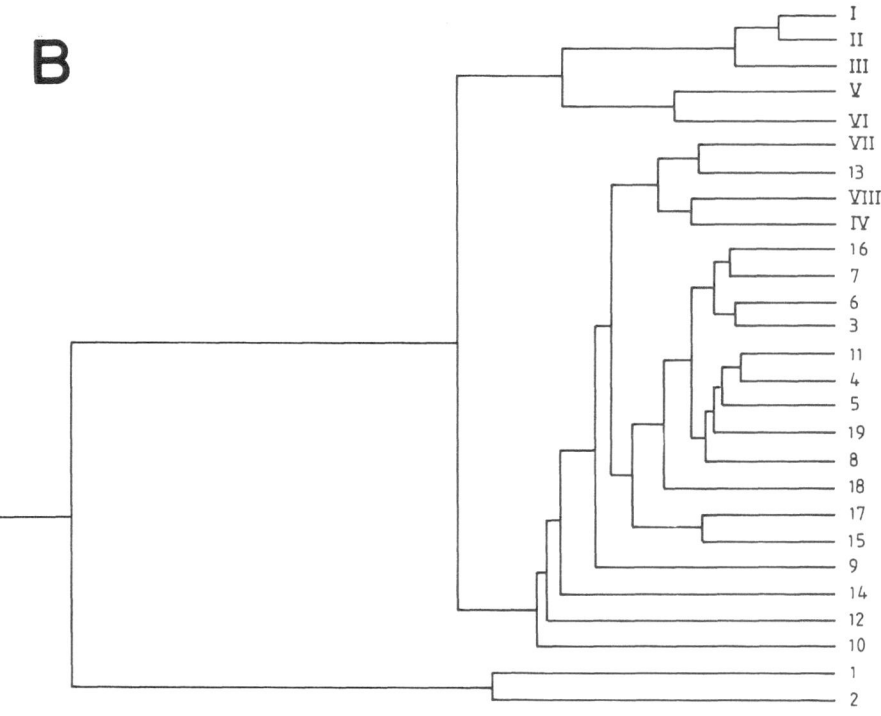

pected to have the same allelic frequencies at each locus. Variation at any locus between sympatric populations will then indicate a restriction to gene flow, and at least partial reproductive isolation. If the measured genetic distance is proportional to the time of divergence from a common ancestor, such a measure will suggest something about the evolutionary history of the populations. Even if it is questionable how well the time of divergence is correlated to the genetic distance (see e.g. Thorpe,

1982 and references therein), most available evidence suggests a good correlation between molecular differences and time of separation (Farris, 1981). It therefore seems reasonable to assume that distance data will bear information that can be the basis for estimating the evolutionary history of a group of populations.

Such an estimate, however, will not be better than the method used; the best data are useless unless properly analyzed. Phenetic clustering has

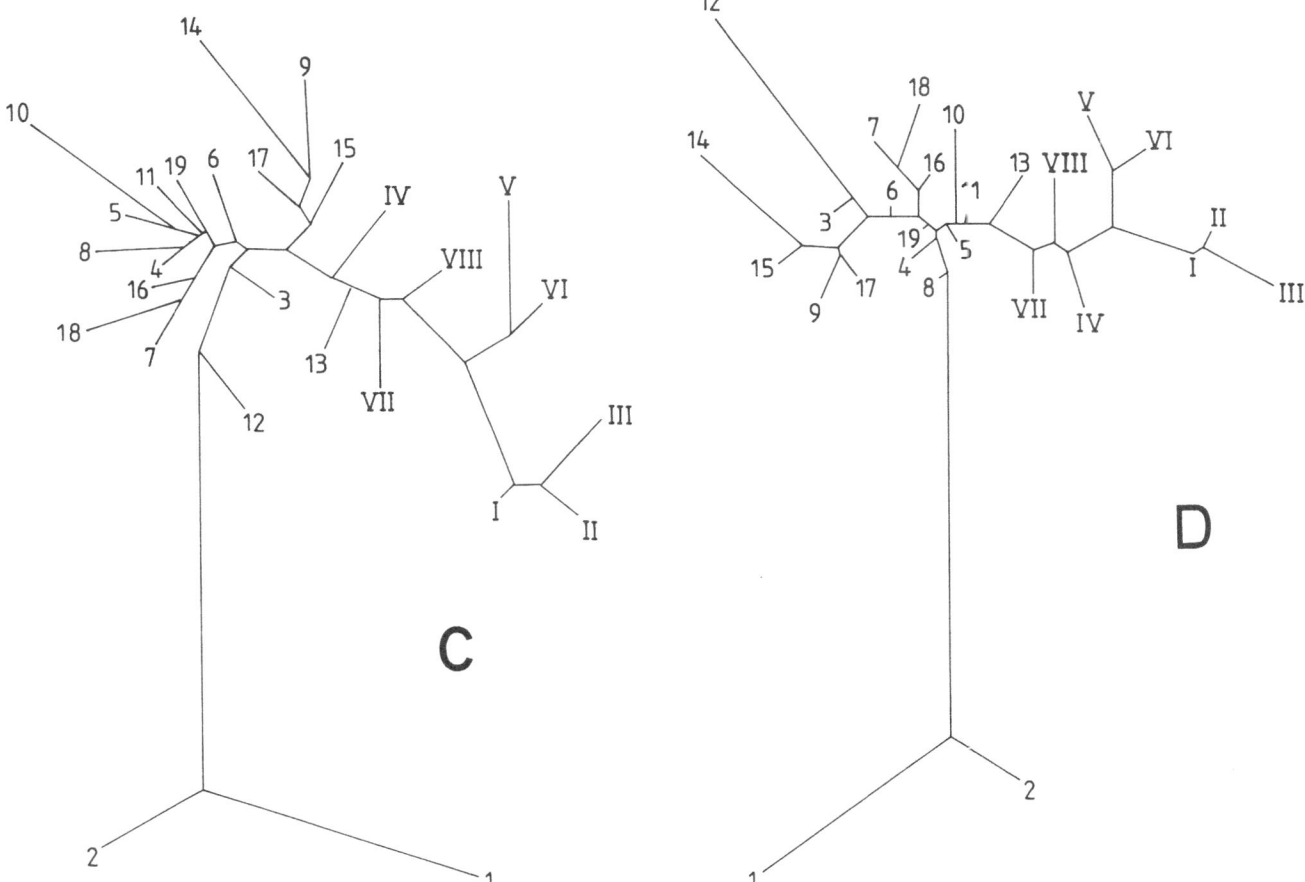

*Fig. 1.* The estimated evolutionary trees for the nineteen populations of *Littorina saxatilis* (Arabic numbers 1–19), together with eight populations of the closely related *L. arcana* (Roman numbers I–VIII). The population numbers refer to Table 1. The phylogenies in A. and B. are rooted due to methodological assumptions, while the trees in C. and D. are unrooted.
A. The phylogeny estimated by average-linkage clustering (UPGMA).
B. Estimated by constrained Fitch-Margoliash method, assuming that the time of divergence is the same from the root to all terminal populations.
C. The estimates obtained by Fitch-Margoliash unconstrained method.
D. The maximum likelihood estimate (Felsenstein, 1981).

become a standard procedure for summarizing a distance matrix and the technique is also commonly applied to estimate the evolutionary history of the group of animals studied. The clustering procedure forms a tree by repeatedly uniting the most reciprocally similar pairs of taxa until a complete hierarchy is produced. Unequal rates of evolution can cause incorrect estimates of the tree topology (Colless, 1970). The assumption of rate constancy is probably unrealistic in many biological applications. This impression is supported by our results where neither of the two methods assuming rate constancy lead to a reasonable evolutionary history of the analysed populations. The complete failure of average clustering to separate the two species, and the placement of the deviant South African populations in the root, makes it clear that this method should not be used in this case. The deviant allele frequencies in the South African populations almost certainly do not represent an ancient origin, rather they are far more likely to result from comparatively recent founder events caused by human introduction (Knight *et al.*, 1987). In such cases, the assumption of rate constancy is not valid, and methods utilizing such assumptions cannot produce 'true' phylogenies.

The two varieties of the Fitch and Margoliash method are likewise based on the distance matrix. The constrained version, where it is assumed that all terminal populations are equidistant from the root, gives a much less good result (Fig. 1B) than the unconstrained version (Fig. 1C). In the Fitch-Margoliash method, a first tree modified by a series of trial-and-error rearrangements and the distances between populations are calculated for each new tree. The obtained lengths in the tree-derived distance matrix are compared with the original, with the statistic described above. Preferences among possible rearrangements are based on the fit between these two matrices. The performance of this method depends on the actual algorithm for doing the analysis. The program FITCH looked at 2135 trees in the Fitch-Margoliash unconstrained analysis, and KITSCH went through 6328 trees in the constrained variant. The algorithm, however,

does not guarantee a global maximum in best fit, and some of these trees may in fact be the same trees. The program does not guarantee that all trees are different trees, it will sometimes wander back to a previous one.

The maximum likelihood estimation technique, on the other hand, does not make the detour over a distance measure but operates upon the differences in allele frequencies. The primary assumption is that each locus evolves independently according to a Brownian motion process. Such a process approximates random genetic drift according to Felsenstein (1981); which is somewhat unrealistic since the diffusion rate will differ from site to site. The program CONTML does, however, contain an internal step to correct for that which works for all except extreme gene frequencies (Felsenstein, pers. comm.). In the Brownian motion model, the net change after a certain number of time units is normally distributed with zero mean and variance determined by the time since divergence under the two assumptions: the changes in two lineages after splitting are assumed independent of each other, and the change of a character during two successive intervals of time is also independent. This distribution of character changes can be capitalized on to develop a maximum likelihood technique to obtain both topology and branch lengths (Felsenstein, 1981). The assumptions underlying this model are too simple and unrealistic for many data sets, but it can be argued to be applicable for problems involving intraspecific populations or closely related species (Swofford & Berlocher, 1987), as in this study. The results of our analysis support this view.

Farris (1981) and Mickevich & Mitter (1981, 1983) have criticized the use of frequency information in phylogenetic analyses. Farris' objection lies primarily in this findings that the 'branch-length fitting' methods (e.g. Fitch-Margoliash method) are logically flawed. Farris (1981) demonstrated that the hypothetical ancestors derived by these methods cannot exist in the original allele frequency space. As a result, the estimated branch lengths cannot be interpreted as genetic distances and the sum of lengths between two populations cannot therefore, according to this view, represent genetic distances. There is thus no logical reason why one should try to minimize the difference between these lengths and the distances in the original population by population matrix. An argument against this is that the lengths in the tree should be considered as expected lengths in a statistical sense (Felsenstein, 1984). Farris (1981) reached the conclusion that some different approach needs to be used when analyzing electrophoretic data, and he advocated a direct phylogenetic analysis of the alleles as characters. However, Swofford & Berlocher (1987) have shown there are ways to reach phylogeny estimates from gene frequencies that can be logically defended. They also argue against the use of alleles as characters in a convincing way, and show how easy it is to miss a particular allele just from sampling error.

The differences in assumptions between the tree-making techniques will show in results, and their ability to recover the correct tree will not be the same. The cluster analysis groups populations after degree of similarity in a straightforward way and does not look for alternative configurations. The methods also contrast in their assumptions. Average-linkage clustering estimates the phylogeny when evolutionary rates are constant, whereas the others do not require this assumption. One might therefore expect that phenetic clustering is inferior to the others in more realistic situations, and our study supports this suspicion. Tateno et al. (1982), however, reached a different conclusion on the basis of computer simulated trees, but their trees were simulated to have constant rates so the result is somewhat limited. For those trees where the rate differed along different branches, a modified version of Farris' distance Wagner method (Farris, 1972) did better in recovering the true tree. Tateno et al. (1982) and Nei et al. (1983) claimed, however, that average-linkage clustering almost always estimated the expected branch lengths best although perhaps not the topology. A similar conclusion was reached by Rohlf & Wooten (1988) in a study where they assessed three methods (UPGMA, Wagner parsimony, and Felsenstein's maximum

likelihood method) by comparing how well they estimated a true, simulated, evolutionary tree. They concluded that UPGMA clustering gives the most accurate estimates of the true phylogeny. Their simulations, however, were based on a model with constant evolutionary rates and the superiority of UPGMA is therefore not unexpected. In another simulation study, Kim & Burgman (1988) compared the maximum likelihood approach (using program CONTML) with a maximum parsimony estimation and average-linkage clustering (UPGMA). They found the maximum likelihood technique to perform considerably and consistently better than the two other methods. Our results point in the same direction and we conclude that methods that do not rely on assumptions of constancy are preferable when estimating the phylogeny in *Littorina saxatilis*.

## Acknowledgements

We thank J. Felsenstein for constructive comments on this paper, and for making the PHYLIP programs available to PS. The preparation of this study was supported by the Swedish Natural Science Research Council (grants to PS and KJ), and by Natural Environment Research Council (grant to RDW). AJK and RDW wish to thank T. Warwick, J. Perry, J. Knight, I. Munro, J. Hope, N. Billington, E. Rodino, R. Hughes, and A. Grenade for their invaluable assistance in the collection of samples.

## References

Buth, D. G., 1984. The application of electrophoretic data in systematic studies. Annu. Rev. Ecol. Syst. 15: 501–522.
Colless, D. H., 1970. The phenogram as an estimate of the phylogeny. Syst. Zool. 19: 352–362.
Eldredge, N. & J. Cracraft, 1980. Phylogenetic patterns and the evolutionary process. Columbia University Press, 349 pp.
Farris, S. J., 1972. Estimating phylogenetic trees from distance data. Am. Nat. 106: 645–668.
Farris, J. S., 1981. Distance data in phylogenetic analysis. In V. A. Funk & D. R. Brooks (eds), Advances in cladistics,

Vol. 1, Proceedings 1st Meeting Willi Hennig Society, New York: 3–23.

Farris, S. J., 1985. Distance data revisited. Cladistics 1: 67–85.

Felsenstein, J., 1973. Maximum-likelihood estimation of evolutionary trees from continuous characters. Am. J. hum. Genet. 25: 471–492.

Felsenstein, J., 1981. Evolutionary trees from gene frequencies and quantitative characters: finding maximum likelihood estimates. Evolution 35: 1229–1242.

Felsenstein, J., 1984. Distance methods for inferring phylogenies: a justification. Evolution 38: 16–24.

Felsenstein, J., 1986. Distance methods: a reply to Farris. Cladistics 2: 130–143.

Fitch, W. M. & E. Margoliash, 1967. Construction of phylogenetic trees. Science 155: 279–284.

Hughes, R. N., 1979. South African populations of *Littorina rudis*. J. linn. Soc., Zool. 65: 119–126.

Janson, K., 1985. A morphologic and genetic analysis of *Littorina saxatilis* (Prosobranchia) from Venice, and on the problem of *saxatilis – rudis* nomenclature. Biol. J. linn. Soc. 24: 51–59.

Johannesson, K., 1988. The paradox of Rockall: why is a brooding gastropod (*Littorina saxatilis*) more widespread than one having a planktonic larval dispersal stage (*L. littorea*)? Mar. Biol. 99: 507–513.

Kim, J. & M. A. Burgman, 1988. Accuracy of phylogenetic-estimation methods under unequal evolutionary rates. Evolution 42: 596–602.

Knight, A. J., R. N. Hughes & R. D. Ward, 1987. A striking example of the founder effect in the mollusc *Littorina saxatilis*. Biol. J. linn. Soc. 32: 417–426.

Mickevich, M. F. & C. Mitter, 1981. Treating polymorphic characters in systematics: A phylogenetic treatment of electrophoretic data. In V. A. Funk & D. R. Brooks (eds), Advances in cladistics, Vol. 1, Proceedings 1st Meeting Willi Hennig Society, New York: 45–60.

Mickevich, M. F. & C. Mitter, 1983. Evolutionary patterns in enzyme data: A systematic approach. In V. A. Funk & D. R. Brooks (eds), Advances in cladistics, Vol. 2, Columbia University Press, New York: 169–176.

Nei, M., 1972. Genetic distance between populations. Am. Nat. 106: 282–292.

Nei, M., 1978. The theory of genetic distance and evolution of human races. Japan J. Hum. Genet. 23: 341–369.

Nei, M., 1987. Molecular evolutionary genetics. Columbia University Press, New York, 512 pp.

Nei, M., F. Tajima & Y. Tateno, 1983. Accuracy of estimated phylogenetic trees from molecular data. II. Gene frequency data. J. Mol. Evol. 19: 153–170.

Nei, M., J. C. Stephens & N. Saitou, 1985. Methods for computing the standard errors of branching points in an evolutionary tree and their application to molecular data from humans and apes. Mol. Biol. Evol. 2: 66–85.

Patton, J. C. & J. C. Avise, 1983. An empirical evaluation of qualitative Hennigian analyses of protein electrophoretic data. J. Mol. Evol. 19: 244–254.

Rogers, J. S., 1972. Measures of genetic similarity and genetic distance. Studies Genet. VII. Univ. Texas Publ. 7213: 145–153.

Rohlf, F. J. & M. C. Wooten, 1988. Evaluation of the restricted maximum-likelihood method for estimating phylogenetic trees using simulated allele-frequency data. Evolution 42: 581–595.

Sneath, P. H. A. & R. R. Sokal, 1973. Numerical taxonomy. The principles and practice of numerical taxonomy. W. H. Freeman, San Francisco, 573 pp.

Sokal, R. R. & C. D. Michener, 1958. A statistical method for evaluating systematic relationships. Kans. Univ. Sci. Bull. 38: 1409–1438.

Swofford, D. L. & S. H. Berlocher, 1987. Inferring evolutionary trees from gene frequency data under the principle of maximum parsimony. Syst. Zool. 36: 293–325.

Tateno, Y., M. Nei & F. Tajima, 1982. Accuracy of estimated phylogenetic trees from molecular data. I. Distantly related species. J. Mol. Evol. 18: 387–404.

Thorpe, J. P., 1982. The molecular clock hypothesis: Biochemical evolution, genetic differentiation and systematics. Annu. Rev. Ecol. Syst. 13: 139–168.

Ward, R. D. & K. Janson, 1985. A genetic analysis of sympatric subpopulations of the sibling species *Littorina saxatilis* (Olivi) and *Littorina arcana* Hannaford Ellis. J. moll. Stud. 51: 86–94.

Ward, R. D. & T. Warwick, 1980. Genetic differentiation in the molluscan species *Littorina rudis* and *Littorina arcana* (Prosobranchia: Littorinidae). Biol. J. linn. Soc. 14: 417–428.

Wiley, E. O., 1981. Phylogenetics. The theory and practice of phylogenetic systematics. John Wiley & Sons, New York, 439 pp.

*Hydrobiologia* **193**: 41–52, 1990.
*K. Johannesson, D. G. Raffaelli and C. J. Hannaford Ellis (eds), Progress in Littorinid and Muricid Biology.*
© 1990 *Kluwer Academic Publishers.*

# Are the opposing selection pressures on exposed and protected shores sufficient to maintain genetic differentiation between gastropod populations with high intermigration rates?

Elizabeth Grace Boulding
*Dept. of Zoology NJ-15, University of Washington, Seattle, WA 98195, USA*

*Key words: Littorina*, selection, polygenic trait, genetic differentiation, phenotypic plasticity, simulation model

## Abstract

Throughout the world intertidal gastropods living on exposed rocky shores differ strikingly in a number of morphological and life history traits from those on protected shores. Where surf is heavy gastropods tend to be smaller and to have thinner and smoother shells with larger apertures than do those from sheltered areas where crab predation is more intense. These morphological differences can occur within a species and there is evidence that they can be partially genetic and partially environmental. In addition the convergence of shell features in each habitat suggests that there are consistent differences between the selective pressures on exposed shores and the selective pressures on protected shores. I constructed a simulation model for a polygenic trait that experiences different selective pressures on exposed and sheltered shores. The results show that genetic differences can be maintained between the two populations despite high intermigration rates. Replacement of a portion of the random environmental variance with adaptive environmental variance reduces the effect of selection and thus the size of the difference maintained between the two populations. Genetic differentiation between exposed and protected populations can persist for significant periods of time and may have sometimes been the first step in speciation.

## Introduction

Throughout the world intertidal gastropods living on exposed rocky shores differ strikingly in a number of morphological and life history traits from closely related gastropods living on sheltered shores. Marine prosobranchs from exposed shores where surf is heavy tend to be smaller and to have thinner and smoother shells with larger apertures than prosobranchs from sheltered areas where crab predation is more intense (Table 1). This pattern can not be attributed to common ancestry as it occurs within families of gastropods as distantly related as the periwinkles (Littorinidae) and the dog-whelks (Thaididae). Nor does it seem correlated with reproductive mode (Table 1) which perhaps has been incorrectly equated with dispersal ability (Johannesson, 1988). Instead, the convergence of shell features in each habitat is best explained by postulating selective pressures characteristic of that habitat type.

Dislodgement by waves is thought to be an important selective agent on intertidal gastropods

*Table 1.* Patterns of shell morphology on exposed and protected shores.

| Species complex | Shell morphology | | Evidence for genetic diff. | Reproduction |
|---|---|---|---|---|
| | Exposed | Sheltered | | |
| *Littorina sitkana*[1] (North. Pacific) | small, smooth | large, thick, ridged | rearing & allozymes | benthic eggs[1] hatch as juveniles |
| *L. saxatilis*[2] (North Atlantic) | small, globose | large, thick, elongate | transplants & allozymes | brooded until juveniles emerge[2] |
| *L. picta*[3] (Hawaii) | small, smooth | large, ridged | rearing | planktonic egg capsules[3] planktonic veligers |
| *L. unifasciata*[4] (Australia) | small wide aperture | large | ? | planktonic egg capsules[11] planktonic veligers |
| *Nucella lapillus*[5] (North Atlantic) | short, squat wide aperture | large, thick, elongate | transplants[6] | benthic eggs[5], hatch as juveniles |
| *Nucella emarginata*[7] (Northeast Pacific) | short, squat wide aperture | large | transplants | benthic eggs[7], hatch as juveniles |
| *Dicathais orbita*[8] (Australia) | short, squat | large | ? | benthic eggs[12] planktonic veligers |
| *Purpura collumellaris*[9] (tropical East. Pacific) | short, squat wide aperture | large, thick | ? | benthic eggs?[9] planktonic veligers? |
| *Lepisella albomarginata*[10] (New Zealand) | short, squat | large, thick | ? | benthic eggs[13] |

[1] Boulding *et al.* in prep., Boulding unpubl.
[2] Janson, 1982a, 1982b; Janson & Ward, 1984 (between subpopulations).
[3] Struhsaker, 1968.
[4] Basingthwaighte & Foulds, 1985.
[5] Crothers, 1983.
[6] Etter, 1988.
[7] Holthuis, in prep.
[8] W. F. Ponder unpubl. data cited in Phillips *et al.*, 1973.
[9] Wellington & Kuris, 1983.
[10] Kitching & Lockwood, 1974.
[11] Pilkington, 1971; Underwood, 1974.
[12] Phillips, 1969; Phillips *et al.*, 1973.
[13] Morton & Miller, 1968.

living in the exposed intertidal (Kitching *et al.*, 1966; Struhsaker, 1968; Behrens, 1972; Raffaelli & Hughes, 1978; Crothers, 1983; Etter, 1988). A smaller body size may increase the number of crevice refuges available (Raffaelli & Hughes, 1978) and a larger foot may decrease the probability of dislodgement. Predation by shore crabs (Heller, 1976; Crothers, 1983; Johannesson, 1986) can be an important source of gastropod mortality on protected shores. Possession of a large, thick shell with a small aperture reduces predation by crabs (Elner & Raffaelli, 1980; Johannesson, 1986) and probably also by fish (Boulding, pers. obs.). Thus the heavy wave action on exposed shores appears to select for shell characters opposite to those selected for by the intense predation of protected shores. In addition a thick shell resists crushing by gravel (Johannesson, 1986) and a small aperture increases resistance to desiccation (Atkinson & Newbury, 1984) both of which can be more serious on protected shores. Although shell shape is included in Table 1 it is correlated with growth rate, and other intertidal prosobranchs may show a different pattern (Kemp & Bertness, 1984).

In the Northeastern Pacific, *Littorina* sp., a new subspecies of *Littorina kurila* Middendorff (Boulding *et al.*, in prep.), lives on shores exposed to surf and in other refuges from crab predation. Electrophoretic and morphological data (Boulding *et al.*, in prep.) suggest that *L.* sp. is closely related to *Littorina sitkana* Philippi.

*L. sitkana* lives mainly on sheltered shores or in high, deep tidepools on moderately exposed shores. *L.* sp. has a relatively larger foot, larger shell aperture, and thinner shell and lacks the deep spiral sculpture common in *L. sitkana*.

*Littorina* sp. lives exposed to extreme surf where the risk of dislodgement is high but where the foraging efficiency of crab and fish predators is low. In an accelerating flow tank individual *L.* sp. were dislodged significantly less often than *L. sitkana* of the same size (Boulding & Van Alstyne, in prep.) but were dislodged by flow rates considerably lower than have been measured during storms on exposed beaches (Denny, 1985). Its small size may be an adaptation allowing it to take refuge in *Iridaea cornucopiae* (Postels & Ruprecht) beds, between barnacles or in crevices where the flow rates are reduced, and may also allow it to reproduce before winter storms wash most of the adults away (Boulding & Van Alstyne, in prep.). Indeed, *L.* sp. decreases its growth rate and becomes sexually mature at a smaller size than *L. sitkana*. (Boulding & Van Alstyne, in prep.).

In contrast, *Littorina sitkana* lives in more protected habitats where the shore crabs *Hemigrapsus nudus* (Dana) and *Hemigrapsus oregonensis* (Dana) are abundant. In an area where *H. nudus* was abundant the recovery of transplanted, marked *L.* sp. was only 1.3% after 6 weeks; significantly less than the 14% recovery of *L. sitkana* and the 70% recovery of a third species, *Littorina scutulata* Gould (Boulding & Van Alstyne, in prep.). These crabs are capable of eating large *L. sitkana* but prefer *L.* sp. or small *L. sitkana* (Boulding & Van Alstyne, in prep.), presumably because of their shorter handling time (Boulding, 1984). Thus *L. sitkana* may greatly reduce its probability of mortality to crabs by growing as rapidly as possible at the cost of postponing reproduction.

I investigated whether opposing selective pressures could cause genetic differentiation between exposed and sheltered populations of the immediate ancestor of *Littorina* sp. and *L. sitkana*. Mechanisms that permit the maintenance of genetic differentiation between populations with high rates of intermigration are of interest because such differentiation could be the first step in non-allopatric speciation (Maynard Smith, 1966).

The morphological and life history characters that differ between exposed and protected shores are continuously distributed within a population and are probably affected by many genes. Quantitative genetic models are appropriate for metric traits that have an unknown molecular genetic basis (Falconer, 1981); in quantitative genetic theory the value of a metric trait measured on an organism is assumed to be the result of the influence of environment and of many genes of small effect. Many polygenic traits that strongly affect fitness are best considered in terms of such models (e.g. Bulmer, 1980; Lande, 1982; Turelli, 1987).

This paper describes a model of a quantitative trait that experiences different selection pressures in each of two habitats. The object is to investigate the genetic difference maintained between the two populations as a function of selection intensity, intermigration rate, and the degree to which the trait is heritable. The magnitude of selection and migration measured for intertidal prosobranchs in the field is then compared with the magnitude necessary to produce genetic differentiation in the model.

## Material and methods

### Model assumptions

Consider a quantitative trait in a diploid gastropod that is determined by 8 loci of small effect that combine additively and by a random normal 'environmental' deviate with mean zero and variance $V_e$ (Fig. 1). The relative contribution of the genetic and the environmental components of the phenotype is determined by the parameter heritability (denoted $h^2$) so that $h^2 = V_g/(V_g + V_e)$ where $V_g$ is the variance of genetic values in the previous generation. When $V_e$ was nonzero (i.e. $h^2 < 1.0$) it was kept at a constant value of 4.0. Thus the heritability decreased with time because the additive genetic variance

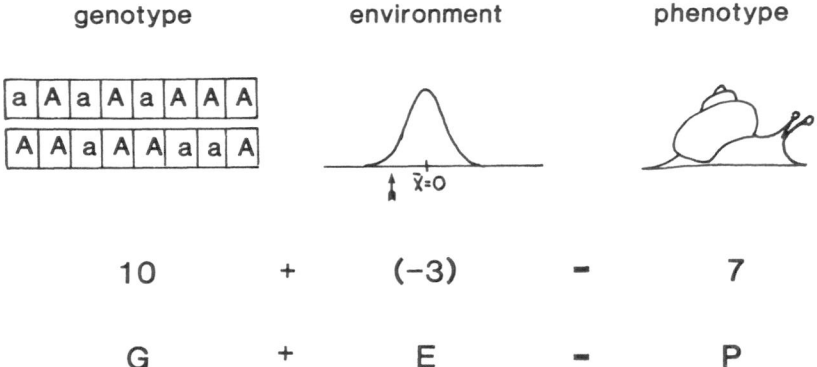

*Fig. 1.* In the model the phenotypic value of the trait results from a genetic and an environmental component. The value of the genetic component is obtained by assuming each A allele contributes 1 and each a allele 0. The environmental component is determined by a random normal deviate with mean zero and variance proportional to the genetic variance. For most simulations the environmental variance was kept constant at 4.0.

decreased under strong selection. At each locus there are two alleles each which are either A or a. Each A contributes 1 to the genetic value of the snail, and each a contributes 0. If there is no dominance or epistasis at any of the 8 loci then the minimum possible genetic value is 0 and the maximum is 16. For any two parents the mean genetic value of their offspring will be equal to the average of their genetic values, but the range their offspring show can not be predicted without knowing the arrangement of alleles in the parental genome (Fig. 2).

*Model parameters*

The parameters included in the model were maximum population size, the optimum phenotypes in each environment, selection intensity, migration rate, environmental variance and the presence or absence of adaptive environmental plasticity. In this paper I will describe the effects of manipulating migration, selection, and environmental variance. I will also show the effect of having some of the environmental variance adaptive instead of random (*i.e.* adaptive phenotypic plasticity). I did

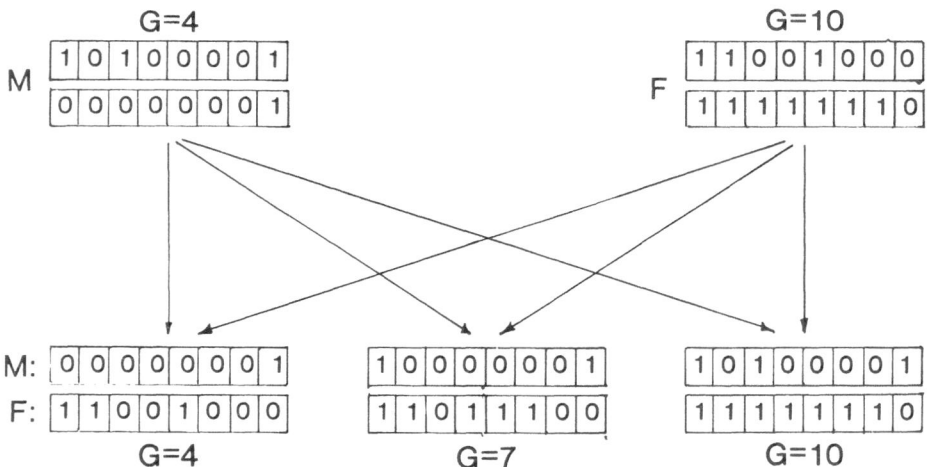

*Fig. 2.* At each locus the offspring receives an allele chosen at random from those at each parental locus. Two parents with genetic values of 4 and 10 will have offspring that have a mean genetic value of 7 but whose range will depend on the arrangement of the parental alleles on the parental chromosomes. The offspring on the extreme left and extreme right represent the highest and lowest genetic values possible for this arrangement of parental alleles.

this by calculating the phenotype of the offspring, then moving it towards the optimum by the amount of the parameter plasticity. For example if the phenotypic value is 3, the optimum 7, and the plasticity = 0.5 then the final phenotypic value will be 3.5.

The model assumes there are two habitats: an exposed habitat in which the optimal phenotype has a value of 7 and a protected habitat in which the optimal phenotype has a value of 9. The positions of these optima were not varied although analytical models suggest that putting the optima further apart would increase the amount of genetic variance maintained (Bulmer, 1971). The initial frequency of A is 0.5 so the initial mean genetic value for all 8 loci averages 8.0 in each population. For the simulations reported here the selection intensities and migration rates are the same in both populations.

*Field estimates of selection and migration*

Field data were obtained from the literature and from field experiments done on Tatoosh Island, Washington State, USA (Boulding & Van Alstyne, in prep.) and transformed so they could be plotted in the units of the model. Where data from reciprocal transplants were available the selection intensity, $C$, was estimated from the equation (the inverse of the fitness function):

$$C = -(\ln W)/(Y - A)^2$$

where $W$ is estimated as the survival of the transplanted phenotype relative to the native phenotype, $Y$ as the mean phenotypic value of the transplanted phenotype in the units of the model and $A$ as the mean phenotypic value of the local population. In calculating $W$ this way I am treating the native phenotype as if it was at the optimum for its habitat. This will be a better approximation when the exposed phenotype and protected phenotype do not interbreed, either because they are separate species or because intermigration rates are very low, than when they do interbreed. In the model the optima of the two habitats are 2 units

apart so for the exposed habitat $Y = 7, A = 9$ and $Y - A = 2$.

The intermigration rates in the field were estimated by comparing the average and maximum net movement of marked snails in their native habitats with the distances between exposed and protected habitats. It was estimated as the proportion of the marked snails from that habitat that moved far enough to move into the other habitat type. Intermigration was considered to be the same in both directions.

*Simulation of a generation*

During each generation (Fig. 3) offspring are produced in each habitat and undergo selection. This continues until a population size of 1000 juveniles is reached. The parents then die. Intermigration of randomly chosen individuals then occurs symmetrically between the two populations. After intermigration, mating occurs; within each habitat two snails are chosen at random to produce one offspring, then are returned to the mating pool. The offspring's genome is constructed by randomly choosing an allele from the two at each parental locus. The offspring's phenotype is constructed from its genetic value as summed over all its loci and from a random environmental component. For a heritability of 1.0 its phenotype will be 100% from the genetic value but most of the runs were done by holding the absolute amount of environmental variance constant at 4.0. The probability of an offspring surviving is determined by first calculating fitness, $w$, in the equation:

$$w = e^{\{-c(y-a)^2\}}$$

where $y$ is the offspring's phenotypic value, $a$ is the optimum for the habitat it is in, and $c$ is the selection intensity constant. Then a random number between zero and one is generated and the offspring survives if the random number is less than the offspring's fitness. When the fitness $w$ is large it has a high probability of being larger than the random number so the offspring has a high probability of surviving. After each generation the difference between the mean genotypic values in the two habitats is calculated and stored.

## The Model
## (exposed population)

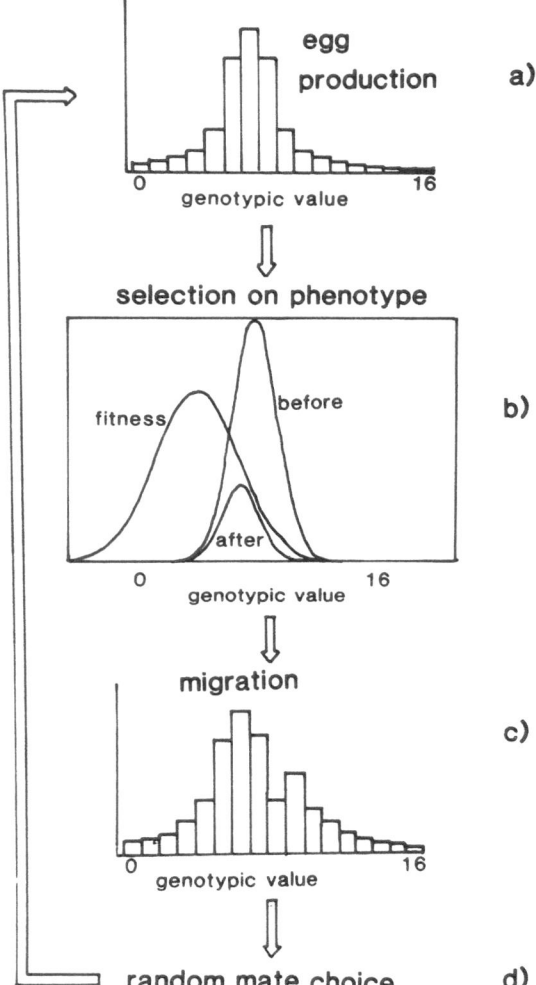

a) egg production

selection on phenotype

b) fitness / before / after

genotypic value

c) migration

genotypic value

d) random mate choice

*Fig. 3.* One generation shown for the exposed population. a) egg production: is shown for the entire exposed population before selection. b) selection on phenotype: note fitness curve in the exposed habitat is to left of population mean. Snails with a high phenotypic value of the trait have a low probability of survival in the exposed habitat but the environmental component of the phenotype dilutes the effect of selection. c) migration: The lower peak represents migrants from protected population with a relatively higher average genetic value for trait. d) In each population mating is random. The offspring receive an allele chosen at random from the two at each locus of the parental chromosomes.

*The program*

The program was written in Pascal with some Assembler subroutines for an IBM clone with 640K of memory. The genome for each snail was stored as an integer and reconstructed with the aid of bit manipulation routines. A driver program runs the desired number of generations of the model for each combination of selection, migration, and heritability that are in the parameter file and saves the average of the difference in the mean genetic values of the two populations for the last three generations. When selection and migration were close to zero, genetic drift caused the simulation results to be substantially different from run to run; this case was not investigated at length as the results for large values of selection and migration were of primary interest.

**Results**

*Simulations*

During a simulation under moderately high selection and migration, there was a correspondence between the allele frequency at a given locus in the two populations. As fixation occurred in the populations it occurred at the same loci (Table 2). The overall results of the simulations show that as selection increases and migration decreases the difference maintained between the mean genetic values of the exposed and protected populations increases (Fig. 4). Increasing $V_e$ from 0 to 4 decreased the ability of selection to maintain differences between the means when migration rates were high; the results for a heritability of 1.0 (*i.e.*, $V_e = 0$, the phenotype entirely genetically determined) maintain larger differences between the means at nonzero migration rates than those for a lower heritability (*i.e.*, $V_e = 4.0$) (Fig. 4). Where selection and migration are both zero, the difference between the means varies from run to run; this is the effect of genetic drift and the difference becomes zero when the results from many runs are averaged. However results from runs with

*Table 2.* Results of 1000 generation iteration of a 8 locus, 2 allele model of a population with 1000 diploid individuals. Values of 0 or 2000 indicate fixation at that locus for the '0' or '1' allele respectively.

| Gen. | Pop. | Number of '1' alleles | | | | | | | | Genotype | |
|---|---|---|---|---|---|---|---|---|---|---|---|
| | | L1 | L2 | L3 | L4 | L5 | L6 | L7 | L8 | Mean | SD |
| 0[1] | Exposed | 1000 | 1000 | 1000 | 1000 | 1000 | 1000 | 1000 | 1000 | 8.00 | |
| 0[1] | Protect. | 1000 | 1000 | 1000 | 1000 | 1000 | 1000 | 1000 | 1000 | 8.00 | |
| 1 | Exposed | 951 | 979 | 965 | 932 | 975 | 1021 | 995 | 948 | 7.77 | 1.68 |
| 1 | Protect. | 1004 | 1041 | 1056 | 1013 | 1026 | 1106 | 1028 | 1060 | 8.37 | 1.67 |
| 51 | Exposed | 702 | 837 | 996 | 845 | 1089 | 1264 | 701 | 918 | 7.35 | 1.56 |
| 51 | Protect. | 847 | 940 | 1269 | 942 | 1265 | 1435 | 877 | 1143 | 8.72 | 1.52 |
| 101 | Exposed | 522 | 573 | 1405 | 528 | 1377 | 1495 | 605 | 1006 | 7.51 | 1.45 |
| 101 | Protect. | 708 | 682 | 1544 | 649 | 1496 | 1544 | 861 | 1159 | 8.64 | 1.47 |
| 201 | Exposed | 142 | 76 | 1900 | 150 | 1984 | 1927 | 87 | 1327 | 7.59 | 0.93 |
| 201 | Protect. | 291 | 122 | 1952 | 256 | 1973 | 1952 | 171 | 1629 | 8.35 | 0.98 |
| 501 | Exposed | 0 | 0 | 1838 | 0 | 2000 | 2000 | 57 | 2000 | 7.89 | 0.45 |
| 501 | Protect. | 0 | 0 | 1925 | 0 | 2000 | 2000 | 206 | 2000 | 8.13 | 0.49 |
| 951 | Exposed | 0 | 0 | 1814 | 0 | 2000 | 2000 | 70 | 2000 | 7.88 | 0.47 |
| 951 | Protect. | 0 | 0 | 1954 | 0 | 2000 | 2000 | 163 | 2000 | 8.12 | 0.45 |

[1] The numbers of '1' alleles will average 1000 at each locus over many runs but will fluctuate within an individual run due to sampling error due to finite population size.

nonzero migration and selection parameters varied little.

Replacing a proportion of the random environmental variance with adaptive environmental variance – which is always in the direction of the optimum for that habitat – reduces the strength of the selection and maintains a smaller difference between the means (Fig. 5). Increasing the population size by an order of magnitude increased the persistence time of the genetic differentiation for given parameter values (Table 3).

*Table 3.* The effect of population size on the difference between the means and on the average genetic variance of the two populations after 50 generations, and the time of persistence of a difference of greater than 0.01 between the means of the two populations.

| Selec. | Migrat. | Pop size | Persistence (generations) | Difference of means (after 50 generations) | Genetic variance (after 50 generations) |
|---|---|---|---|---|---|
| 0.5 | 0.1 | 10 | 37 | 0.00 | 0.00 |
| | | 100 | 300 | 1.49 | 1.19 |
| | | 1000 | 400 | 0.98 | 1.36 |
| 0.1 | 0.1 | 10 | 48 | 0.00 | 1.28 |
| | | 100 | 407 | 1.34 | 1.49 |
| | | 1000 | >1000 | 1.44 | 1.35 |
| 0.3 | 0.1 | 10 | 36 | 0.00 | 0.54 |
| | | 100 | 417 | 1.32 | 1.40 |
| | | 1000 | >1000 | 0.79 | 1.54 |
| 0.1 | 0.2 | 10 | 39 | 0.00 | 1.30 |
| | | 100 | 320 | 0.57 | 1.37 |
| | | 1000 | 551 | 0.95 | 1.37 |

a.

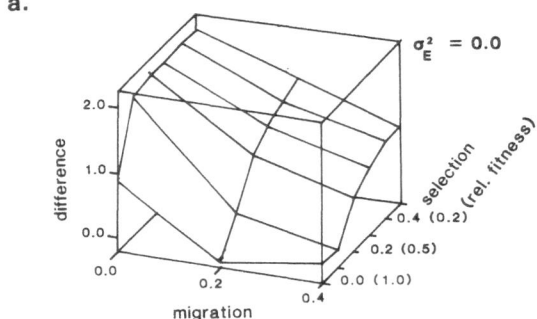

b.

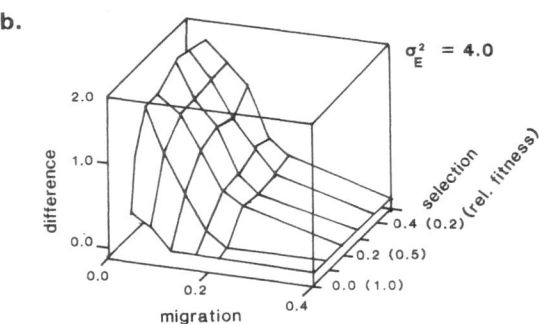

*Fig. 4.* The mean difference between the exposed and protected genetic means as a function of selection intensity and migration rate for a population size of 1000 after 1000 generations. In a) the heritability is 1.0 so the environmental variance is 0 and in b) The environmental variance is held at 4.0; for selection = 0.5 and migration = 0.1 the heritability is about 0.1 after 1000 generations. $\sigma_E^2$ is equal to $V_e$ in the text.

a.

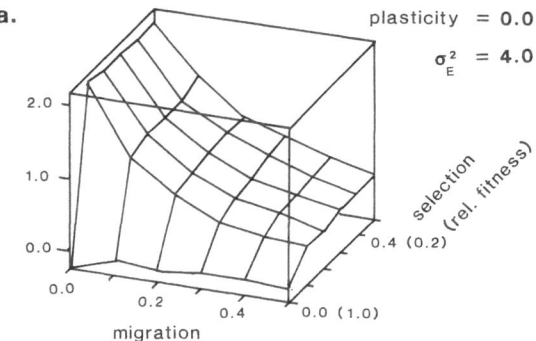

b.

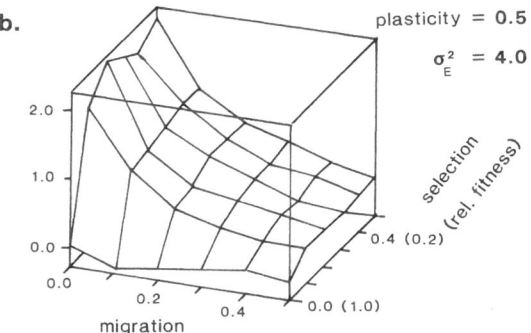

*Fig. 5.* The mean difference between the mean genetic values in the exposed and the protected populations as a function of selection intensity and migration rate after 20 generations. In a) the environmental variance is 4.0 and adaptive plasticity = 0.5. In b) the environmental variance is 4.0 and there is no plasticity. The relative fitness differences corresponding to the selection intensities are also given for reference. $\sigma_E^2$ is equal to $V_e$ in the text.

## Field estimates of selection and migration

Exposed gastropods (*Littorina* sp.) transplanted into a protected area with shore crabs had a survival of 0.09 for 6 weeks in 1986 and 0.13 for 6 weeks in 1987 relative to protected snails (*L. sitkana*) of the same size (Boulding & Van Alstyne, in prep.). This translates into lifetime selection intensities of 0.6 and 0.5 respectively in the units of the model (Table 4). Published estimates of selection intensity on snails transplanted from their native to the opposite habitat range from 0.3 to 0.5 in the units of the model (Table 4).

In one experiment with 300 snails of each type 22% of the snails showed a displacement of more than one meter in the first few days. After 6 weeks two snails were found more than 15 meters away

along the shore. Had this been in the right direction it would have taken them into a moderately exposed area. Therefore the annual migration rate is estimated at $(2/300) \times (52$ weeks/6 weeks) $= 0.06$ but it will obviously vary considerably in different areas and different seasons. In some areas of intermediate exposure both *L. sitkana* and *L.* sp. occur together in the same bed of *I. cornucopiae* which is equivalent to a migration rate of 0.5 in the units of the model.

Many published estimates of the distances travelled by intertidal snails are not useful in this context since the distance between the exposed and protected habitat types is not given nor the proportion of the snails which travelled that far. It is easy to underestimate the amount of migration between two populations unless the *Littorina*

*Table 4.* Field measurements of selection and migration translated into units of the model.

| Species[1] | Habitat | Fitness[2] | Intensity[3] | Migration[4] | Reference |
|---|---|---|---|---|---|
| *L.* sp. (1986) | exposed | 0.09 | 0.6 | 0.7 (0–15) m | Boulding & Van Alstyne, in prep. |
| *L.* sp. (1987) | exposed | 0.13 | 0.51 | | Boulding & Van Alstyne, in prep. |
| *L. saxatilis* | shelter | 0.14 | 0.5 | 2.2 (0–8) m | Janson, 1983; Table 7 |
| *L. saxatilis* | exposed | 0.27 | 0.33 | 1.8 (0–10) m | Janson & Sundberg, 1986 |
| *N. lapillus* | exposed | 0.2 | 0.4 | | Kitching *et al.*, 1966; Table 1 |
| *N. lamellosa* | shelter | | | 5.5%[5] | Spight, 1974 |

[1] See Table 1 for full species name.

[2] Relative survival when transplanted to opposite habitat type if the survival of the native species or morph is given a value of 1.

[3] Selection intensity constant, $c$, in the units of the model.

[4] Average and ranges of movement of snails in their native habitat over duration of experiment.

[5] Proportion of snails breeding in a different breeding aggregation than previously (25 m away).

are very conspicuously marked; snails that are still close to the point of release are more likely to be detected (Boulding, unpub. data). More than five percent of tagged *Nucella lamellosa* were found in breeding aggregations 25 meters along the shore from those where they had previously bred (Spight, 1974). On the outer coast of Washington this distance can result in transition from very exposed to sheltered.

## Discussion

### Interpretation of results

The results of the simulations show that genetic differences in the mean value of a trait can be maintained between two populations under biologically realistic selection and migration rates. Thus my results support the view of Barton & Charlesworth (1984) that rapid evolutionary divergence does not require small populations nor complete geographical isolation. Even though I did not include factors regenerating genetic variance such as mutation or immigration from other populations, I obtained maintenance of genetic differentiation and additive genetic variance for over 1000 generations for a population of 1000 individuals. Felsenstein (1975) has pointed out that fixation of an allele in a cline can long be delayed if selection is not too weak. This

delay in fixation could have sometimes been sufficient for reproductive isolation to develop, something previously ignored by theoreticians concerned with equilibrium solutions.

Although many models assume selection is weak in natural populations (Bulmer, 1971; Lande, 1982), Endler (1986) reviewed field studies of selection and concluded there was often evidence for strong selection. The selection data summarized in Table 4 agree with this conclusion. The results of the model show that strong selection can overcome very high migration rates.

Estimation of migration rates in the field is difficult and migration of individuals is not directly equivalent to migration of genes. Unless there are private alleles or some other genetic markers it is more accurate to report a range rather than a single rate. Accurate estimates of migration rates require marking a high proportion of the population and carefully monitoring who they breed with; Spight (1974) may be the only study where this has been done.

My model considers only one polygenic, quantitative trait whereas intertidal gastropods on exposed shores differ in several characters from those on protected shores (Table 1). I plan to extend this model to more traits and to genetically correlated traits and to incorporate my heritability estimates. However I do not expect my conclusions to alter markedly; selection intensities of the magnitude observed in field populations should result in differences in the genetic means of the modelled populations despite the presence of such genetic constraints.

*Comparison with similar models*

An analytical two habitat model of a quantitative trait was first published by Bulmer (1971) and then extended to a cline by Felsenstein (1977) and Slatkin (1978). My results are similar to those of Bulmer (1971) in that differences in the means of the two habitats are maintained for high inter-migration rates given sufficient selection. However Bulmer's model did not consider genetic drift or linkage disequilibrium and is appropriate only for very large populations under weak, symmetrical selection.

Felsenstein (1979) did numerical iterations for a four locus, two allele model of a population under disruptive selection for two optima and found that a small amount of heterozygousity was maintained at equilibrium when the genotype of highest fitness is not homozygous at all loci. His result is of interest because the position of the optima in my model insure that nor all loci are homozygous in the most fit genotypes. My model is more likely to maintain genetic variance over time than his model because the population size is separately regulated in the exposed and in the protected population (Maynard Smith, 1966).

Janson & Sundberg (1986) built a model of speciation in littorinids as a function of two un-linked loci; one locus had alleles with different fitnesses in exposed and protected habitats and one locus controlled assortative mating. Their model has an intermediate habitat as Janson (1982b) has documented in her field populations of *Littorina saxatilis*. My model differs from their model in considering only the first stage of speciation (Maynard Smith, 1966) and in having traits determined by many loci and by environment.

The results from models of one and two locus systems can be different from models of a normally distributed, polygenic character because changes in the mean are effectively decoupled from changes in the phenotypic variance (Slatkin, 1978).

Rice (1984) constructed a polygenic model of how disruptive selection on habitat preference can result in sympatric speciation. He found that moderate levels of habitat-based assortative mating combined with moderate levels of disruptive selection produced the highest level of reproductive isolation.

Kondrashov (1986) modelled sympatric speciation driven by disruptive selection and assortative mating and concluded that reproductive isolation would be more likely to develop for a polygenic character than for a character determined by one or two loci.

*Phenotypic plasticity vs speciation*

Some of the phenotypic differentiation between exposed and protected populations of intertidal snails can be attributed to environmentally induced variation. Indeed if snails were able to recognize early the type of environment they were in they could change their morphology and thus increase their fitness. Adaptive phenotypic plasticity is known in barnacles (Lively, 1986), corals (De Weert, 1981), bryozoans (Harvell, 1984), sponges (Palumbi, 1984), and for snail shell thickness (Appleton & Palmer, 1988) and snail foot size (Etter, 1988). The results from the simulations show that the presence of environmental variation biased towards the direction of the optima both increases individual fitness and reduces the likelihood of genetic divergence between the two habitats. However the available data (Table 1) suggest the presence of genetic differentiation between exposed and sheltered snail populations which in turn suggests constraints on the evolution of adaptive phenotypic plasticity.

**Conclusion**

The results of the simulation model of a polygenic, quantitative trait under different selection pressures in each of two habitats suggest that genetic differentiation could be maintained between exposed and sheltered gastropod populations under biologically realistic selection and migration rates. Thus opposing selective pressures between exposed and protected shores can provide a mechanism to explain the often observed genetic differentiation of exposed and protected snail populations. In addition, such differentiation in populations of the immediate ancestor of *Littorina sitkana* and *L.* sp. could have been the first step in their speciation.

## Acknowledgements

I thank A. R. Palmer for reading my paper at this meeting, T. K. Hay for helping me with the programming and the statistics, and J. Felsenstein for answering my questions on quantitative genetics and modelling. Reviews by N. H. Barton, J. Felsenstein, T. K. Hay, B. V. Holthuis, K. Johannesson, A. J. Kohn, D. G. Raffaelli, and an anonymous reviewer greatly improved this manuscript. Paul Kemp allowed me to use his computer. Financial support was obtained from NSF grant No. BSR-8700523 to A. J. Kohn and from an NSERC Canada postgraduate scholarship, a Sigma Xi Grant-in-Aid of Research, a Lerner-Gray Fund for Marine Research award, and a Pacific Northwest Shell Club award to E.G. Boulding.

## References

Appleton, R. D. & A. R. Palmer, 1988. Water-borne stimuli released by predatory crabs and damaged prey induce more predator-resistant shells in a marine gastropod. Proc. natn. Acad. Sci. USA 85: 4387–4391.

Atkinson, W. D. & S. F. Newbury, 1984. The adaptations of the rough winkle, *Littorina rudis*, to desiccation and to dislodgement by wind and waves. J. anim. Ecol. 53: 93–105.

Barton, N. H. & B. Charlesworth, 1984. Genetic revolutions, founder effects, and speciation. Annu. Rev. Ecol. Syst. 15: 133–164.

Basingthwaighte, G. & W. Foulds, 1985. The effect of wave action on the shell morphology of *Littorina unifasciata* Gray. J. r. Soc. West. Aust. 68: 9–12.

Behrens, S., 1972. The role of wave impact and desiccation on the distribution of *Littorina sitkana* Philippi, 1885. Veliger 15: 129–132.

Boulding, E. G., 1984. Crab-resistant features of shells of burrowing bivalves: decreasing vulnerability by increasing handling time. J. exp. mar. Biol. Ecol. 76: 201–223.

Bulmer, M. G., 1971. Stable equilibria under the two-island model. Heredity 27: 321–330.

Bulmer, M. G., 1980. The Mathematical Theory of Quantitative Genetics. Clarendon Press, Oxford, 255 pp.

Crothers, J. H., 1983. Variation in dog-whelk shells in relation to wave action and crab predation. Biol. J. linn. Soc. 20: 85–102.

Denny, M. W., 1985. Wave forces on intertidal organisms: a case study. Limnol. Oceanogr. 30: 1171–1187.

De Weert, W. H., 1981. Transplantation experiments with Caribbean *Millepora* species (Hydrozoa, Coelenterata), including some ecological observations on growth forms. Bijdragen tot de Dierkunde 51: 1–19.

Elner, R. W. & D. G. Raffaelli, 1980. Interactions between two marine snails, *Littorina rudis* Manton and *Littorina nigrolineata* Gray, a predator, *Carcinus maenas* (L.), and a parasite, *Microphallis similis* Jagerskiold. J. exp. mar. Biol. Ecol. 43: 151–160.

Endler, J. A., 1986. Natural selection in the wild. Princeton Univ. Press, Princeton. 337 pp.

Etter, R. J., 1988. Asymmetrical developmental plasticity in an intertidal snail. Evolution 42: 322–334.

Falconer, D. S., 1981. Introduction to Quantitative Genetics 2nd Ed. Longman, New York. 340 pp.

Felsenstein, J., 1975. Genetic drift in clines which are maintained by migration and natural selection. Genetics 81: 191–207.

Felsenstein, J., 1977. Multivariate normal genetic models with a finite number of loci. In E. Pollak, O. Kempthorne & T. B. Bailey (eds), Proceedings of the International Conference on Quantitative Genetics. Iowa State University Press, Ames: 227–246.

Felsenstein, J., 1979. Excursions along the interface between disruptive and stabilizing selection. Genetics 93: 773–795.

Harvell, C. D., 1984. Predator-induced defense in a marine bryozoan. Science 224: 1357–1359.

Heller, J., 1976. The effects of exposure and predation on the shell morphology of two British winkles. J. Zool., Lond. 179: 201–213.

Janson, K., 1982a. Genetic and environmental effects on the growth rate of *Littorina saxatilis*. Mar. Biol. 69: 73–78.

Janson, K., 1982b. Phenotypic differentiation in *Littorina saxatilis* Olivi (Mollusca, Prosobranchia) in a small area on the Swedish west coast. J. moll. Stud. 48: 167–173.

Janson, K., 1983. Selection and migration in two distinct phenotypes of *Littorina saxatilis* in Sweden. Oecologia 59: 58–61.

Janson, K. & P. Sundberg, 1986. Speciation in *Littorina saxatilis*? – A one-dimensional selection-migration model. In K. Janson (ed.), Polymorphisms, Causes and Evolutionary Consequences in a Marine Prosobranch Species, *Littorina saxatilis*. Ph. D. thesis, Goteborgs University.

Janson, K. & R. D. Ward, 1984. Microgeographic variation in allozyme and shell characters in *Littorina saxatilis* Olivi (Prosobranchia: Littorinidae). Biol. J. linn. Soc. 22: 289–307.

Johannesson, B., 1986. Shell morphology of *Littorina saxatilis* Olivi: the relative importance of physical factors and predation. J. exp. mar. Biol. Ecol. 102: 183–195.

Johannesson, K., 1988. The paradox of Rockall: why is a brooding gastropod (*Littorina saxatilis*) more widespread than one having a planktonic larval dispersal stage (*L. littorea*)? Mar. Biol. 99: 507–513.

Kemp, P. & M. D. Bertness, 1984. Snail shapes and growth rates: evidence for plastic shell allometry in *Littorina littorea*. Proc. natn. Acad. Sci. USA 81: 811–813.

Kitching, J. A. & J. Lockwood, 1974. Observations on shell form and its ecological significance in thaisid gastropods of the genus *Lepisella* in New Zealand. Mar. Biol. 28: 131–144.

Kitching, J. A., L. Muntz & F. J. Ebling, 1966. The ecology of Lough Ine XV. The ecological significance of shell and body forms in *Nucella* J. anim. Ecol. 35: 113–126.

Kondrashov, A. S., 1986. Multilocus model of sympatric speciation. III. Computer simulations. Theor. pop. Biol. 29: 1–15.

Lande, R., 1982. A quantitative genetic theory of life history evolution. Ecology 63: 607–615.

Lively, C. M., 1986. Predator-induced shell dimorphism in the acorn barnacle *Chthamalus anispoma*. Evolution 40: 232–242.

Maynard Smith, J., 1966. Sympatric speciation. Am. Nat. 100: 637–650.

Morton, J. & M. Miller, 1968. The New Zealand Sea Shore. Collins, London.

Raffaelli, D. G. & R. N. Hughes, 1978. The effect of crevice size and availability on populations of *Littorina rudis* and *Littorina neritoides*. J. anim. Ecol. 47: 71–83.

Palumbi, S. R., 1984. Tactics of acclimation: Morphological changes of sponges in an unpredictable environment. Science 225: 1478–1480.

Phillips, B. F., 1969. The population of the whelk *Dicathais aegrota* in western Australia. Aust. J. mar. Freshwat. Res. 20: 225–265.

Phillips, B. F., Campbell, N. A. & B. R. Wilson, 1973. A multivariate study of geographic variation in the whelk *Dicathais*. J. exp. mar. Biol. Ecol. 11: 27–69.

Pilkington, M. C., 1971. Eggs, larvae, and spawning in *Melarapha cincta* (Quoy & Gaimard) and *M. oliveri* Finlay (Littorina, Gastropoda). Aust. J. mar. Freshwat. Res. 22: 79–90.

Rice, W. R., 1984. Disruptive selection on habitat preference and the evolution of reproductive isolation. Evolution 38: 1251–1260.

Slatkin, M., 1978. Spatial patterns in the distribution of polygenic characters. J. Theoret. Biol. 70: 213–228.

Spight, T. M., 1974. Sizes of populations of a marine snail. Ecology 55: 712–729.

Struhsaker, J. W., 1968. Selection mechanisms associated with intraspecific shell variation in *Littorina picta* (Prosobranchia: mesogastropoda). Evolution 22: 459–480.

Turelli, M., 1987. Effects of pleiotropy on predictions concerning mutation-selection balance for polygenic traits. Genetics 111: 165–195.

Underwood, A. J., 1974. The reproductive cycles and geographical distribution of some common eastern australian prosobranchs. Aust. J. mar. Freshwat. Res. 25: 63–88.

Wellington, G. M. & A. M. Kuris, 1983. Growth and variation in the tropical Eastern Pacific in the gastropod genus *Purpura*: ecological and evolutionary implications. Biol. Bull. 164: 518–535.

Hydrobiologia **193**: 53–69, 1990.
K. Johannesson, D. G. Raffaelli and C. J. Hannaford Ellis (eds), Progress in Littorinid and Muricid Biology.
© 1990 Kluwer Academic Publishers.

# Biochemical genetic variation in the genus *Littorina* (Prosobranchia:Mollusca)

Robert D. Ward
*Environmental Biology Unit, Department of Human Sciences, University of Technology, Loughborough, Leicestershire LE11 3TU, England, UK (present address: CSIRO Division of Fisheries, GPO Box 1538, Hobart, Tasmania 7001, Australia)*

*Key words:* heterozygosity, genetic differentiation, biochemical systematics

## Abstract

The genus *Littorina* has been subject to many studies of electrophoretically detectable variation, mostly aimed either at clarifying questions concerned with population structure, or at clarifying difficult taxonomic/systematic problems. This paper reviews many of these studies. Topics covered include Hardy-Weinberg deviations, the extent of genetic differentiation among populations within species, founder effects and the effects of human introductions on genetic variation, the biological significance of allozyme variation, and the uses of allozyme variation in *Littorina* systematics.

## Introduction

Since the original demonstrations by Harris (1966) and Lewontin & Hubby (1966) of high levels of genetic variation in man and *Drosophila pseudoobscura* respectively, variation has been exposed through the use of gel electrophoresis and histochemical enzyme staining, and several hundred animal and plant species have been examined by these or related techniques. The initial demonstration of such enzyme variation prompted questions concerning its biological significance, that is, whether or not it affected fitness, and contrasting views on this subject persist today. This debate enlivened the field of population genetics, but the adoption of these techniques derived from molecular biology also had significant effects in other disciplines. Taxonomists and systematists began to appreciate the benefits of such studies, since the enzyme variation revealed

by gel electrophoresis almost invariably reflects variation in DNA, whereas morphological variation, the traditional mainstay of systematists, is influenced by both environmental and genetic factors. The use of enzyme electrophoresis enabled quantification of the genetic similarities between populations and species, and permitted the derivation of genetically based evolutionary trees.

The genus *Littorina* has been subjected to a considerable number of genetic investigations using electrophoretic techniques. This interest reflects not only the abundance and importance of this genus in intertidal ecology but also the uncertain taxonomic status of some groups. In this paper I aim to describe levels of genetic variation in different species of *Littorina*, examine the relationship between population structure and dispersal ability, discuss the possible adaptive nature of enzyme variation, and show how taxonomic controversies have benefitted from molecular approaches.

*Levels of genetic variation within species*

Sixteen species of *Littorina* have been investigated for electrophoretically detectable variation, with varying degrees of precision. Clearly, the more loci or enzymes that have been screened, the more reliable will be the estimates of parameters such as average heterozygosity per locus ($\overline{H}$) or proportion of polymorphic loci ($P$). The $\overline{H}$ values of different species, presented in Table 1, range from 0.008 to 0.381, with a mean of 0.150. This interspecies variation in $\overline{H}$ may be partly attributable to real differences in genomic heterozygosity and partly attributable to variation in the types of enzymes screened and to the different resolution capabilities of different laboratories. *Littorina mandshurica* Schrenck probably does have lower than average genomic heterozygosity, since a substantial number of the 13 loci screened were of enzymes that normally show higher than average degrees of variation (such as esterases and leucine aminopeptidases), although it should be noted that so far only a single population of this species has been screened.

The *Littorina* assayed show broadly comparable levels of genetic variation to other molluscs. Data on 46 such species (including two littorinids) summarised by Nevo *et al.* (1984) give an average $\overline{H}$ of 0.148 (SD = 0.170), compared with 0.150 (SD = 0.071) of the 16 *Littorina*. Nevo *et al*'s figure is based on observed heterozygosities, whereas the figures compiled for *Littorina* are based on Hardy-Weinberg expected heterozygosities. These two estimates are generally very similar in *Littorina* (see next section). The proportion of polymorphic loci, a less sensitive measure of genetic variation since a highly polymorphic locus is weighted equally with a weakly polymorphic locus, averaged 0.468 for 44 molluscs and 0.356 for the *Littorina*. This apparent difference is probably the consequence of Nevo *et al.* using the less stringent 0.99 criterion for polymorphism rather than the 0.95 level adopted in Table 1.

*Testing for Hardy-Weinberg equilibrium within populations*

When screening populations for enzyme polymorphisms, one of the first tests to be applied to a variable locus is to determine whether the genotypes at that locus are in Hardy-Weinberg equilibrium. Deviations from equilibrium denote a situation of particular evolutionary interest. Significant heterozygote excesses may result from heterozygote advantage (heterosis) or, very rarely, from negative assortative mating. Heterozygote deficiencies may result from negative heterosis, positive assortative mating, the presence of null or inactive alleles, or population admixture.

In general, subpopulations of *Littorina*, usually with between 15 and 100 individuals assayed per locus, show genotype deviations that accord with Hardy-Weinberg expectations. Where deviations do exist, they are usually in the direction of heterozygote deficiencies. For example, Janson & Ward (1984) found that only ten out of 100 tests in *L. saxatilis* (Olivi) showed significant departures from expectations, eight attributable to heterozygote deficiencies, two to heterozygote excesses. In that study, out of 15 polymorphic loci examined in 11 subpopulations, ten loci showed an overall tendency to heterozygote deficiencies (three significantly so) and five showed overall heterozygote excesses (none significantly). Fits to Hardy-Weinberg expectations were also recorded for *L. saxatilis* by Snyder & Gooch (1973) and Ward & Warwick (1980). Newkirk & Doyle (1979) observed heterozygote deficiencies at an esterase locus in both *L. saxatilis* and *L. obtusata* (L.), which they speculated may be attributable to a null allele. Fevolden & Garner (1987) found populations of *L. littorea* (L.) to show general agreement with Hardy-Weinberg expectations: the only instances of deviations were two cases of heterozygote excesses for 6-phosphogluconate dehydrogenase. Two Caribbean species of *Littorina*, *L. lineolata* (d'Orbigny) and *L. ziczac* (Gmelin) showed no significant deviations from equilibrium, a third species, *L. angustior* (Mörch), showed some significant heterozygote deficiencies (Janson, 1985a). One of 20 populations of the

*Table 1.* Levels of genetic variation in *Littorina* species. Only surveys screening ten or more loci for 15 or more specimens per locus are listed.

| Species | Number of | | $\overline{H}$ | P | Reference |
|---|---|---|---|---|---|
| | loci | populations | | | |
| European and North Atlantic species | | | | | |
| *L. saxatilis* | 25 | 2–15 | 0.128 | 0.359 | Morris, 1979 |
| | 21 | 6 | 0.153 | 0.381 | Ward & Warwick, 1980 |
| | 23 | 11 | 0.127 | 0.324 | Janson & Ward, 1984 |
| | 16 | 15 | 0.163 | 0.396 | Knight *et al.*, 1987 |
| *L. s. tenebrosa* | 23 | 2 | 0.101 | 0.283 | Janson & Ward, 1985 |
| *L. arcana* | 21 | 9 | 0.132 | 0.365 | Ward & Warwick, 1980 |
| | 16 | 6 | 0.179 | 0.490 | Knight *et al.*, 1987 |
| *L. nigrolineata* | 19 | 1–4 | 0.070 | 0.281 | Morris, 1979 |
| | 16 | 12 | 0.130 | 0.304 | Knight & Ward, unpublished |
| *L. obtusata* | 25 | 1–6 | 0.108 | 0.394 | Morris, 1979 |
| | 15 | 2 | 0.060 | 0.234 | Kemp & Ward, unpublished |
| | 18 | 2 | 0.088 | 0.263 | Janson, 1987a |
| *L. mariae* | 17 | 1 | 0.099 | 0.294 | Morris, 1979 |
| | 18 | 6 | 0.064 | 0.213 | Janson, 1987a |
| *L. littorea* | 24 | 1–14 | 0.093 | 0.290 | Morris, 1979 |
| | 31 | 4 | 0.043 | 0.065 | Fevolden & Garner, 1987 |
| | 15 | 2 | 0.022 | 0.133 | Kemp & Ward, unpublished |
| | 16 | 9 | 0.037 | ? | Janson, 1987a |
| *L. neritoides* | 17 | 3 | 0.381 | 0.806 | Noy *et al.*, 1987 |
| | 13 | 1 | 0.019 | 0.077 | Kemp & Ward, unpublished |
| *L. punctata* | 17 | 3 | 0.167 | 0.481 | Noy *et al.*, 1987 |
| West Coast USA species | | | | | |
| *L. scutulata* | 10 | 5 | 0.229 | 0.500 | Mastro *et al.*, 1982 |
| *L. plena* | 10 | 5 | 0.305 | 0.720 | Mastro *et al.*, 1982 |
| Caribbean species | | | | | |
| *L. angustior* | 12 | 4 | 0.195 | 0.417 | Janson, 1985a |
| *L. lineolata* | 12 | 3 | 0.098 | 0.278 | Janson, 1985a |
| *L. ziczac* | 12 | 2 | 0.082 | 0.335 | Janson, 1985a |
| *L. anguilifera* | 12 | 4 | 0.226 | 0.417 | Janson, 1985b |
| Sea of Japan species | | | | | |
| *L. brevicula* | 13 | 1 | 0.129 | 0.230 | Kartavtsev & Ephremov, 1981 |
| *L. mandshurica* | 13 | 1 | 0.008 | 0.077 | Kartavtsev & Ephremov, 1981 |

$\overline{H}$ = Mean Hardy–Weinberg expected heterozygosity per locus.

P = Proportion polymorphic loci (frequency of most common allele < 0.95).

*Note:* the species in this table (and in the text of this paper) are commonly referred to as members of the genus *Littorina*. However, following various taxonomic revisions, *L. neritoides* is more properly placed in the genus *Melarhaphe*, *L. punctata*, *L. angustior*, *L. angustior*, *L. lineolata* and *L. ziczac* in the genus *Nodilittorina*, and *L. angulifera* in the genus *Littoraria* (Bandel & Kadolsky, 1982; Reid, 1989).

mangrove periwinkle *L. angulifera* (Lamarck) showed a significant deficiency of heterozygotes at an esterase locus, and 15 further populations showed slight but non-significant deficiencies (Gaines *et al.*, 1974). Surveys in the same species of five polymorphic loci in four populations, making 20 tests in all, revealed three instances of heterozygote deficiencies and one of heterozygote excess (Janson, 1985b). Exceptions to this general finding of Hardy-Weinberg equilibrium in littorinids are provided by *L. neritoides* (L.) and *L. punctata* (Gmelin) (Noy *et al.*, 1987), where average observed heterozygosities per locus were only about half those expected (*L. neritoides* (L.), $\overline{H}_{obs} = 0.177$, $\overline{H}_{exp} = 0.381$; *L. punctata*, $\overline{H}_{obs} = 0.073$, $\overline{H}_{exp} = 0.167$). Noy *et al.* consider that these heterozygote deficiencies may have been caused by larval immigration from areas outside the sites they studied, thus generating the Wahlund (1928) effect.

These generally reasonable fits to Hardy-Weinberg equilibrium observed for most *Littorina* species contrast with what is commonly observed in marine bivalves. These species, perhaps because of their commercial significance, have been particularly well studied. Here, heterozygote deficiencies seem to be a general phenomenon: Zouros & Foltz (1984) list 27 species showing such deficiencies. A concensus explanation for these deficiencies is not yet forthcoming. Inbreeding seems unlikely since bivalves have external fertilisation and a pelagic larval phase typically of some weeks duration. In *Mytilus edulis* (L.), the common mussel, some observations are consistent with population admixture (Koehn *et al.*, 1976), whereas in other studies the observed degree of genetic differentiation between populations is less than that required to explain the heterozygote deficiencies (Gartner-Kepkay *et al.*, 1983; Skibinski *et al.*, 1983). Interestingly, breeding tests demonstrated an overall heterozygote deficiency in sib groups of *Mytilus edulis*, and it appears that different larval genotypes have different mortality rates (Mallet *et al.*, 1985).

Why are significant Hardy-Weinberg deviations commonly observed in bivalves but not littorinids? Since the heterozygote deficiencies in bivalves are not well understood, it is not surprising that this contrast also cannot be easily explained. However, sample sizes in littorinids tend to be smaller than in bivalves, since many littorinid studies seek to clarify the taxonomy or systematics of particular groups rather than examining genetic variation *per se*, and small non-significant heterozygote deficiencies (which, it has already been stated, are commoner than heterozygote excesses in *Littorina*) may attain statistical significance in larger surveys. But this seems unlikely to be the complete explanation. Indeed, those littorinids with limited dispersal ability, such as the ovoviviparous *L. saxatilis*, might have been expected to be more inbred and more likely to exhibit heterozygote deficiencies than bivalves. Could it be that bivalves, which produce very large numbers of zygotes, have more scope for selectively determined deviations from Hardy-Weinberg equilibrium through genotype-dependent larval mortality than littorinids which produce fewer zygotes? The more *K*-selected *Littorina* such as *L. saxatilis* and *L. arcana* Hannaford Ellis will have fecundities some orders of magnitude less than the bivalves. Indeed, breeding experiments in *L. saxatilis* (Ward *et al.*, 1986) have revealed no evidence of the genotype-dependent mortalities revealed in *M. edulis* (Mallet *et al.*, 1985). However, *L. littorea* also produces huge numbers of planktonic larvae but shows no significant heterozygote deficiencies; indeed in this species the rare deviations from equilibrium were cases of heterozygote excess (Fevolden & Garner, 1987). It is clear that we still have much to learn concerning the causes of Hardy-Weinberg deviations in natural populations, and both further fieldwork and further laboratory work is called for.

*Genetic differentiation among populations within species*

The extent of genetic differentiation among populations is expected to be related to the extent of gene flow. Species with a pelagic larval stage and thus increased levels of gene flow are expected to show less inter-population differentiation than those with direct development, regardless of whether such differentiation arises from stochastic or deterministic forces. This is because gene flow among populations will reduce differentiation due to the genetic drift of neutral or nearly neutral alleles, and will also tend to prevent or retard the evolution of specialised genotypes in each micro-environment.

This topic has attracted the interest of several investigators. Berger (1973) examined esterase variation in a number of North American Atlantic populations of *L. saxatilis*, *L. obtusata* (both with direct development) and *L. littorea* (pelagic larvae) and observed higher genetic differentiation with the first two species, a result in line with predictions. Similar results were observed for the same species in Irish and French populations screened for two further loci (Wilkins & O'Regan, 1980). The most thorough study of the effects of gene flow on population differentiation in *Littorina* is Janson's (1987a) study of *L. littorea* and *L. saxatilis* from the Swedish west coast, and the

results are unequivocal. Far more local differentiation is observed among *L. saxatilis* populations than those of *L. littorea*. However, one observation appears to be inconsistent with these expectations. A European population of *L. littorea* was found to be far more differentiated from a North American population than were samples of *L. obtusata* from the same two sites. This indicates that there is unlikely to be significant gene flow across the Atlantic for *Littorina*, and it was suggested that the profound *L. littorea* differentiation reflected ancient population divergence and a severe population bottleneck in North America (Berger, 1977).

Table 2 summarises information on genetic differentiation in populations of species screened for 10 or more loci. Here genetic diversity has been partitioned using Nei's methods (1973, 1987). Total heterozygosity is given by $h_t$, average sample or subpopulation heterozygosity by $h_s$, the coefficient of gene differentiation among subpopulations is $g_{st}$ (and is given by $1 - (h_s/h_t)$), and $d_m$ is the average minimum genetic distance

*Table 2.* Analysis of gene diversity among local populations of various *Littorina* species.

| Species | Area | Number of | | $h_t$ | $h_s$ | $g_{st}$ | $d_m$ | Ref. |
|---|---|---|---|---|---|---|---|---|
| | | pops. | loci | | | | | |
| Pelagic larvae | | | | | | | | |
| *littorea* | Swedish west coast | 9 | 14 | 0.042 | 0.041 | 0.021 | 0.001 | a |
| *scutulata* | California | 5 | 10 | 0.251 | 0.230 | 0.085 | 0.027 | b |
| *plena* | California | 5 | 10 | 0.330 | 0.305 | 0.076 | 0.031 | b |
| *angustior* | Florida | 4 | 12 | 0.206 | 0.195 | 0.049 | 0.014 | c |
| *lineolata* | Florida | 3 | 12 | 0.105 | 0.098 | 0.065 | 0.010 | c |
| *ziczac* | Florida | 2 | 13 | 0.083 | 0.082 | 0.011 | 0.002 | c |
| *angulifera* | Florida | 4 | 12 | 0.241 | 0.235 | 0.025 | 0.008 | d |
| Non-pelagic larvae | | | | | | | | |
| *saxatilis* | Saltö, Sweden | 11 | 23 | 0.137 | 0.127 | 0.070 | 0.011 | e |
| | Swedish west coast | 10 | 15 | 0.152 | 0.140 | 0.078 | 0.012 | a |
| | Britain | 6 | 21 | 0.176 | 0.153 | 0.131 | 0.028 | f |
| | 'worldwide' | 15 | 16 | 0.223 | 0.159 | 0.287 | 0.069 | h |
| *arcana* | Britain | 6 | 16 | 0.208 | 0.173 | 0.167 | 0.042 | h |
| *nigrolineata* | Britain | 12 | 16 | 0.130 | 0.105 | 0.188 | 0.027 | i |

'Worldwide' includes populations from Britain, Italy, Canada, USA and South Africa.
a Janson, 1987a; b Mastro *et al.*, 1982; c Janson, 1985a; d Janson, 1985b; e Janson & Ward, 1984; f Ward & Warwick, 1980; h Knight *et al.*, 1987; i Knight & Ward, unpublished.

between subpopulations (given by $s(h_t - h_s)/(s - 1)$ where $s$ is the number of subpopulations). This latter figure is independent of the gene diversity or heterozygosity within subpopulations and can therefore be used to compare the degrees of gene differentiation in different species.

Examination of this table confirms the relative lack of population differentiation in *L. littorea* when compared with *L. saxatilis*. Somewhat surprisingly, *L. scutulata* Gould and *L. plena* Gould both have quite high levels of interpopulation differentiation, despite having veliger larvae. However, the mean extent of genetic differentiation among the seven species with pelagic larvae, whether $g_{st}$ or $d_m$ is considered, is about one-third that of the three species with crawling young (using the *L. saxatilis* figures from the British survey). Clearly more data are required to confirm the validity of this difference, but it is in the predicted direction.

Genetic differentiation among populations of *L. saxatilis* has been studied more intensively than for any other *Littorina*. Here populations only metres apart, subject to similar environmental conditions, can show significant allele frequency heterogeneity (Janson & Ward, 1984). Indeed, a survey of 11 populations along a 1 km stretch of Swedish coastline (the island of Saltö) showed levels of genetic differentiation ($d_m = 0.011$, 23 loci) not that much lower than that observed between the three major races of man ($d_m = 0.019$, 62 loci, Nei & Roychoudhury, 1982). This reflects the low dispersal powers of this littorinid, although the paradox remains that despite low mobility and presumably small local population sizes, genetic variation within subpopulations remains high.

The *L. saxatilis* data (Table 2) also show that as the area of enquiry widens, from a 1 km stretch of coastline of the island of Saltö, through the Swedish west coast, to populations around the coast of Britain, and finally to populations many of which are thousands of miles apart, both $g_{st}$ and $d_m$ rise. Increasing physical distance will decrease rates of gene flow, and this effect will be more marked in species like *L. saxatilis* with low

mobility than *L. littorea* with planktonic larvae. Nearly 30% (0.287) of the gene diversity in the survey of *L. saxatilis* populations from Europe, North America and South Africa arises from differentiation between subpopulations, and $d_m$ attains a relatively high value of 0.069.

In the survey of Knight *et al.* (1987), three of the 15 populations of *saxatilis* samples probably originated from accidental introductions by man. These three populations, two from South Africa and one from Italy, are compared with the 'native' populations in Table 3. The Venice, Italy, sample shows a somewhat decreased heterozygosity (this was also recorded by Janson (1985c)), but the South African populations show a substantial reduction in heterozygosity and polymorphism. The mean genetic distance (Nei, 1972) between the Venice population and native populations is about twice that of the average of comparisons among native populations, while the South African samples are even more divergent from native populations. The South African and to a lesser extent the Venice populations appear to exhibit the effects of past bottlenecks in population size, caused by their accidental introduction. However, these founder effects seen in *L. saxatilis* are not observed in introduced populations of three other species of marine molluscs (one bivalve and two slipper limpets, Table 3). These populations show only small reductions in heterozygosity and rather little divergence in terms of genetic distance from native populations. While it may be that different numbers of founders were involved, it is possible that the explanation of this contrast lies in the differing reproductive modes of the species. *L. saxatilis* is ovoviviparous while the other three species all produce large numbers of planktonic larvae, and thus the latter species have the potential to expand in population size much more rapidly than *L. saxatilis*. Nei *et al.* (1975) and Chakraborty & Nei (1977) have shown that, following a bottleneck, the reduction in heterozygosity depends not only on the size of the bottleneck but also on the rate of recovery from the bottleneck. Hence bottleneck effects in species with high fecundity and high intrinsic rates of increase may be expected to be less

*Table 3.* The effects of human introductions on genetic variation in marine molluscs.

| Population | Status | Number of loci | $\overline{H}$ | $P$ | Genetic distance to 'native' populations | | | Ref. |
|---|---|---|---|---|---|---|---|---|
| | | | | | Range | $n$ | Mean | |
| *Littorina saxatilis* | | | | | | | | |
| Venice, Italy | introduced | 16 | 0.131 | 0.313 | 0.044–0.126 | 12 | 0.076 | a |
| Langebaan, S. Africa | introduced | 16 | 0.049 | 0.125 | 0.149–0.353 | 12 | 0.225 | a |
| Knysna, S. Africa | introduced | 16 | 0.055 | 0.125 | 0.141–0.270 | 12 | 0.187 | a |
| North Atlantic (12) | native | 16 | 0.184 | 0.448 | 0.006–0.125 | 66 | 0.039 | a |
| *Mytilus galloprovincialis* | | | | | | | | |
| Cape, S. Africa | introduced | 23 | 0.22 | 0.478 | 0.010 | 1 | – | b |
| S. Spain (1) | native | 23 | 0.24 | 0.478 | | | | b |
| *Crepidula fornicata* | | | | | | | | |
| Portsmouth, UK | introduced | 24 | 0.030 | 0.33 | 0.002–0.012 | 6 | ? | c |
| New England, USA (6) | native | 24 | 0.045 | 0.57 | 0.003–0.016 | 15 | ? | c |
| *Crepidula onyx* | | | | | | | | |
| Hong Kong | introduced | 23 | 0.141 | 0.565 | 0.049–0.058 | 3 | 0.053 | d |
| California, USA (3) | native | 23 | 0.167 | 0.579 | 0.020–0.027 | 3 | 0.023 | d |

$P$ = proportion of loci polymorphic (0.95 criterion except *D. fornicata* which uses 0.98). Single populations assayed except where indicated.
a Knight *et al.*, 1987; b Grant & Cherry, 1985; c Hoagland, 1985; d Woodruff *et al.*, 1986.

marked than in ovoviviparous species. Interestingly, *saxatilis* populations inhabiting small rocks (skerries) off the Swedish west coast had heterozygosities averaging around 0.128 compared with 0.162 for Swedish mainland populations (13 loci screened), despite probable founding groups of only one or a few females (Janson, 1987b). The reduction in heterozygosity of the very isolated South African populations was substantially greater than for these skerries, indicating either that the rate of increase subsequent to colonisation was lower in the South African sites, or that the Swedish skerries received a few migrants each year from neighbouring populations.

## The biological significance of allozyme variation in littorinids

There has been very little critical work on littorinids which has attempted to assess the bio-logical significance of allozyme variation, that is, whether or not it affects fitness parameters. The majority of studies has been concerned with systematics or population structure, and have not specifically addressed the adaptive nature of enzyme polymorphism.

A number of hypotheses attempt to relate average heterozygosity $(\overline{H})$ of species to particular models. For example, under neutral theory, species with high $N_e$ (effective population size) are expected to have higher $\overline{H}$ than species with lower $N_e$ (Kimura & Crow, 1964); and in general terms such expectations are fulfilled (Nei & Graur, 1984). One selection model predicts that a positive correlation may be expected between niche width and level of genetic diversity (Van Valen, 1965), the allozyme variation being regarded as an adaptive strategy in a heterogeneous environment. An alternative (and somewhat contradictory) selection theory suggests that ecological generalists (with broad niche specificity) may possess 'general purpose' genotypes and hence show reduced variation (Hochachka & Somero, 1973). Nevo *et al.* (1984) have amassed data from many different species, and they believe that the niche width/heterozygosity theory of Van Valen is supported by this data. However, it is clear that numerous problems exist in attempting to discriminate between these types of theories. For example, quantifying niche width is itself difficult. Furthermore, even if a positive correlation is generally found between niche width and heterozygosity, this may in fact result from a correlation between effective population size and heterozygosity (as predicted by neutral theory), since species with broader niches are likely also to have larger $N_e$.

Not withstanding these reservations, data from *Littorina* can be examined with respect to these various theories. With respect to *L. neritoides* and *L. punctata* from Israeli coasts, Noy *et al.* (1987) ascribe the relatively high genetic variation of the former species (see Table 1) to differences in niche width, suggesting that because *L. neritoides* is encountered higher up on rocks and walls, it is expected to be more tolerant of environmental fluctuations in, for example, oxygen and temperature, and hence has a broader niche width than *L. punctata*. However, *L. neritoides* may well be more abundant throughout the Mediterranean than *L. punctata*, and hence may have a larger $N_e$. On northern European shores, *L. punctata* is absent and *L. neritoides* is not common, being restricted to the very high tidal zone of exposed rocky shores. Here it appears to have a narrow niche width, and a limited survey of a single population of *L. neritoides* from the Isle of Man, Britain (Table 1) showed low levels of heterozygosity, consistent with the expectations of the niche width hypothesis (but also with models of neutral variation, since its $N_e$ values will be relatively low). The difference in heterozygosity between the Mediterranean and British populations is striking, but these surveys were carried out in different laboratories and using different enzymes: the difference may be more apparent than real.

Perhaps the most extensively studied species are *L. saxatilis*, *L. arcana* and *L. littorea*. *L. saxatilis* is ovoviviparous (giving birth directly to juvenile snails). *L. arcana* is oviparous (laying egg masses attached to the substrate from which juvenile snails emerge), and *L. littorea* produces planktonic eggs and larvae. Thus the dispersal abilities of *L. littorea* are expected to be considerably greater than either *L. saxatilis* or *L. arcana*. Both *L. saxatilis* and *L. littorea* are abundant on a wide range of habitats on the coastlines of northern Europe, whereas *L. arcana* is only locally abundant on British coastlines and is absent, for example, from Sweden. Consideration of these features leads to the conclusion that the effective population sizes ($N_e$) of these species must be in the order *L. littorea* > *L. saxatilis* > *L. arcana*. Hence under neutral theory for the species considered here, $\overline{H}$ is expected to be in the same order of *L. littorea* > *L. saxatilis* > *L. arcana*. Under the niche width theory, the order might be expected to be *L. littorea* = *L. saxatilis* > *L. arcana*. Yet the observed data clearly do not fit either of these predictions. A number of independent surveys have found relatively low levels of variation in *L. littorea*, with $\overline{H}$ averaging around 0.049, and relatively high levels in both *L. saxatilis* and *L. arcana* ($\overline{H}$ = 0.143 and 0.155 respectively) (see Table 1).

Surveys of the extent of population differentiation for different enzymes have sometimes been used in attempts to distinguish between selectionist and neutralist models, although in the great majority of cases, plausible hypotheses can be erected to explain the observed variation on either stochastic or deterministic grounds.

Janson & Ward (1984) screened 23 loci (15 polymorphic) in 11 subpopulations of *L. saxatilis* occupying different habitats (four exposed, four intermediate and three sheltered) along a 1 km stretch of coastline. They showed that shell morphology varied considerably and consistently with respect to degree of exposure, and since there is evidence that such morphology is at least partly under genetic control (Newkirk & Doyle, 1975; Janson, 1982), it is likely that natural selection selects for particular genotypes affecting shell shape at particular locations. There was significant allozyme heterogeneity between neighbouring populations, but as little of this could be related to environmental pressures, it seems likely that the differences arose from genetic drift in populations with low migration rates. The exception was at the octanol dehydrogenase locus, where virtually all the differentiation between subpopulations was attributable to habitat type. The conclusion was that with the exception of the *Odh* polymorphism, there was no evidence of natural selection influencing allozyme differentiation in this species. Johannesson & Johannesson (1989) observed a vertical cline in a Swedish population of *saxatilis* in the frequency of alleles at the aspartate aminotransferase locus (*Aat*) but no differentiation at five other polymorphic loci: they ascribe the *Aat* cline to natural selection.

Lavie & Nevo (1987) have examined the genotype dependent mortality of samples of *L. punctata* and *L. neritoides* in the laboratory when stressed with mercury and/or cadmium. The response patterns were complex, sometimes favouring heterozygotes, sometimes homozygotes. In these experiments, selection appeared to be operating on the two polymorphic loci screened, but replicate tests would be worthwhile to confirm this.

Other marine molluscs have been more intensively studied from the standpoint of adaptation than littorinids. The bivalves are one such group. A particularly noteworthy investigation is that of Koehn and his colleagues, who have carried out in depth studies of a leucine aminopeptidase cline in *Mytilus edulis* from Long Island Sound, USA, studying the polymorphism from the standpoints of both population genetics and physiology. They conclude that this salinity-related cline is maintained by selective forces (Koehn, 1978; Hilbish & Koehn, 1985). No such detailed work exists in littorinids. Another topic that has in recent years attracted considerable interest is the correlation between growth rate and heterozygosity observed in *Mytilus* and many other bivalves, although it remains unclear whether this correlation is due directly to the genes detected or to linked genes (Zouros, 1987). Again, this is a topic that has not been studied in littorinids. Indeed, it would be rather difficult to study in species such

as *L. saxatilis* which breed throughout the year and do not have discreet cohorts.

## Allozymes and Littorina systematics

The potential of allozyme electrophoresis for assisting in the resolution of difficult taxonomic questions has long been recognised, and in the genus *Littorina* has been applied most intensively in the study of the *saxatilis* complex. This complex was originally described by Dautzenberg & Fischer (1912) as comprising seven subspecies. Later James (1968) described six British subspecies, but recognised that some could coexist without interbreeding. Heller (1975) recognised four distinct *L. saxatilis* species in Wales, viz. *L. rudis* (Maton), *L. patula* (Jefferey), *L. nigrolineata* Gray and *L. neglecta* Bean. Subsequently Raffaelli (1979) concluded that *L. patula* was a form of *L. rudis*, and in 1978 Hannaford Ellis described a new species, *L. arcana*. The concensus of opinion today is that four saxatilid species exist on British shores, *L. saxatilis* ( = *L. rudis*), *L. neglecta*, *L. nigrolineata* and *L. arcana* (e.g. Hannaford Ellis, 1979, 1983; Fretter & Graham, 1980; Raffaelli, 1982). The first two of these species are ovoviviparous, the latter two oviparous. However, there are still some controversies. For example, it has been argued that *L. saxatilis* and *L. arcana*, which can thus far only be unambiguously separated with respect to the structure of the female reproductive trait, are not true species, and that *L. arcana* may only be a reproductive morph of *L. saxatilis* (Caugant & Bergerard, 1980; Hughes & Roberts, 1981). J.E. Smith (1981) considers on morphological grounds that the species referred to above as *L. saxatilis* ( = *L. rudis*) is truly two species, *L. saxatilis* and *L. rudis*, found respectively on rocky shores and on boulder and stone beaches. A small ovoviviparous form, living permanently submerged in lagoons, *L. tenebrosa* Montagu, has been suggested to be a separate species within the *saxatilis* complex (Fretter & Graham, 1980; S.M. Smith, 1982).

How has allozyme electrophoresis contributed

to a resolution of some of these questions? Initial genetic investigations of sympatric and allopatric *L. saxatilis* and *L. arcana* indicated that the two taxa did represent distinct species, although they are clearly very closely related to one another (see Table 4). This conclusion was supported because

the single strictly sympatric population of two species screened in that survey showed highly significant differences in allele frequencies at a number of polymorphic loci, although no diagnostic locus was identified (Ward & Warwick, 1980). A detailed analysis at another site where the two

*Table 4.* Genetic distances (Nei, 1972) between pairs of *Littorina* species.

| Species pair | Number of loci | Genetic distance values | | | Ref. |
| --- | --- | --- | --- | --- | --- |
| | | Range | *n* | Mean | |
| Sibling species pairs | | | | | |
| *L. saxatilis* complex | | | | | |
| *saxatilis/arcana* | 21 | 0.022–0.079 | 18 | 0.044 | a |
| *saxatilis/arcana* | 16 | 0.012–0.200 | 75 | 0.072 | b |
| *saxatilis*/nigrolineata* | 19 | ? | ? | 0.098 | c |
| *saxatiliis/nigrolineata* | 16 | 0.042–0.237 | 120 | 0.106 | d |
| *saxatilis/tenebrosa* | 19–23 | 0.004–0.017 | 4 | 0.010 | e |
| *arcana/nigrolineata* | 16 | 0.052–0.227 | 72 | 0.140 | d |
| *L. obtusata* complex | | | | | |
| *obtusata/mariae* | 17 | ? | ? | 0.087 | c |
| *obtusata/mariae* | 17 | – | 1 | 0.501 | f |
| *obtusata/aestuarii* | 17 | – | 1 | 0.003 | f |
| *mariae/aestuarii* | 17 | – | 1 | 0.473 | f |
| *L. ziczac* complex | | | | | |
| *ziczac/lineolata* | 12 | 0.423–0.510 | 6 | 0.474 | g |
| *ziczac/angustior* | 12 | 0.256–0.313 | 8 | 0.281 | g |
| *lineolata/angustior* | 12 | 0.309–0.406 | 12 | 0.352 | g |
| *L. scutulata* complex | | | | | |
| *scutulata/plena* | 10 | 0.276–0.477 | 25 | 0.348 | h |
| Non-sibling species pairs | | | | | |
| *saxatilis*/obtusata* | 25 | ? | ? | 0.137 | c |
| *saxatilis*/mariae* | 17 | ? | ? | 0.296 | c |
| *nigrolineata/obtusata* | 19 | ? | ? | 0.203 | c |
| *nigrolineata/mariae* | 16 | ? | ? | 0.169 | c |
| *littorea/saxatilis*$ | 23 | ? | ? | 0.528 | c |
| *littorea/nigrolineata* | 17 | ? | ? | 0.535 | c |
| *littorea/obtusata* | 23 | ? | ? | 0.472 | c |
| *littorea/mariae* | 15 | ? | ? | 0.886 | c |
| *angulifera/ziczac* | 11 | – | 1 | 2.38 | i |
| *angulifera/lineolata* | 11 | – | 1 | 1.92 | i |
| *angulifera/angustior* | 11 | – | 1 | 1.72 | i |

*saxatilis**: some of these specimens may have included unrecognised *arcana*.
a Ward & Warwick, 1980; b Knight *et al.*, 1987; c Morris, 1979; d Knight & Ward, unpublished; e Janson & Ward, 1985; f Moyse *et al.*, 1982; g Janson, 1985a; h Mastro *et al.*, 1982; i Janson, 1985b.

species were sympatric supported that conclusion (Ward & Janson, 1985). A broader survey revealed that North Atlantic populations of *L. saxatilis* and *L. arcana* are genetically more similar to one another than either is to the isolated South African populations of *L. saxatilis* (Knight *et al.*, 1987). A somewhat similar phenomenon has been recorded for the *Partula* land snails found on Moorea, French Polynesia (Johnson *et al.*, 1986). Species of this genus, like *L. saxatilis* and *L. arcana*, show rather little inter-specific differentiation and rather high intra-specific differentiation. Genetic relationships between populations of *L. saxatilis* and *L. arcana* are discussed further in Sundberg *et al.* (this volume).

A third saxatilid, *L. nigrolineata*, is somewhat less closely related to *L. saxatilis* than is *L. arcana* (Table 4). Figure 1 shows the genetic relationships between these species based on a survey of 16 loci in ten populations of *L. saxatilis*, six of *L. arcana*, and 12 of *nigrolineata*. All populations were located on British coasts. These relationships are consistent with the ovoviviparous form of reproduction, as found in *L. saxatilis*, being the derived or advanced form compared with the oviparous type displayed by *L. arcana* and *L. nigrolineata*.

Surveys of *L. tenebrosa* and *L. saxatilis* from Scotland and Sweden showed that specimens of *L. tenebrosa* were genetically almost identical to neighbouring samples of *L. saxatilis*, and that furthermore over a micro-environmental cline the Swedish snails showed a gradual morphological transition from typical *tenebrosa*-like animals to phenotypes resembling *L. saxatilis* from sheltered boulder shores. It was concluded that *L. tenebrosa* is an ecotype of *L. saxatilis*, and not a valid species (Janson & Ward, 1985).

A morphological and electrophoretic examination of ovoviviparous animals from exposed and sheltered populations along a 1 km shoreline revealed that animals from the exposed populations were small, thin-shelled and wide apertures (E forms), whereas those from sheltered boulder shores were large, thick-shelled and small apertured (S forms), resembling respectively the *L. patula* and *L. rudis* species of Heller (1975) and

the *L. saxatilis* and *L. rudis* of J. E. Smith (1981). However, sites subject to intermediate degrees of exposure contained both E and S morphs, and morphs of intermediate form. Twenty-three loci were screened in these samples, and in nearly all cases the differentiation at polymorphic loci was not related to environmental type. At only a single locus (octanol dehydrogenase) was the genetic differentiation related to habitat type. The conclusion of this study was that the E and S morphs did not represent different species (Janson & Ward, 1984). A similar conclusion was reached by Mill & Grahame (1988) in their study of esterase variation in British *L. saxatilis* and *L. arcana*.

The species status of *L. neglecta* is currently being examined by Johannesson & Johannesson (this volume), but it is too early to say whether genetic data confirm its distinction as a separate species.

A further species complex that has provoked biochemical genetic investigation in Europe is the *L. obtusata* complex. These are the flat periwinkles, for which until fairly recently the names *L. littoralis* and *L. obtusata* were used more or less synonymously. Sacchi & Rastelli (1966) showed that some of the apparent variability within this species was caused by the confusion of two similar but distinct species, *L. obtusata* and *L. mariae*. Genetic comparisons between species indicate that they are viable taxa, although the two currently available genetic distance values are a low 0.087 (Morris, 1979) and a much higher 0.501 (Moyse *et al.*, 1982). A third 'species', *L. aestuarii* Jeffreys, appears to be genetically identical to *L. obtusata* (Table 4), and is not considered a valid species (Moyse *et al.*, 1982). On the coast of New England, high-spired thin-walled obtusates have given way this century to low-spired thick walled morphs, apparently due to an increase in the range of the predatory inter-tidal crab *Carcinus maenas* (L.). At some localities, the high-spired form can still be found: electrophoresis has shown that these two forms represent different morphs of *L. obtusata* rather than different species (Seeley, 1986).

Outside Europe, electrophoresis has confirmed

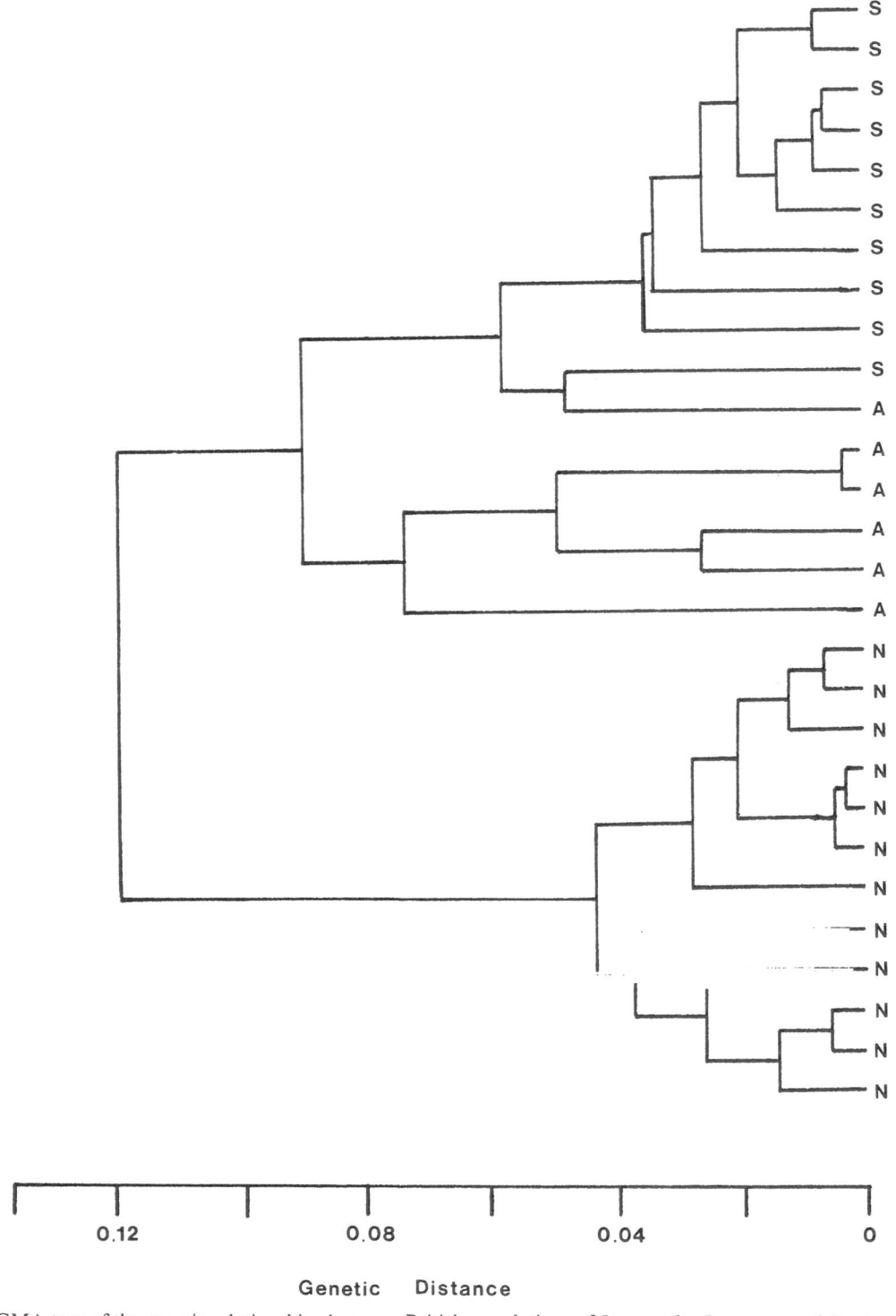

*Fig. 1.* UPGMA tree of the genetic relationships between British populations of *L. saxatilis*, *L. arcana* and *L. nigrolineata*. Tree derived from a matrix of Nei's (1972) genetic distance based on the allele frequencies of 16 loci screened in ten populations of *L. saxatilis*, six populations of *L. arcana*, and 12 populations of *L. nigrolineata*.

the existence of a new species of *Littorina*. Murray (1979) showed that a dichotomy existed in the reproductive biology of *L. scutulata* Gould from California, such that male genitalia was dimorphic and females produced two different types of planktonic egg capsules. Genetic study of these morphs indicated that they represented two species showing considerable genetic differentiation (Table 4) and Mastro *et al.* (1982) suggested that the name *L. plena* Gould be revived for one of these species.

Similarly, species of the morphologically confusing *L. ziczac* (Gmelin) complex (Borkowski, 1975; Bandel & Kadolsky, 1982) from the Caribbean show substantial differences at a number of polymorphic loci, inter-specific distances far exceeding intra-specific distances (Janson, 1985a).

Comparison of the genetic distance values among species of the *saxatilis* complex with those of the *ziczac* and *scutulata* complexes (Table 4) show that the saxatilids are far more closely related to one another than are the component species of the *ziczac* or *scutulata* complexes. Since genetic distance values are roughly proportional to evolutionary time, this indicates that the saxatilid radiation has occurred substantially more recently than that of the other two sibling groups. Genetic distances for most congeneric comparisons in animal species range from $D = 0.02$ to $D = 1.5$ (Thorpe, 1982), again confirming the close genetic relationships of the saxatilids where inter-specific distance are of the order 0.05–0.10.

Morris (1979) has electrophoretically compared with one another the European species *L. saxatilis* (although his sample of *L. saxatilis* may have inadvertently included some *L. arcana*). *L. nigrolineata*, *L. obtusata*, *L. mariae* and *L. littorea*. Genetic distances between all pairwise comparisons are given in Table 4. If this information is combined with the *L. arcana*, *L. saxatilis* and *L. nigrolineata* data of Ward & Warwick (1980), Knight *et al.* (1987), and Knight & Ward (unpublished), an approximate evolutionary tree of all these species can be drawn up (Fig. 2). This indicates that the obtusates and the

saxatilids are more closely related to one another than either is to *L. littorea*. This accords with their reproductive biology, since *L. littorea* produces planktonic larvae while both obtusates and saxatilids have direct development of juveniles. A further species of *Littorina* in Europe is *L. neritoides*. This species is considered as quite distinct from the others, and is often placed in the subgenus *Melarhaphe* (Bandel & Kadolsky. 1982; Reid, 1986): electrophoretic comparisons at 13 loci confirm that this species is only distantly related to the *littorea/saxatilis/obtusata* grouping (Kemp & Ward, unpublished). Warmoes (1986) came to similar conclusions based on his electrophoretic studies of the species *L. neritoides*, *L. littorea*, *L. saxatilis*, *L. arcana*, *L. obtusata* and *L. mariae*.

*Conclusions and prospects for future work*

There is still much to be gained from the further application of molecular techniques to problems in littorinid biology. Although it is probably over-optimistic to think that we will ever have a complete understanding of the biological significance of allozyme variation in littorinids, or indeed in any other group of species, the genetic markers revealed have already proved their worth in illuminating various aspects of population structure and systematics.

The relationship between allozyme variation and morphological variation has not yet been studied very intensively in littorinids. A particular finding from some other groups of animals is that enzyme heterozygosity is frequently associated with a reduced morphological variance (Mitton & Grant, 1984). As mentioned earlier, looking for relationships between heterozygosity and growth rate, a fruitful area of research in bivalves, may be less promising in *Littorina*, since identifying cohort members is likely to be difficult.

Particular areas of research in the future might include further examination of the genetic relationships of European and Pacific *Littorina*, and their inter-relationships. Comparative morphology indicates that the northern Atlantic species are relatively recent immigrants (late Pliocene)

66

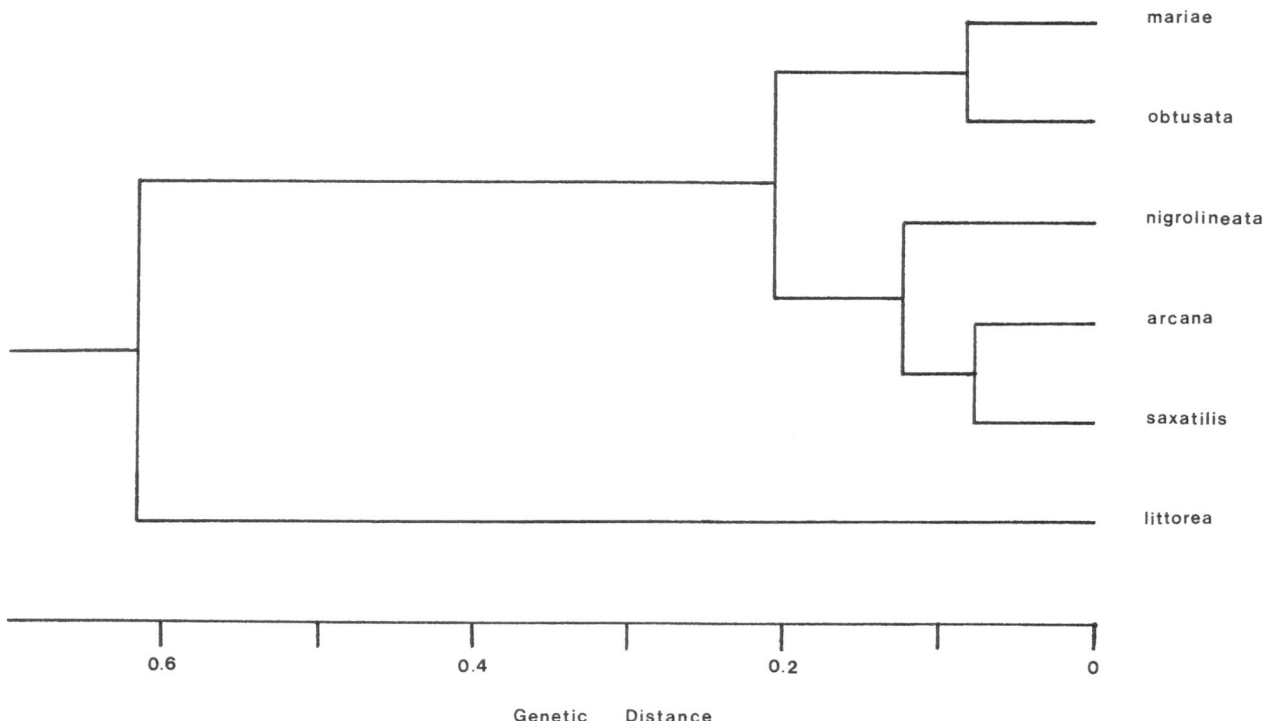

*Fig. 2.* UPGMA tree of the genetic relationships between six species of the genus *Littorina* from British waters. This is compiled from Morris (1979) and Knight & Ward (unpublished), and is based on between 16 and 25 loci per species comparison. See Table 4 for additional information.

from the northern Pacific (Reid, this volume), and it would be most interesting to see whether electrophoretic data support this hypothesis.

Another topic, as yet rather little studied, concerns the genetic dynamics of species when sympatric. In particular, it is unclear whether in such circumstances some species hybridise. It is known that in the laboratory certain species do hybridise and produce fertile offspring (e.g. female *L. arcana* × male *L. saxatilis*; Ward *et al.*, 1986; Warwick *et al.*, this volume), although it is not yet clear whether this happens under natural conditions. The likelihood is that this is possible, since there seems to be little ethological isolation between species of *Littorina* (e.g. Gallagher & Reid, 1974; Raffaelli, 1977).

One technique that has still not yet been used in the furtherance of littorinid studies is that of restriction enzyme analysis of mitochondrial and/or nuclear DNA. Mitochondrial DNA is small (about 16 kb long), evolves rapidly, is non-recombining and maternally inherited. These peculiarities compared with nuclear DNA offer substantial advantages for phylogenetic analysis (Avise *et al.*, 1987). Furthermore, the fact that it is maternally inherited offers the prospect of identifying, for a given hybrid animal (perhaps recognised through the use of standard enzyme electrophoresis), the maternal and paternal parental species. such information can not be gained from the study of nuclear DNA alone. Mitochondrial DNA variation is currently being studied in some marine molluscs, notably *Mytilus* (Skibinski, 1985; Edwards & Skibinski, 1987). However, most littorinids are rather small animals, and the analysis of mtDNA variation in them may thus be more difficult than in some other species, and may have to await the development of suitable probes for detecting the small quantities of mtDNA likely to be extracted.

## Acknowledgements

I would like to thank the Natural Environment Research Council for their support in the form of research grants, and Andrew Knight and Tom Warwick for their help in various aspects of the work, David Reid kindly commented on an earlier version of the paper.

## References

Avise, J. C., J. Arnold, R. M. Ball, E. Bermingham, T. Lamb, J. E. Neigel, C. A. Reeb & N. C. Saunders, 1987. Intraspecific phylogeography: the mitochondrial DNA bridge between population genetics and systematics. Annu. Rev. Ecol. Syst. 18: 489–522.

Bandel, K. & D. Kadolsky, 1982. Western Atlantic species of *Nodilittorina* (Gastropoda: Prosobranchia): comparative morphology and its functional, ecological, phylogenetic and taxonomic implications. Veliger 25: 1–42.

Berger, E. M., 1973. Gene-enzyme variation in three sympatric species of *Littorina*. Biol. Bull. 145: 83–90.

Berger, E. M., 1977. Gene-enzyme variation in three sympatric species of *Littorina*. II. The Roscoff population, with a note on the origin of North American *L. littorea*. Biol. Bull. 153: 255–264.

Borkowski, T. V., 1975. Variability among Caribbean Littorinidae. Veliger 17: 369–377.

Caugant, D. & J. Bergerard, 1980. The sexual cycle and reproductive modality in *Littorina saxatilis* Olivi (Mollusca: Gastropoda). Veliger 23: 107–111.

Chakraborty, R. & M. Nei, 1977. Bottleneck effects on average heterozygosity and genetic distance with the stepwise mutation model. Evolution 31: 347–356.

Dautzenberg, P. H. & H. Fischer, 1912. Mollusques provenant des campagnes de 'Hirondelle' et de la 'Princesse Alice' dans les Mers du Nord. Result. Camp. scient. Prince Albert 1, 37: 187–201.

Edwards, C. A. & D. O. F. Skibinski, 1987. Genetic variation of mitochondrial DNA in mussel (*Mytilus edulis* and *M. galloprovincialis*) populations from South West England and South Wales. Mar. Biol. 94: 547–556.

Fevolden, S. E. & S. P. Garner, 1987. Environmental stress and allozyme variation in *Littorina littorea* (Prosobranchia) Mar. Ecol. prog. Ser. 39: 129–136.

Fretter, V. & A. Graham, 1980. The prosobranch molluscs of Britain and Denmark. Part V – Marine Littorinacea. J. moll. Stud., suppl. 7.

Gaines, M. S., J. Caldwell & A. M. Vivas, 1974. Genetic variation in the mangrove periwinkle *Littorina angulifera*. Mar. Biol. 27: 327–332.

Gallagher, S. B. & G. K. Reid, 1974. Reproductive behaviour and early development in *Littorina scabra angulifera* and

*Littorina irrorata* (Gastropoda: Prosobranchia) in the Tampa Bay region of Florida. Malac. Rev. 7: 105–125.

Gartner-Kepkay, K. E., E. Zouros, L. M. Dickie & K. R. Freeman, 1983. Genetic differentiation in the face of gene flow: a study of mussel populations from a single Nova Scotian embayment. Can. J. Fish. aquat. Sci. 40: 443–451.

Grant, W. S. & M. I. Cherry, 1985. *Mytilus galloprovincialis* Lmk. in southern Africa. J. exp. mar. Biol. Ecol. 90: 179–191.

Hannaford Ellis, C. J., 1978. *Littorina arcana* sp. nov.: a new species of winkle (Gastropoda: Prosobranchia: Littorinidae). J. Conch. 29: 304.

Hannaford Ellis, C. J., 1979. Morphology of the oviparous rough winkle, *Littorina arcana* Hannaford Ellis, 1978, with notes on the taxonomy of the *L. saxatilis* species-complex (Prosobranchia: Littorinidae). J. Conch. 30: 43–56.

Hannaford Ellis, C. J., 1983. Patterns of reproduction in four *Littorina* species. J. moll. Stud. 49: 98–106.

Harris, H., 1966. Enzyme polymorphisms in man. Proc. r. Soc. Ser. B 164: 298–319.

Heller, J., 1975. The taxonomy of some British *Littorina* species, with notes on their reproduction (Mollusca: Prosobranchia). Zool. J. linn. Soc. 56: 131–151.

Hilbish, T. J. & R. K. Koehn, 1985. Dominance in physiological phenotypes and fitness at an enzyme locus. Science 229: 52–54.

Hoagland, K. E., 1985. Genetic relationships between one British and several North American populations of *Crepidula fornicata* based on allozyme studies (Gastropoda: Calyptraeidae). J. moll. Stud. 51: 177–182.

Hochachka, P. & G. Somero, 1973. Strategies of Biochemical Adaptation, Saunders, Philadelphia.

Hughes, R. N. & D. J. Roberts, 1981. Comparative demography of *Littorina rudis, L. nigrolineata* and *L. neritoides* on three contrasted shores in North Wales. J. anim. Ecol. 50: 251–268.

James, B. L., 1968. The characters and distribution of the subspecies and varieties of *Littorina saxatilis* (Olivi, 1872) in Britain. Cah. Biol. mar. 9: 143–165.

Janson, K., 1982. Genetic and environmental effects on the growth rate of *Littorina saxatilis*. Mar. Biol. 69: 73–78.

Janson, K., 1985a. Genetic variation in three species of Caribbean periwinkles, *Littorina angustior, L. lineolata,* and *L. ziczac* (Gastropoda: Prosobranchia). Bull. mar. Sci. 37: 871–879.

Janson, K., 1985b. Genetic and morphologic variation within and between populations of *Littorina angulifera* from Florida. Ophelia 24: 125–134.

Janson, K., 1985c. A morphologic and genetic analysis of *Littorina saxatilis* (Prosobranchia) from Venice, and on the problem of *saxatilis-rudis* nomenclature. Biol. J. linn. Soc. 24: 51–59.

Janson, K., 1987a. Allozyme and shell variation in two marine snails (*Littorina*, Prosobranchia) with different dispersal abilities. Biol. J. linn. Soc. 30: 245–256.

Janson, K., 1987b. Genetic drift in small and recently

founded populations of the marine snail *Littorina saxatilis*. Heredity 58: 31–37.

Janson, K. & R. D. Ward, 1984. Microgeographic variation in allozyme and shell characters in *Littorina saxatilis* Olivi (Prosobranchia: Littorinidae). Biol. J. linn. Soc. 22: 289–307.

Janson, K. & R. D. Ward, 1985. The taxonomic status of *Littorina tenebrosa* Montagu as assessed by morphological and genetic analyses. J. Conch. 32: 9–15.

Johannesson, K. & B. Johannesson, 1989. Differences in allele frequencies of *Aat* between high and mid rocky shore populations of *Littorina saxatilis* (Olivi) suggest selection in this enzyme locus. Genet. Res. 54: 7–11.

Johnson, M. S., J. Murray & B. Clarke, 1986. Allozymic similarities among species of *Partula* on Moorea. Heredity 56: 319–327.

Kartavtsev, Y. P. & V. V. Ephremov, 1981. Genetic similarity and variability of two Littorinid species (Mollusca: Littorinidae). Genetika 17: 1029–1033.

Kimura, M. & J. F. Crow, 1964. The number of alleles that can be maintained in a finite population. Genetics 49: 725–738.

Knight, A. J., R. N. Hughes & R. D. Ward, 1987. A striking example of the founder effect in the mollusc *Littorina saxatilis*. Biol. J. linn. Soc. 32: 417–426.

Koehn, R. K., 1978. Physiology and biochemistry of enzyme variation: the interface of ecology and population genetics. In P. Brussard (ed.), Ecological Genetics: the Interface. Springer-Verlag, New York.

Koehn, R. K., R. Milkman & J. B. Mitton, 1976. Population genetics of marine pelecypods. IV. Selection, migration and genetic differentiation in the blue mussel *Mytilus edulis*. Evolution 30: 2–32.

Lavie, B. & E. Nevo, 1987. Differential fitness of allelic isozymes in the marine gastropods *Littorina punctata* and *Littorina neritoides*, exposed to the environmental stress of the combined effects of cadmium and mercury pollution. Envir. Manage. 11: 345–349.

Lewontin, R. C. & J. L. Hubby, 1966. A molecular approach to the study of genic heterozygosity in natural populations of *Drosophila pseudoobscura*. Genetics 54: 595–609.

Mallet, A. L., E. Zouros, K. E. Gartner-Kepkay, K. R. Freeman & L. M. Dickie, 1985. Larval viability and heterozygote deficiency in populations of marine bivalves: evidence from pair matings of mussels. Mar. Biol. 87: 165–172.

Mastro, E., V. Chow & D. Hedgecock, 1982. *Littorina scutulata* and *Littorina plena*; sibling species status of two prosobranch gastropod species confirmed by electrophoresis. Veliger 24: 239–246.

Mill, P. & J. Grahame, 1988. Esterase variability in the gastropod *Littorina saxatilis* (Olivi) and *L. arcana* Ellis. J. moll. Stud. 54: 347–355.

Mitton, J. B. & M. C. Grant, 1984. Associations among protein heterozygosity, growth rate, and developmental homeostasis. Annu. Rev. Ecol. Syst. 15: 479–499.

Morris, S. R., 1979. Genetic Variation in the Genus *Littorina*. Unpublished thesis, University of Wales.

Moyse, J., J. P. Thorpe & E. Al-Hamadani, 1982. The status of *Littorina aestuarii* Jeffreys. An approach using morphology and biochemical genetics. J. Conch. 31: 7–15.

Murray, T. E., 1979. Evidence for an additional *Littorina* species and a summary of the reproductive biology of *Littorina* from California. Veliger 21: 469–474.

Nei, M., 1972. Genetic distance between populations. Am. Nat. 106: 283–292.

Nei, M., 1973. Analysis of gene diversity in subdivided populations. Proc. natn. Acad. Sci. USA 70: 3321–3323.

Nei, M., 1987. Molecular Evolutionary Genetics. Columbia University Press, New York.

Nei, M. & D. Graur, 1984. The extent of protein polymorphism and the neutral mutation theory. Evol. Biol. 17: 73–118.

Nei, M. & A. K. Roychoudhury, 1982. Genetic relationship and evolution of human races. Evol. Biol. 14: 1–59.

Nei, M., T. Maruyama & R. Chakraborty, 1975. The bottleneck effect and genetic variability in populations. Evolution 29: 1–10.

Nevo, E., A. Beiles & R. Ben-Shlomo, 1984. The evolutionary significance of genetic diversity: ecological, demographic and life history correlates. Lect. Notes Biomathematics 53: 13–213.

Newkirk, G. F. & R. W. Doyle, 1975. Genetic analysis of shell shape variation in *Littorina saxatilis* on an environmental cline. Mar. Biol. 30: 227–237.

Newkirk, G. F. & R. W. Doyle, 1979. Clinal variation at an esterase locus in *Littorina saxatilis* and *L. obtusata*. Can. J. Genet. Cytol. 21: 505–513.

Noy, R., B. Lavie & E. Nevo, 1987. The niche-width variation hypothesis revisited: genetic diversity in the marine gastropods *Littorina punctata* (Gmelin) and *L. neritoides* (L.). J. exp. mar. Biol. Ecol. 109: 109–116.

Raffaelli, D. G., 1977. Observations on the copulatory behaviour of *Littorina rudis* (Maton) and *Littorina nigrolineata* Gray. Veliger 20: 75–77.

Raffaelli, D. G., 1979. The taxonomy of the *Littorina saxatilis* species-complex, with particular reference to the systematic status of *Littorina patula* Jeffrys. Zool. J. linn. Soc. 65: 219–232.

Raffaelli, D. G., 1982. Recent ecological research on some European species of *Littorina*. J. moll. Stud. 48: 342–354.

Reid, D. G., 1986. The Littorinid Molluscs of Mangrove Forests in the Indo-Pacific Region: the Genus *Littoraria*. British Museum (Natural History), London.

Reid, D. G., 1989. The comparative morphology, phylogeny and evolution of the gastropod family Littorinidae. Phil. Trans. r. Soc., London. Ser. B 324: 1–110.

Sacchi, C. F. & M. Rastelli, 1966. *Littorina mariae* n. sp.: les differences morphologique et ecologique entre 'nairns' et 'normanaux' chez l'espece *L. obtusa* (L.) (Gastropoda: Prosobranchia) et leur signification adaptive et evolutive. Atti Soc. ital. Sci. nat. 105: 351–369.

Seeley, R. H., 1986. Intense natural selection caused a rapid morphological transition in a living marine snail. Proc. natn. Acad. Sci. USA 83: 6897–6901.

Skibinski, D. O. F., 1985. Mitochondrial DNA variation in *Mytilus edulis* L. and the Padstow mussel. J. exp. mar. Biol. Ecol. 92: 251–258.

Skibinski, D. O. F., J. A. Beardmore & T. F. Cross, 1983. Aspects of the population genetics of *Mytilus* (Mytilidae: mollusca) in the British Isles. Biol. J. linn. Soc. 19: 137–183.

Smith, J. E., 1981. The natural history and taxonomy of shell variation in the periwinkles *Littorina saxatilis* and *Littorina rudis*. J. mar. biol. Ass. UK 61: 215–241.

Smith, S. M., 1982. A review of the genus *Littorina* in British and Atlantic waters (Gastropoda: Prosobranchia). Malacologia 22: 535–539.

Snyder, T. P. & J. L. Gooch, 1973. Genetic differentiation in *Littorina saxatilis* (Gastropoda). Mar. Biol. 22: 177–182.

Thorpe, J. P., 1982. The molecular clock hypothesis: biochemical evolution, genetic differentiation, and systematics. Annu. Rev. Ecol. Syst. 13: 139–168.

Van Valen, L., 1965. Morphological variation and width of ecological niche. Am. Nat. 99: 377–390.

Wahlund, S., 1928. Zuzammensetzung von Populationen und Korrelationserscheinungen vom Standpunkt der Vererbungslehre aus betrachtet. Hereditas 11: 65–106.

Ward, R. D. & K. Janson, 1985. A genetic analysis of sympatric subpopulations of the sibling species *Littorina saxatilis* (Olivi) and *Littorina arcana* Hannaford Ellis. J. moll. Stud. 51: 86–94.

Ward, R. D. & T. Warwick, 1980. Genetic differentiation in the molluscan species *Littorina rudis* and *Littorina arcana* (Prosobranchia: Littorinidae). Biol. J. linn. Soc. 14: 417–428.

Ward, R. D., T. Warwick & A. J. Knight, 1986. Genetic analysis of ten polymorphic enzyme loci in *Littorina saxatilis* (Prosobranchia: Mollusca). Heredity 57: 233–241.

Warmoes, T., 1986. Een inleidende systematische en taxonomische studie van het genus *Littorina* (Gastropoda, Prosobranchia). Licentiaatsthesis, Universitaire Instelling Antwerpen.

Wilkins, N. P. & D. O'Regan, 1980. Generic variation in sympatric sibling species of *Littorina*. Veliger 22: 355–359.

Woodruff, D. S., L. L. McMeekin, M. Mulvey & M. P. Carpenter, 1986. Population genetics of *Crepidula onyx*: variation in a Californian slipper snail recently established in China. Veliger 29: 53–63.

Zouros, E., 1987. On the relation between heterozygosity and heterosis: an evaluation of the evidence from marine molluscs. In M. C. Rattazzi, J. G. Scandalios & G. S. Whitt (eds), Isozymes: Current Topics in Biological and Medical Research Vol. 15 Alan R. Liss, New York.

Zouros, E. & D. W. Foltz, 1984. Possible explanations of heterozygote deficiency in bivalve molluscs. Malacologia 25: 583–591.

*Hydrobiologia* **193**: 71–87, 1990.
*K. Johannesson, D. G. Raffaelli and C. J. Hannaford Ellis (eds), Progress in Littorinid and Muricid Biology.*
© 1990 *Kluwer Academic Publishers.*

# *Littorina neglecta* Bean, a morphological form within the variable species *Littorina saxatilis* (Olivi)?

Bo Johannesson[1] & Kerstin Johannesson[1]
[1] *Tjärnö Marine Biological Laboratory, Pl 2781, S-452 00 Strömstad, Sweden*

*Key words:* prosobranchia, population differentiation, shell shape, Northeastern Atlantic, principal component analysis

## Abstract

The morphological variation of *Littorina saxatilis* and *L. neglecta* on a microscale (vertical transects down a shore) and on a geographical scale (Northeastern Atlantic) was examined to see if the sampled snails could be consistently separated into two groups on morphological criteria. Size, shell form and shell banding pattern, subopercular pattern and size at maturity were recorded for 21 samples from different shore levels (barnacle zone – usual habitat of *L. neglecta*, low littoral fringe and high littoral fringe – usual habitat of *L. saxatilis*) in Iceland, Norway, Sweden, Isle of Man and Wales. A multivariate approach (principal component analysis) was applied to analyse 14 quantitative shell characters. Between-shore variation in shell shape of samples from similar levels was generally larger than within-shore variation of samples from different levels, while size of snails was consistently smaller in barnacle zone compared to littoral fringe samples. The frequency of snails with qualitative characters, used in earlier studies to define *L. neglecta*, differed between samples from the littoral fringe and the barnacle zone at each site (except at the Swedish site), but the differences between geographical areas were generally larger. The results indicate that snails fitting earlier descriptions of *L. neglecta* are present in the barnacle zone in Iceland, Norway, Isle of Man and Wales, but that this form is not clearly distinguishable from *L. saxatilis*. Snails with intermediate shapes were common and, furthermore, the qualitative and the quantitative characters used to define *L. neglecta* were not closely coupled. No snails of *L. neglecta* form were found at the Swedish site, yet shape differences over the intertidal gradient were greatest at this site.

The extensive morphological variation found in *L. saxatilis*, between shores and between microhabitats on the same shore, is likely to be due to low gene-flow within the species and a heterogeneous environment. The area over which mating can be random is probably small, thus a subpopulation will behave as a semi-isolate with more or less unique ecological and evolutionary factors influencing its detailed morphology.

## Introduction

The morphological variation among different habitat types is well documented in *Littorina saxatilis* (Olivi) (see Raffaelli, 1982 and Faller-Fritsch & Emson, 1985 for a review of some of the literature). Boulder shores, rocky cliffs and eel-grass meadows are, for example, inhabited by different morphological forms (Janson & Ward, 1984, 1985). Little interest has, however, been paid to variation among populations from different sites of similar habitat type, or to the variation among different subpopulations within the same site.

Presence of geographic and micro-scale variation in rocky cliff populations of *L. saxatilis* may be important to the discussion of the taxonomic status of *L. neglecta* Bean. This taxon is a member of the *L. saxatilis* species complex, and it has been separated as a distinct species mainly on shell morphology, e.g. shell form, size at maturity, shell ornamentation and colour pattern, and the morphology of the penis (Heller, 1975; Fish & Sharp, 1985).

Most studies of *L. neglecta* mention problems of classifying individual snails. Fish & Sharp (1985), for example, remark that it is difficult to separate individuals of *L. neglecta* from those of *L. saxatilis* on shell characters, and that the morphological description they give in their paper only applies to snails from Wales, although *L. neglecta* from all over the British Isles were examined. Hannaford Ellis (1984) states that it is hard to recognize those individuals of *L. neglecta*, which do not have the shell colour pattern usually typical of the species (a dark band running into the aperture). She also states that the penis characteristics used by Heller (1975), are invalid for distinguishing between *L. neglecta* and *L. rudis*, ( = *L. saxatilis* see Janson, 1985), at least outside Anglesey.

In this study we examine the morphological variation in a number of shell characters in *L. saxatilis*-like and *L. neglecta*-like snails to see if it is compatible with the hypothesis of two distinct species being present. Our investigation includes samples from three geographical areas from which *L. neglecta* have been reported to occur (Norway; Sneli & Marion, 1979; Isle of Man, and Wales; Fretter & Graham, 1980; Fish & Sharp, 1985), as well as from two areas where it has not been previously recorded (Iceland and Sweden). As no character has so far been reported which is diagnostic of *L. neglecta*, our study is based on a comparison of samples from the barnacle zone (mid-shore) and the littoral fringe (high-shore), and we assume, following earlier descriptions of

the ecology of the species (e.g. Heller, 1975; Hannaford Ellis, 1980; Fish & Sharp, 1985) that, if present, *L. neglecta* will be more or less confined to the barnacle zone, while *L. saxatilis* sensu stricto will dominate in the littoral fringe.

In addition to scoring the presence or absence of qualitative characters described as being useful for distinguishing *L. neglecta* from other littorinids, we have applied multivariate analyses to quantitative characters in order to allow an objective study of shape by separating effects of size. Obviously both size and shape are biologically interesting and both have in earlier studies been used to distinguish between *L. neglecta* and *L. saxatilis*. Corruchini (1987) and Sundberg (1988) present sound arguments for the use of multivariate methods instead of ratios or regressions to separate differences due to shape and size, and stress the importance of considering both.

Snails of *L. neglecta* form are usually described as concentrated in the barnacle belt of rocky shores, while 'typical' *L. saxatilis* are predominant in the littoral fringe. If both *L. neglecta* and *L. saxatilis* are widespread, and if they also are reproductively isolated, one would expect discontinuities in snail morphology along the intertidal gradient from mid- to high-shore levels, and conformity in the morphology of snails from similar shore levels from different sites. This hypothesis was tested by comparison of between-shore and between-shore level variation in shell morphology of *L. saxatilis* and *L. neglecta*. The possibility that there may be two sibling species, with a large morphological overlap, present in our material was examined in an accompanying paper (K. Johannesson & B. Johannesson, this volume), in which we used electrophoresis to study genetic variation within and between populations.

## Materials and methods

### Study sites

We sampled vertical transects at five sites: Sandgerði in Iceland (S-, 64° 02′ N; 22° 43′ W,

September 1986), Herdla in Norway (H-, 60° 34′ N; 04° 56′ E, September 1987), Ursholmen in Sweden (U-, 58° 52′ N; 11° 00′ E, August 1987), Port St. Mary in Isle of Man (P-, 54° 03′ N; 04° 46′ W, November 1986) and Cable Bay on Anglesey, UK (C-, 53° 12′ N; 04° 28′ W, January 1988). All these sites are exposed rocky cliffs, but the Swedish site differs from the others in its lack of significant tides (10–30 cm) and being exposed to ice during cold winters. We located transects from mid- to high-shore at each locality, and each transect at least included one sample from the barnacle zone (-B), one from the low littoral fringe (-LF), and one from the high littoral fringe (-HF). The vertical distance between the barnacle zone and the high littoral fringe samples was about 2 m except at Ursholmen (see below). For consistency, we use the same designation of each sample in this study, as in an accompanying genetic study (K. Johannesson & B. Johannesson this volume; Fig. 1 & 2).

In the present study we analyse three data sets:
(1) Pooled samples from all five sites
(2) Separate samples from the Swedish site (Ursholmen)
(3) Separate samples from the Norwegian site (Herdla)

The first data set includes the three samples from each transect, (one transect per site), except for Cable Bay, as we could only use the barnacle zone sample from this site, the remaining samples containing an unresolvable mixture of *L. saxatilis* and its sibling species, *L. arcana* Hannaford Ellis 1978 (Hannaford Ellis, 1979).

The second data set consists of eight samples from two transects, U1 and U2, 10 m apart on Ursholmen. U1B and U2B are from the barnacle zone, U1LF and U2LF are from 2 m above in the low littoral fringe, and U1HF and U2HF are from 3 m above the low fringe samples. In this set we also included for comparative purposes barnacle zone sample from between U1B and U2B designated U1B*, and a rockpool sample from 10 m above the barnacle zone designated U2R. The barnacle zone was patchy at Ursholmen and U1B* was actually the only

sample taken from within barnacles at this site.

The third data set comprised six samples from two parallel transects 10 m apart on Herdla. The positions of the samples along the transects were the same as at the other sites (except Ursholmen).

*Sampling procedure*

We collected each sample of snails from within 1 m² and included all snails larger than about 2 mm shell height. The snails were then kept alive or deep frozen before examination. In the samples from Herdla, Ursholmen and Cable Bay morphometrics and qualitative characters were recorded for the same individuals as screened enzymatically (K. Johannesson & B. Johannesson, this volume), while in the study of snails from Sandgerði and Port St. Mary, different random subsamples from the same original samples were used.

*Quantitative characters*

We measured the shell dimensions shown in Fig. 1, except shell thickness, on the magnified picture of each shell displayed on a video monitor. The picture was produced either directly through a dissecting microscope or from taped records. We positioned all the shells in the same way and so that each character was measured perpendicular to the optical axis and at the largest possible magnification. The precision of these measurements ranges from 2.6 $\mu$m to 44 $\mu$m depending on the dimension. Shell thickness was measured with a precision of 1 $\mu$m about 1 mm from the edge of the aperture with a digital dial indicator (Sony, U30). We did not measure whorl heights, whorl breadths and whorl diameter above $WB_2$, $WH_2$ or $WD_3$ (Fig. 1) because the top of the shell was often eroded.

*Qualitative characters*

We scored four qualitative characters: the presence of a subopercular pattern similar in form to

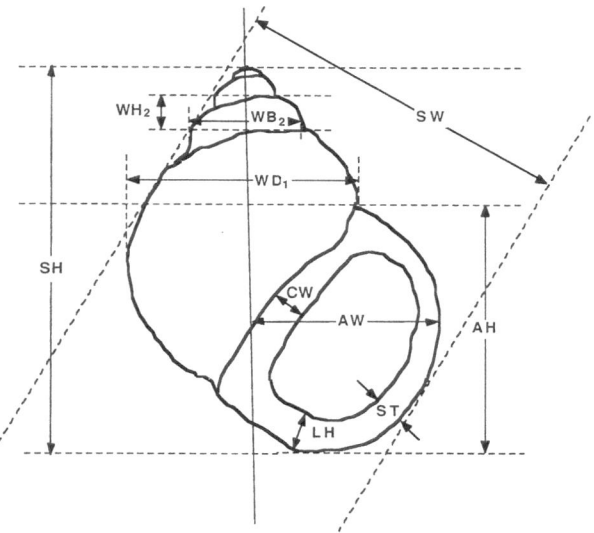

*Fig. 1.* The fourteen shell measurements used in the PCA. AH = aperture height, AW = aperture width, CW = columella width, LH = lip height, SH = shell height, ST = shell thickness, SW = shell width, WB = whorl breadth (the diameter at the middle of whorls), WD = whorl diameter (at the suture of whorls) and WH = whorl height. Subscript number indicates the number of the whorl. $WB_1$, $WD_2$, $WD_3$ and $WH_1$ were also measured, but are not indicated in the figure.

the character 'g'; a dark band running into the aperture; the degree of convexity of the profile of the body whorl and whether the shell was smooth. These characters have in earlier reports been used to distinguish *L. neglecta* from other species, although none is diagnostic (Hannaford Ellis, 1980; Fish & Sharp, 1985). We scored the characters present, if they matched the descriptions given by Fish & Sharp (1985).

*Size at maturity*

We grouped the individuals of all samples into juveniles and adults by inspecting their reproductive organs. Males were scored as mature if the penis was well-formed and the penial glands turgid; females were scored as mature if the brood pouch contained embryos. We scored all individuals with regressed reproductive systems but infected with digenean parasites as adults, since

we did not find any small individuals with parasites. The proportion in different size-classes of females with embryos and males with well-formed penes can change with the season (Hannaford Ellis, 1983). However, the high proportion, in our study, of snails judged as adults in each sample indicated that there were few mature snails which were out of their reproductive phase. Thus, what can be deduced from our samples taken once from each population is at which smallest size the snails at least can be mature.

*Numerical analyses*

The quantitative data were evaluated using principal component analyses (PCA). (See Neff & Marcus, 1980 and Manly, 1986 for a description of this method.) It aims to summarize the original data and make them easier to understand by transforming the observed variables into a new set of variables which are uncorrelated and arranged in decreasing order of importance, reflected by the amount of variation that they explain. This technique can also separate shape from size if the PCA is based on log-transformed observation values and the correlation matrix (Reyment *et al.*, 1984). The first eigenvector may be interpreted as an index of size if all elements (one for each character) are positive and of about equal size, and if they are equal to $(n)^{-1/2}$, where n is the number of characters (Jolicœur, 1963). In addition, the principal component scores of the first axis should be correlated to the original values of all characters. If the first eigenvector can be interpreted to represent size the remaining uncorrelated vectors can be viewed as different aspects of shape (Reyment *et al.*, 1984).

Heterogeneity among samples in each principal component was analysed by one way ANOVA (Sokal & Rolf, 1969). In addition, a Student-Newman-Keuls procedure (SNK, *op. cit.*) was applied to the first two or three principal components from each analysis, to reveal which samples differ from each other.

By pooling samples from different sites one loses information on within-site differentiation

(Pimentel, 1979; Thorpe, 1983b). We therefore supplemented the pooled analysis of samples from the five geographic regions by separate analyses of within-site variation in size and shape in the site showing the largest range of variation (Ursholmen, Sweden), and the one with the smallest range of variation (Herdla, Norway). In the analysis of all sites pooled, and in the separate analysis of samples from Herdla, each sample was represented by 36 individuals, while in the Ursholmen analysis sample sizes were 39.

## Results

### All sites pooled

All the shell dimensions are highly correlated with each other, and the PCA was successful in reducing the original values into a small number of transformed variables which explain much of the variation. In fact, the first component accounts for more than 81.4% of the total variation, and when the three first components have been considered less than 4% of the variance remains (Table 1).

### Size

The elements of the first vector are all positive and most of them are close to the predicted value (number of characters)$^{-1/2}$ = 0.27. Also, original values for all characters are significantly correlated ($P < 0.002$) with the scores of the first principal component. Therefore the scores of the first component mainly express variation due to snail size.

Size distributions of snails in the samples overlap considerably within and among sites. However, there is a trend of increasing size up the

*Table 1.* Results from the principal component analysis of all the sites pooled. Samples from three shore levels were taken at: Sandgerdi in Iceland; Herdla in Norway; Ursholmen in Sweden; Port St. Mary in Isle of Man and from the barnacle belt at Cable Bay on Anglesey in UK. The analysis was based on the correlation matrix for the fourteen characters shown in Fig. 1. Only the first eight eigenvectors with corresponding eigenvalues are reported; the remaining six eigenvalues account for less than 0.5% of the total variance. Significant differences in component scores among samples were found for all components.

| Component: | 1 | 2 | 3 | 4 | 5 | 6 | 7 | 8 |
|---|---|---|---|---|---|---|---|---|
| Eigenvalue: | 12.1 | 0.90 | 0.48 | 0.17 | 0.11 | 0.09 | 0.05 | 0.03 |
| Cumulative %: | 81.4 | 92.8 | 96.2 | 97.4 | 98.2 | 98.8 | 99.2 | 99.5 |
| Eigenvectors: | | | | | | | | |
| character SH | 0.285 | − 0.068 | 0.004 | − 0.026 | − 0.162 | − 0.086 | 0.018 | 0.024 |
| SW | 0.284 | − 0.106 | − 0.082 | − 0.019 | − 0.204 | − 0.117 | 0.018 | 0.057 |
| LH | 0.242 | 0.356 | − 0.422 | 0.697 | 0.112 | 0.347 | 0.104 | − 0.001 |
| WD1 | 0.285 | − 0.049 | − 0.011 | 0.001 | − 0.218 | 0.019 | − 0.103 | − 0.318 |
| WB1 | 0.283 | − 0.073 | 0.066 | − 0.054 | − 0.194 | 0.100 | − 0.113 | − 0.625 |
| WD2 | 0.283 | 0.068 | 0.179 | 0.018 | 0.037 | − 0.086 | 0.000 | − 0.076 |
| WB2 | 0.279 | 0.064 | 0.267 | 0.055 | 0.117 | − 0.135 | 0.097 | − 0.280 |
| WD3 | 0.269 | 0.197 | 0.308 | 0.124 | 0.344 | − 0.166 | − 0.714 | 0.263 |
| WH1 | 0.271 | − 0.091 | 0.267 | − 0.249 | − 0.151 | 0.796 | − 0.006 | 0.328 |
| WH2 | 0.270 | 0.065 | 0.383 | 0.032 | 0.342 | − 0.149 | − 0.663 | 0.130 |
| AH | 0.281 | − 0.082 | − 0.183 | 0.078 | − 0.243 | − 0.232 | 0.005 | 0.166 |
| AW | 0.275 | − 0.200 | − 0.206 | 0.006 | − 0.315 | − 0.275 | 0.051 | 0.439 |
| CW | 0.209 | − 0.626 | − 0.387 | − 0.122 | 0.616 | 0.092 | − 0.050 | − 0.074 |
| ST | 0.210 | 0.593 | − 0.410 | − 0.639 | 0.154 | − 0.036 | − 0.037 | − 0.009 |
| F-values*: | 21.9 | 132.8 | 32.8 | 7.49 | 8.34 | 7.10 | 3.61 | 5.25 |

* From ANOVA of principal component scores for the thirteen samples pooled. All values are significant for $P < 0.001$.

shore (Fig. 2 and 3). The smallest snails are found in the sample from the barnacle zone at Cable Bay and all other barnacle zone samples, with the exception of that from Ursholmen, contain on average smaller snails than all the littoral fringe samples (Fig. 3).

*Shape*

The second principal component, which mainly reflects aspects of snail shape, reveals differences among sites (Figs. 2 & 3). Columella width (CW), shell thickness (ST), and lip height (LH) are the characters which contribute most to the variation

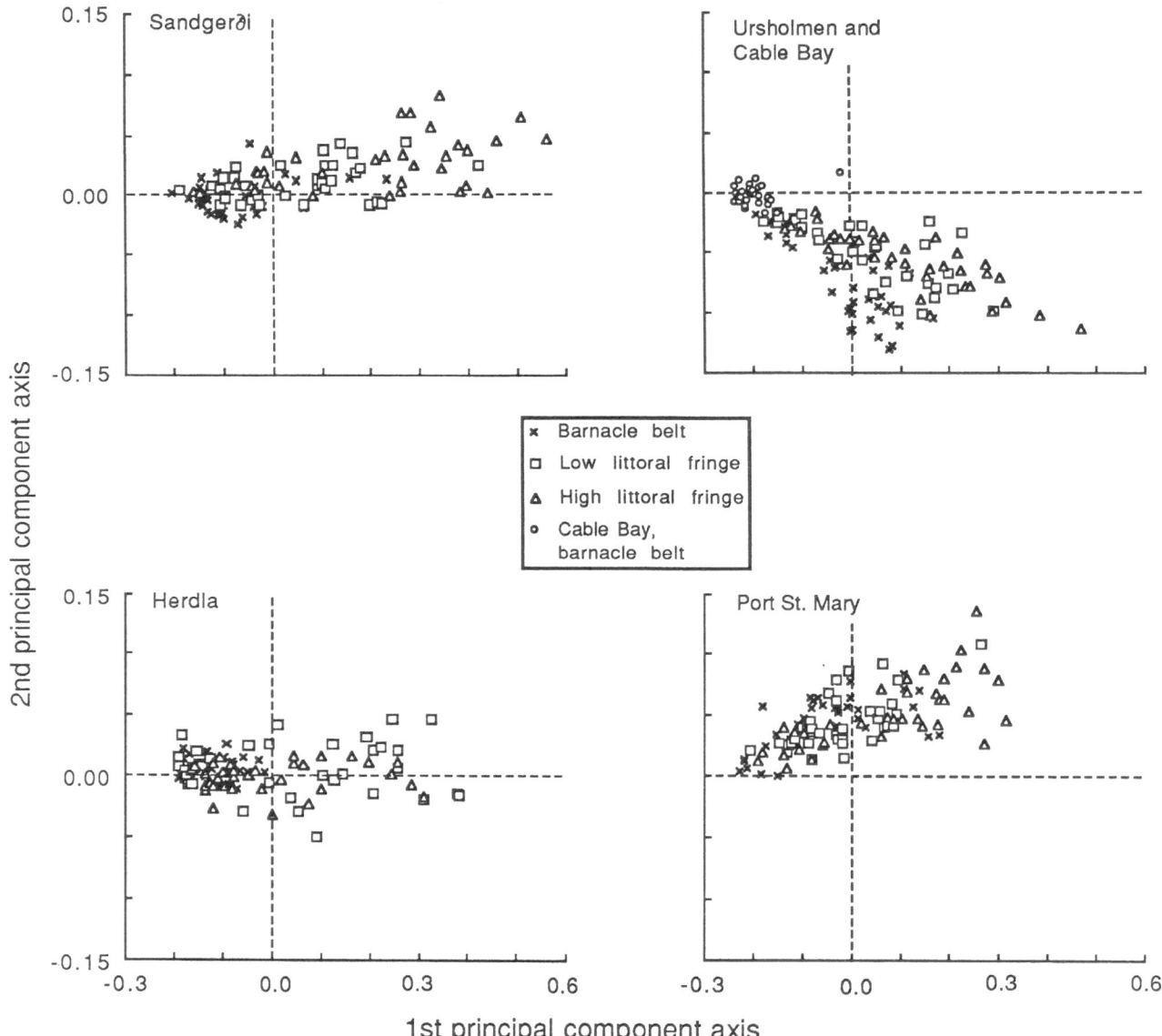

*Fig. 2.* Principal component scores for the first two axes in the analysis of all sites pooled ($n = 36$ in each sample). Each site is represented by samples from three shore levels except Cable Bay which is only represented by a sample from the barnacle belt. The plot is partitioned into four parts only to improve presentation. The first component explains 81.4% of the total variation, which is mainly due to differences in size among individuals (Table 1). The second component explains 11.4% of the total variation, which is mainly due to differences in columella width and shell thickness.

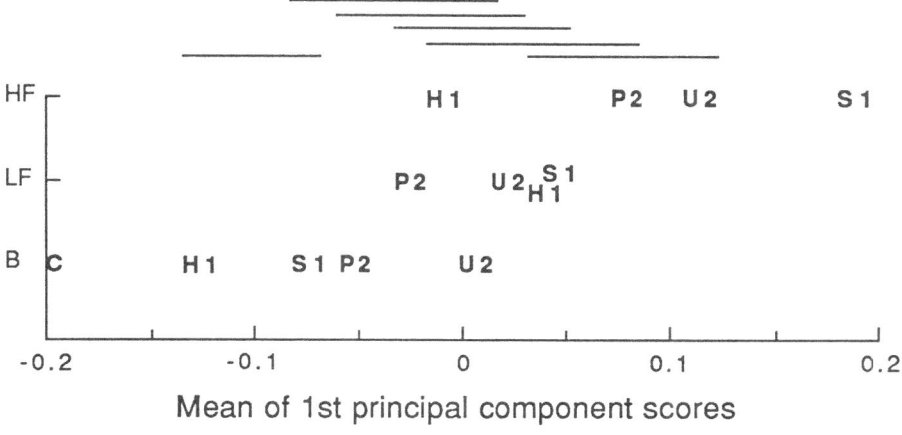

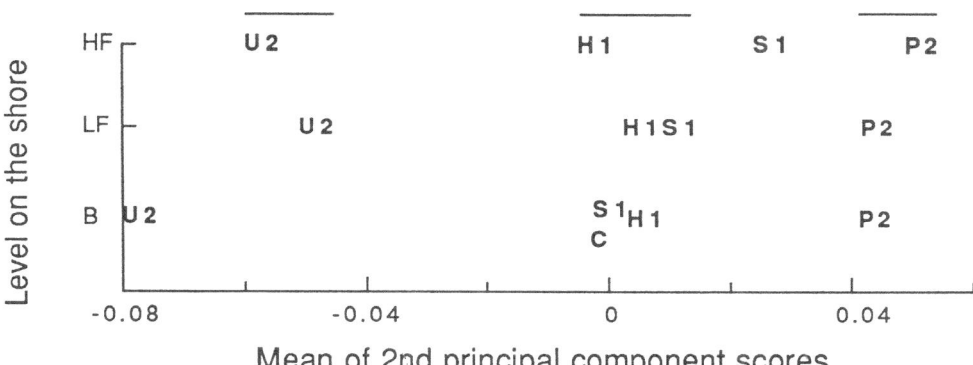

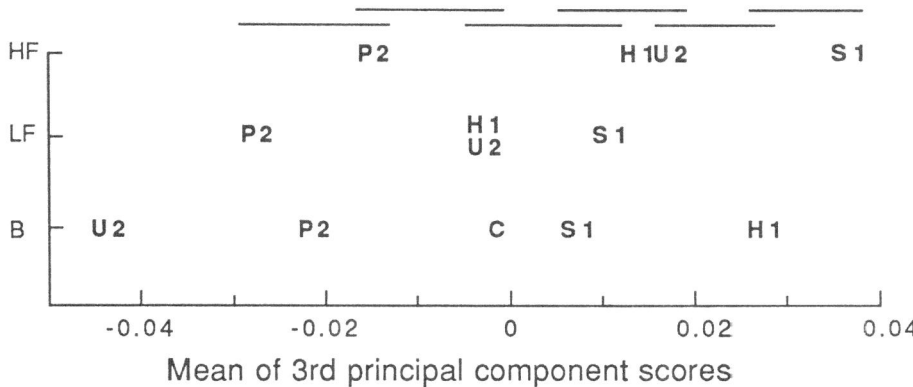

*Fig. 3.* Student-Newman-Keuls test for differences among samples in scores for the three first components from the analysis of all sites pooled. Each site, except Cable Bay, is represented by samples from the barnacle belt (B), the low littoral fringe (LF) and the high littoral fringe (HF), and the mean of the component scores is indicated by the position of the site abbreviation; C = Cable Bay; H1 = Herdla, transect 1, S1 = Sandgerði 1, P2 = Port St. Mary 2, U2 = Ursholmen 2. Horizontal lines join those samples which are not significantly different ($P > 0.05$).

along the second axis (Table 1). For example, snails from Port St. Mary have thicker shells and apertures with more narrow columellae and higher lips than the snails from the other sites, while snails from Ursholmen have thinner shells, wider columellae, and lower lips than the others (Figs. 2 & 3, Table 1). Although the differences along the second axis are most pronounced between different sites significant differences

among samples from the same site are found at Ursholmen and Sandgerði (Fig. 3).

The third principal component reveals some differences among sites as well as among samples within sites (Figs. 3 & 4). The within-site differences are again most pronounced at Ursholmen, while no significant differences are found among the three samples from Port St. Mary.

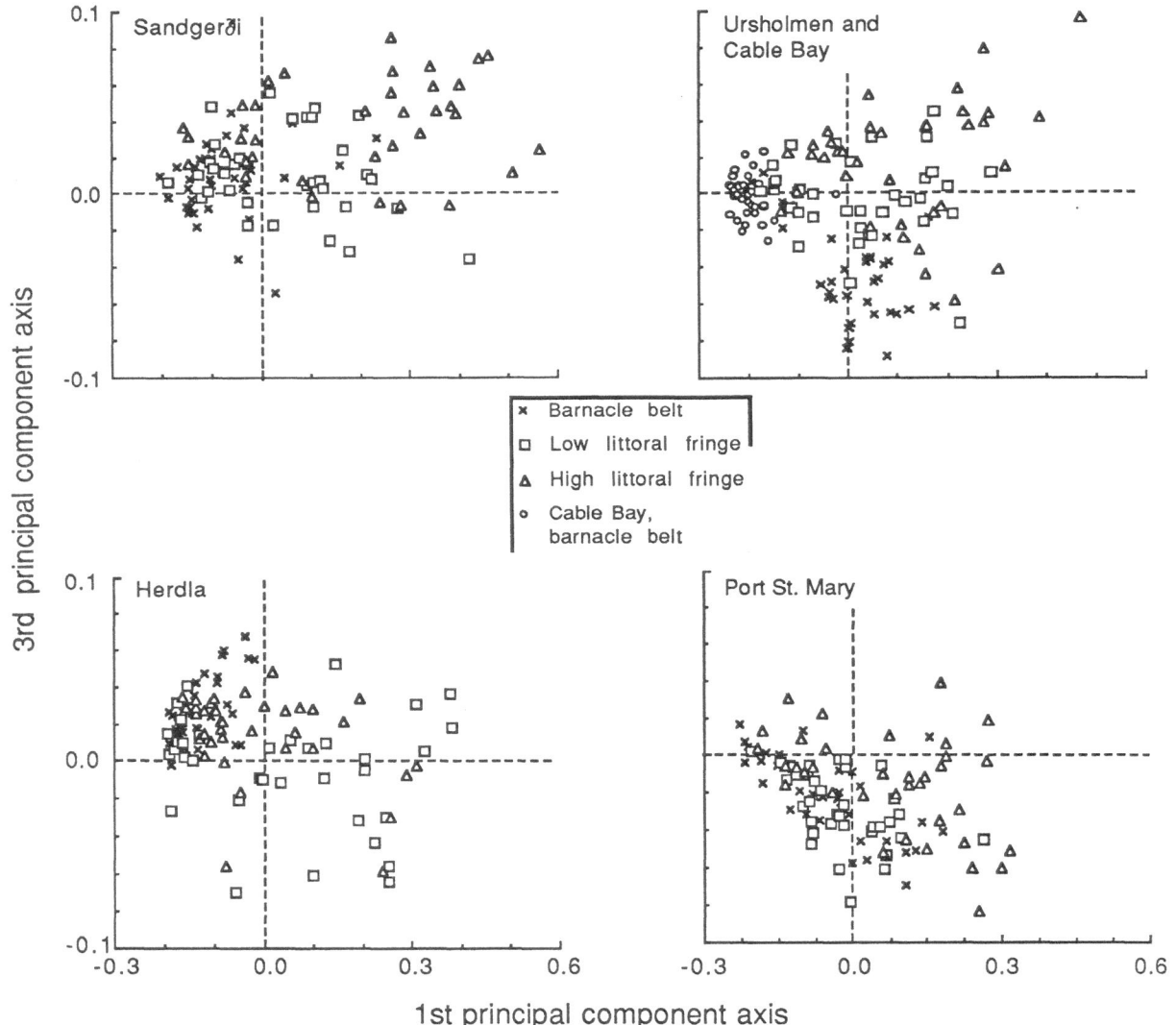

*Fig. 4.* Principal component scores for the first and third axes in the analysis of all sites pooled. Each site is represented by samples from three shore levels except Cable Bay which is only represented by a sample from the barnacle belt. The third component explains 3.4% of the total variation, with separation of individuals mainly because of differences in lip height, shell thickness, columella width and height of whorl number two (Table 1).

Although the ANOVA indicates significant differences among samples along all the remaining axes (Table 1), none of these axis explains more than 1.2% of the total variation in the material, and they are not considered further.

## Qualitative characters

The four qualitative characters described as usual in *L. neglecta* from Anglesey (Fish & Sharp, 1985), vary among shore levels within the same site, but vary even more between different sites (Table 2). The samples from Ursholmen are clearly different from the others with respect to these qualitative characters as, for example, the subopercular pattern (SP) and a dark band running into the aperture (DB) which are usually found in *L. neglecta* from Anglesey (*op. cit.*), are infrequent and often completely absent in snails from Ursholmen. No individual from Ursholmen has all the *neglecta*-characters combined. Two characters also frequent in snails from Anglesey (*op. cit.*) are common in all samples including those from Ursholmen. These are a convex profile of the body whorl (CP) and a smooth shell (SS). On the other hand, all the qualitative characters typical of *L. neglecta* are found in at least some of the snails from the barnacle zone at Cable Bay and many of these snails have the characters combined. Although the *neglecta*-characters are not that frequent in the samples from Sandgerði, Herdla and Port St. Mary there is still a difference between barnacle zone and littoral fringe samples with *neglecta*-characters more common in the former.

## Size at maturity

At all sites we found more small mature individuals in the samples from the barnacle zone than in the samples from the littoral fringe (Fig. 5). Especially high numbers of small mature individuals were found in the sample from Cable Bay, while all mature individuals were large in the high fringe samples from Sandgerði, Herdla and Port St. Mary.

## Within-site variation at Ursholmen

The first four eigenvalues from the PCA of the eight samples from Ursholmen are: $1 = 12.1$ (86.6%), $2 = 0.846$ (6.04%), $3 = 0.420$ (3.00%) and $4 = 0.193$ (1.38%). These eigenvalues account for more than 97% of the total variation. Component scores differ significantly among samples (ANOVA; $P < 0.05$) for all axes except numbers 8, 9, and 14. As in the above analysis of all the sites, the first eigenvector of the Ursholmen analysis fulfils all the requirements of a size vector.

The snails in all samples from Ursholmen span almost the same size range (Fig. 6), although the average size tends to increase up the shore (Fig. 7).

We judge the second eigenvector to mainly represent shape differences and its elements are these: $SH = -2.89 \times 10^{-4}$, $SW = -0.123$, $LH = -0.263$, $WD_1 = -0.0157$, $WB_1 = 6.18 \times 10^{-3}$, $WD_2 = 0.243$, $WB_2 = 0.294$, $WD_3 = 0.426$, $WH_1 = 0.147$, $WH_2 = 0.426$, $AH = -0.236$, $AW = -0.273$, $CW = -0.420$, $ST = -0.276$. While small snails are generally similar in shape, there are pronounced differences between large snails from different levels on the shore (Fig. 6). Large snails from the barnacle zone are different in shape from those from the littoral fringe at Ursholmen mainly because of their: relatively low spires, as the element for $WH_2$ has a large value and the individuals from the barnacle zone have low component scores along the second axis; high whorl expansion rates, as the elements for $WD_{2-3}$ have large values and the value for the element $WD_1$ is considerably smaller; wide aperture margins, as both CW and LH are small; large apertures, as both AH and AW are small; and thick shells, as ST is small (Fig. 6). Snails from the barnacle zone are significantly different ($P < 0.05$ SNK procedure) from the littoral fringe and the rock-pool snails, with the snails from the U_B* sample also different from all other samples (Fig. 7).

*Table 2.* Percentage of individuals in each sample (N = sample size) that have qualitative characters described as common in *Littorina neglecta*: SP = subopercular pattern, DB = dark band running into the aperture, CP = convex profile of the body whorl, SS = smooth shell.

| | Cable Bay | Sandgerdi 1 | | | Herdla 1 | | | Herdla 2 | | | Ursholmen 1 | | | | Ursholmen 2 | | | | Port St. Mary 2 | | |
|---|---|---|---|---|---|---|---|---|---|---|---|---|---|---|---|---|---|---|---|---|---|
| | B | B | LF | HF | B | LF | HF | B | IF | HF | B* | B | LF | HF | B | LF | HF | R | B | LF | HF |
| SP | 92 | † | † | † | 0 | 3 | 0 | † | † | † | 5 | 3 | 3 | 0 | 0 | 0 | 0 | 2 | 15 | 3 | 3 |
| DB | 71 | 13 | 3 | 5 | 38 | 8 | 11 | 28 | 18 | 13 | 0 | 0 | 0 | 0 | 0 | 0 | 0 | 5 | 22 | 10 | 10 |
| CP | 95 | 100 | 100 | 100 | 100 | 74 | 68 | 90 | 88 | 95 | 100 | 100 | 100 | 100 | 100 | 100 | 100 | 100 | 100 | 100 | 100 |
| SS | 40 | 60 | 30 | 23 | 90 | 20 | 22 | 85 | 28 | 21 | 93 | 95 | 92 | 100 | 95 | 100 | 95 | 98 | 20 | 7 | 2 |
| N | 36 | 36 | 36 | 36 | 36 | 36 | 36 | 36 | 36 | 36 | 39 | 39 | 39 | 39 | 39 | 39 | 39 | 39 | 36 | 36 | 36 |

B   = barnacle belt
B*  = additional sample from the barnacle belt
LF = low littoral fringe
HF = high littoral fringe
R   = rock pool

1 = transect 1
2 = transect 2
† = not scored due to technical problems

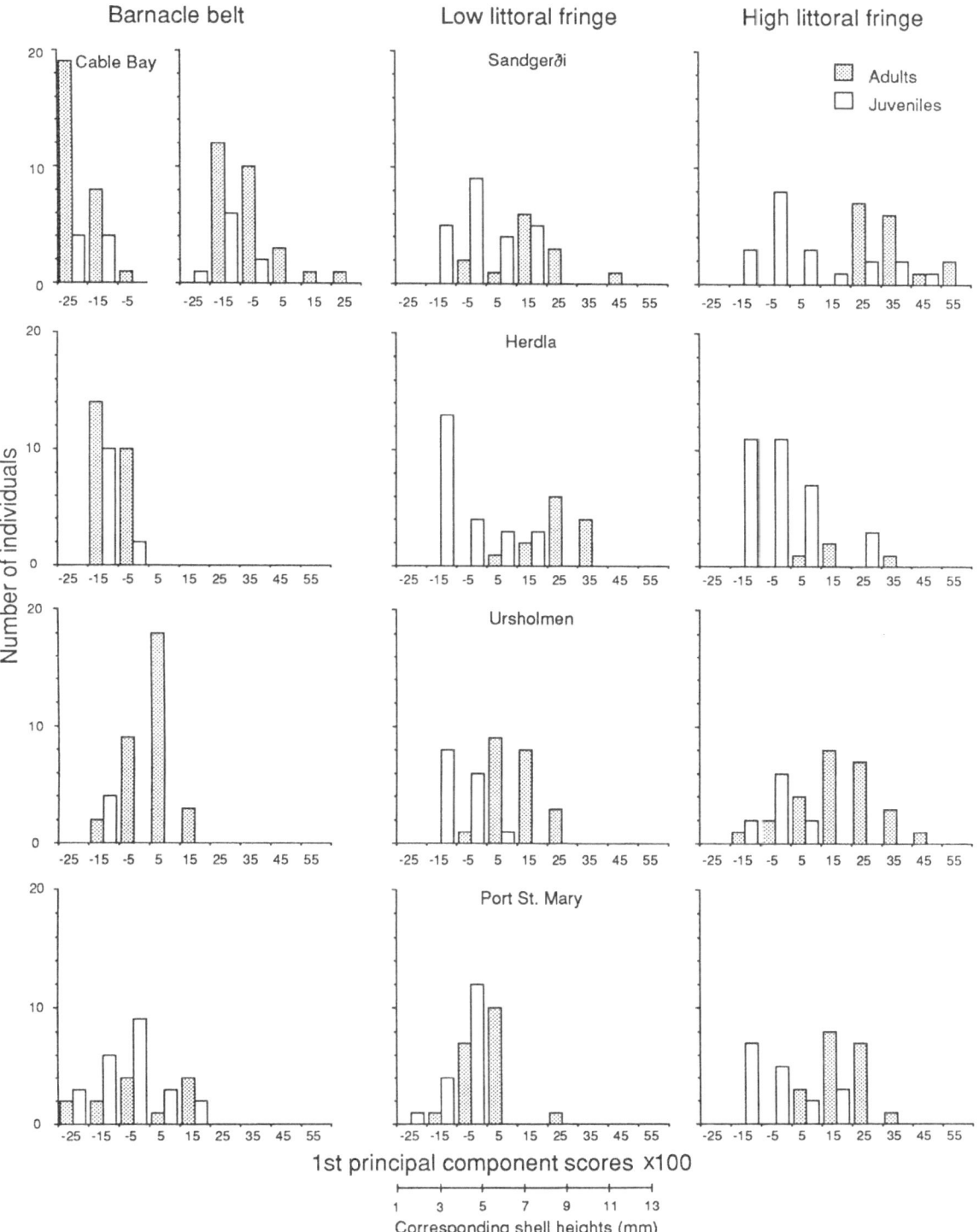

*Fig. 5.* Histograms showing the number of juvenile and adult individuals in nine size classes for the samples used in the analysis of all sites. Each site is represented by samples from three shore levels except Cable Bay which is only represented by a sample from the barnacle belt. The first principal component (PC 1) can be interpreted as showing differences in size (see Table 1). For comparison the relationship between shell height and size (PC $1 = 0.0706 \times$ shell height $- 0.374$, $R^2 = 0.985$) is shown at the bottom of the figure.

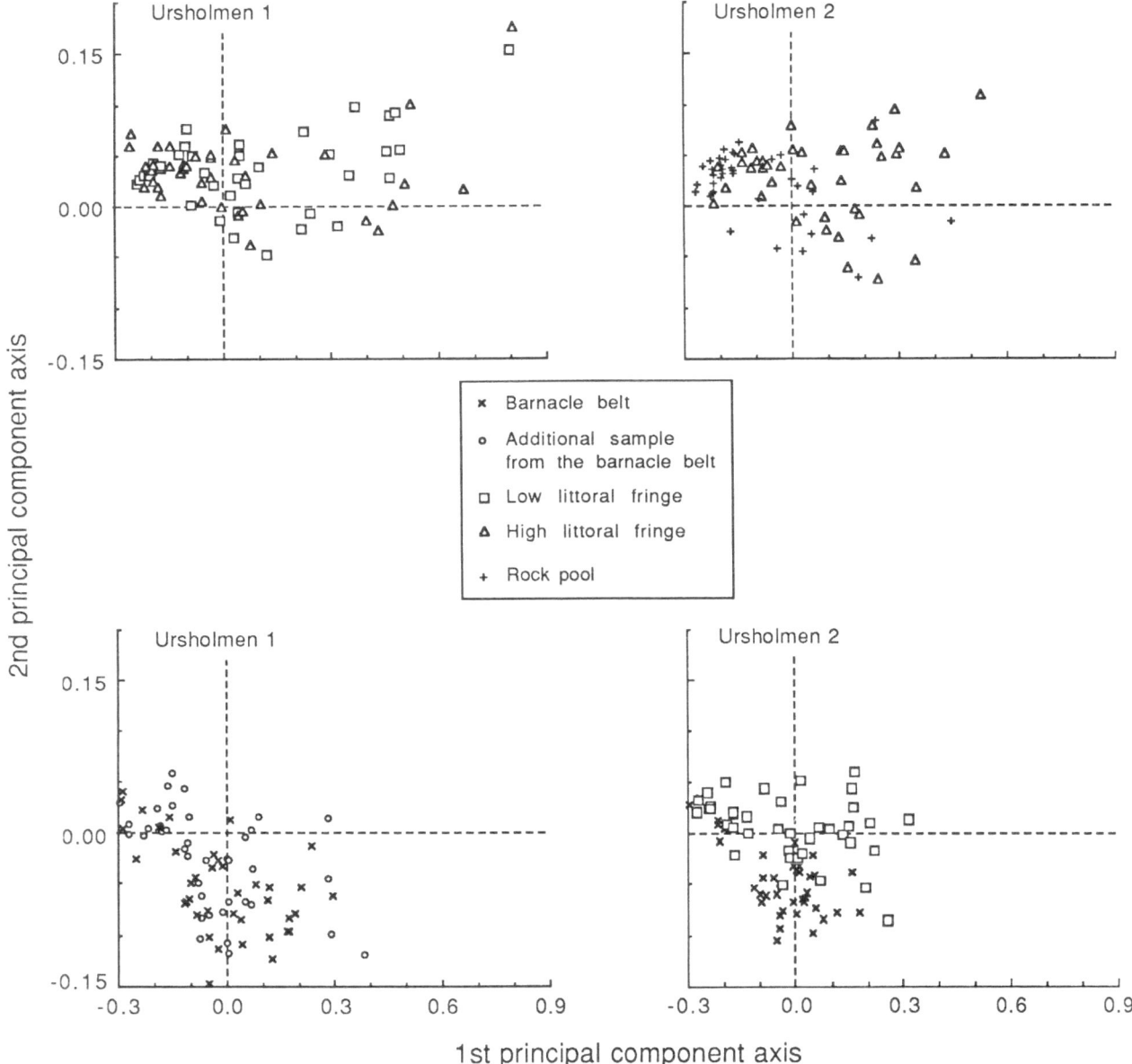

*Fig. 6.* Principal component scores for the first two axes in the analysis of samples from Ursholmen ($n = 39$ in each sample). The plot is partitioned into four parts only to improve presentation. The first component explains 86.6% of the total variation, which is mainly due to differences in size among individuals. The second component explains 6.0% of the total variation, which is mainly due to differences in the diameter of the third whorl, height of the second whorl and columella width (see text).

*Within-site variation at Herdla*

The four first eigenvalues from the PCA of the six samples from Herdla are: 1 = 13.0 (92.9%), 2 = 0.342 (2.44%), 3 = 0.199 (1.42%) and

4 = 0.120 (0.856%). The remaining eigenvalues accounts for less than 2.4% of the variance, and significant differences in scores among samples (ANOVA; $P < 0.05$) were found along all axes except numbers 4, 6, 8, 9, 10, and 13. Also in this

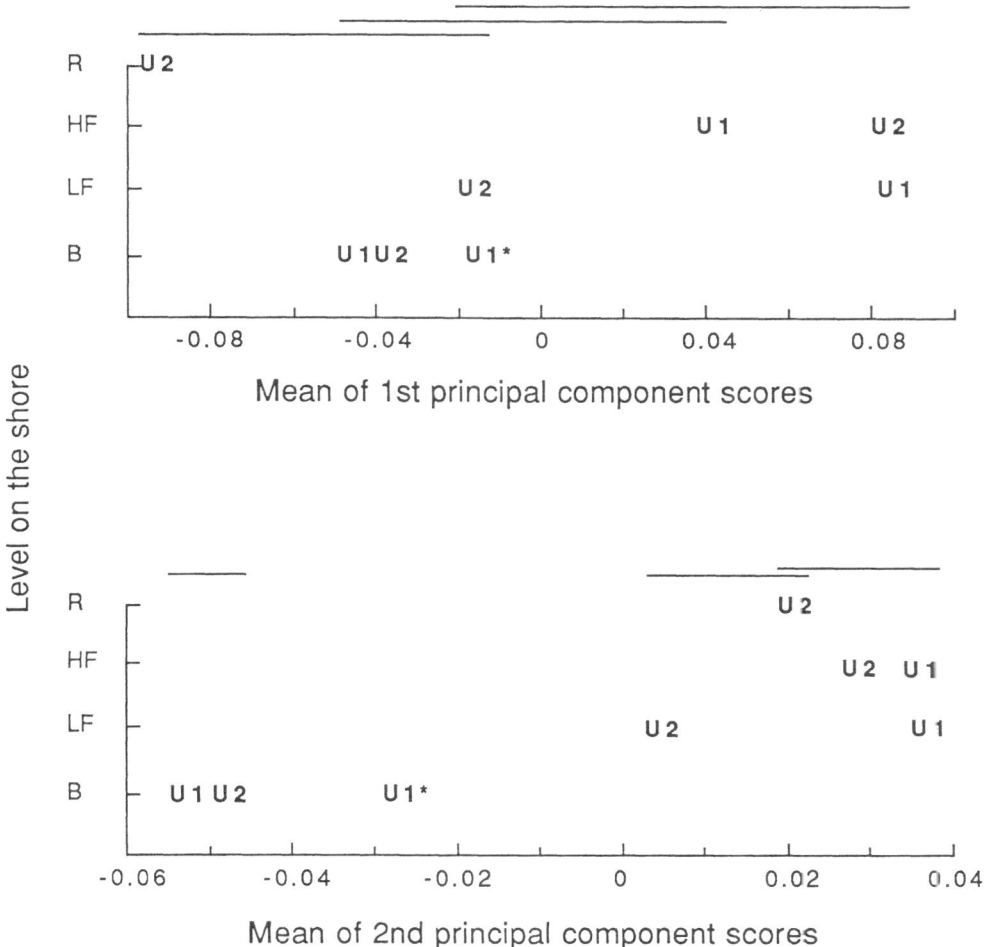

*Fig. 7.* Student-Newman-Keuls test for differences among samples in scores for the two first components from the Ursholmen analysis. Samples from two transects (U1 and U2) were taken from different shore levels; the barnacle belt (**B**), the low littoral (LF), the high littoral fringe (HF), and a high shore rockpool (R). Mean of component scores is indicated by the position of the transect abbreviation. U1* is an additional sample taken 2 m from U1. Horizontal lines join those samples which are not significantly different ($P > 0.05$).

analysis we judge that the first eigenvector represents size while the second axis mainly reflects shape differences.

A plot of the scores along the first axis reveals two groups of samples with significantly different mean sizes of shells ($P < 0.05$ SNK procedure). Snails from the H2B, H1B, and H2LF samples are smaller than those from the H1HF, H2HF, and H1LF samples. On the other hand, there are no significant shape differences among the samples, except for one high fringe sample (H2HF, $P < 0.05$ SNK procedure) which separ-

ation is mainly due to thinner shells, higher aperture lips and lower spires of these snails.

Thus, at Ursholmen there are pronounced differences in shell shape between barnacle zone and littoral fringe samples, but all snails are similar in qualitative characters at this site (Table 2). At Herdla, on the other hand, the differences in qualitative characters found among barnacle zone and littoral fringe samples are not accompanied by large differences in shell shape

## Discussion

Although *L. neglecta* has been separated from *L. saxatilis* on morphological criteria, the morphological descriptions given in the literature are far from consistent. For example, some authors describe *L. neglecta* as having a short blunt spire (Hannaford Ellis, 1980; Fretter & Graham, 1980), while James (1968) describes the subspecies *L. saxatilis neglecta* as having 'a tall to very tall spire'. *L. neglecta* is generally said to have a smooth shell (James, 1968; Raffaelli, 1982; Hannaford Ellis, 1980; Fish & Sharp, 1985), but Heller (1975) states that the shell may range from smooth to heavily ridged. The characters which seem to be the most generally accepted as those of *L. neglecta* (see e.g. James, 1968; Heller, 1975; Raffaelli, 1979; Fretter & Graham, 1980; Hannaford Ellis, 1980; Fish & Sharp, 1985) are its size (up to between 4 and 5 mm shell height), a dark band running into the aperture, and habitat, the species being most often confined to the barnacle zone of rocky shores. (We have re-examined what Robertson & Mann, 1982 called *L. neglecta*, living in an eelgrass bed in Nova Scotia, and we found that this population is *L. saxatilis* of the 'tenebrosa'-form. James' (1968) statement that *L. neglecta* occurs in salt marshes seems questionable to us). The most extensive description of *L. neglecta* including an examination of material from the type locality on the east coast of England, is given by Fish & Sharp (1985). Their description of *L. neglecta*, as well as most other morphological studies of the species (e.g. Heller, 1975; Raffaelli, 1979; Hannaford Ellis, 1980), is, however, based on samples from Wales.

Our barnacle zone sample from Cable Bay on Anglesey in Wales is the one which has the greatest number of snails with qualitative characters and size at maturity most closely fitting the description of *L. neglecta* by Fish & Sharp (1985). Also in Herdla, Port St. Mary and Sandgerði, snails with these characters may be found in the barnacle zone. However, the morphometric analysis of all the sites pooled indicates that there are differences in shape among barnacle zone samples from different sites while samples from different levels on the same shore are more similar in shape (Figs. 2 & 3). Thus although at all of our sites (except Ursholmen, see below) it is possible to pick out snails similar in qualitative characters and size at maturity to those that have been described as *L. neglecta*, there is no consistent difference in shape among the snails living in the barnacle zone and those living in the littoral fringe. Furthermore, we find gradual transitions or no essential differences at all, in the occurrence of the qualitative characters described as typical for *L. neglecta* between mid- and high-shore samples, and the differences between sites may be larger than differences between different shore levels within sites (Table 2).

The barnacle zone samples from Ursholmen were the least similar to the standard descriptions of *L. neglecta*. For example, all snails at this site lack a dark band running into the aperture, and size at maturity does not differ much among mid- and high-shore snails. The shapes of the barnacle zone snails of Ursholmen are also furthest away from that of the Cable Bay snails (Fig. 3). Thus, in contrast to the situation at the other sites, snails of the *neglecta*-form are absent from Ursholmen.

The analysis of morphological variation in samples from Ursholmen has demonstrated that variation in shell shape of *L. saxatilis* along the intertidal gradient may be as great as the variation along a transect which includes both *L. neglecta* and *L. saxatilis*. On the other hand, the Herdla analysis shows that large differences in qualitative characters are not necessarily accompanied by large differences in shape as assessed by the PCA. This implies that the qualitative and the quantitative characters used to define *L. neglecta* are not closely coupled.

The high proportion of intermediate morphs, the large between-site variation recorded in the present study, and the lack of a relationship between differences in qualitative and quantitative characters, do not support the case for the specific status of *L. neglecta*. If there are two reproductively isolated populations within our material they could not be distinguished by our analysis of a range of morphological characters, many of which have been used by other authors to distinguish between *L. neglecta* and *L. saxatilis*. Based on a genetic study of partly the same material (K. Johannesson & B. Johannesson, this volume) we have reached the same conclusion that there are no reproductive barriers between the different morphs.

Although, as we argue in the introduction, principal component analysis is far better than regressions or ratios to compare size and shape differences among snails, the result of the PCA may be somewhat difficult to interpret if one pools samples of different origin within the same analysis. This difficulty can be exaggerated if there are differences in shape with size (e.g. Pimentel, 1979; Thorpe, 1983a, 1983b; Reyment *et al.*, 1984; Corruccini, 1987). We may have allometric tendencies in those cases where small individuals have shapes different to those of large individuals in the same samples, as in Figs. 2, 4 and 6. If there are allometry effects, size and shape may not be completely separated in our analyses. The results of the PCA are, however, verified by direct examination of the size and shape variation among the samples.

Morphological variation among populations of direct developing species of *Littorina* living in different habitats, have been reported several times before (e.g. Raffaelli, 1979; Reimchen, 1981; Naylor & Begon, 1982; Janson & Sundberg, 1983). Our investigation illustrates that morphological variation may be found among units of populations that are smaller than previously assumed, and may be found not only between different shore types (e.g. boulder and cliff shores) but also within the same type of habitat.

The seashore is heterogeneous and many environmental parameters change rapidly over the intertidal gradient. There may also be substantial differences among different sites, even if they are considered as being of the same general type, e.g. exposed rocky cliff. In *L. saxatilis*, migration rates (Janson, 1983), and gene flow (Janson & Ward, 1984, K. Johannesson & B. Johannesson, this volume) are small over distances of a few metres or less, and this may result in both phenotypic and genotypic variation at a scale of only a few metres. Dramatic shifts from one phenotype to another have also been shown to take place over distances of 5 to 50 m where a brackish water seagrass bed borders a boulder shore (Janson & Ward, 1985), or where a boulder shore borders a rocky cliff face (Janson & Sundberg, 1983; Janson & Ward, 1984).

It seems likely that on a rocky shore the barnacle belt and the open rock surface of the littoral fringe have different sets of selective and environmental factors which are strong enough to induce a steep phenotypic cline within a relatively sedentary species like *L. saxatilis*. In Sweden, the barnacle belt is temporary, due to ice scouring, and narrow, due to small tides. As a consequence, factors promoting a *neglecta* form are not present all the time, although there are still large differences in environmental factors between mid- and high-shore. The lack of *L. neglecta*-like snails on Ursholmen might be due to this.

Even if it is easy to understand that *L. saxatilis* may be polymorphic in many characters, a detailed understanding of the reasons why specific characters are present in some snails but not in others demands the study of, heritability, selection, stochastic processes, genetic and mechanical correlations among characters, and phenotypic plasticity for each subpopulation of interest (e.g. Gould & Lewontin, 1979; Falconer, 1981; Endler, 1986). So although we might speculate about the adaptive significance of certain morphological characters in different forms of *L. saxatilis* (see e.g. Raffaelli, 1982; Faller-Fritsch & Emson, 1985; Janson, 1986; Sundberg, 1988) these hypotheses will remain speculative as long as we do not know the relative importance of each of the factors mentioned above. Before general conclusions can be drawn in any study of *L. saxatilis*, it is important to consider the micro-scale structure and semi-isolation of the subpopulations within this species.

Besides the micro-scale structure of populations, a complicating factor of the analysis of phenotypic variation in *L. saxatilis* is the possibility of temporal variation in, for example, selection and environmental influences. This makes the time and method of sampling most critical to the outcome of the study. The way samples are taken and also measured are at least as important as their later numerical analyses. To obtain a deeper understanding of the causes and significances of the extensive morphological variation in *L. saxatilis* it seems therefore necessary to follow individuals within subpopulations throughout their lives.

## Acknowledgements

The trip to Iceland was funded by the Nordic Council of Marine Biology, and we thank Prof. Agnar Ingolfsson for his hospitality during the stay at the Biology Institute of Reykjavik. We are also grateful to Eeva Furman for samples from Anglesey, Friederike Schneider for 'L. neglecta' from Nova Scotia, Jan Sundberg and Hans G. Hansson for their help with computer programs, and to Per Sundberg and Dave Raffaelli for their comments on the manuscript. This study was supported by a research grant to one of us (K.J.) from the Swedish Natural Sciences Research Council.

## References

Corruccini, R. S., 1987. Shape in morphometrics: comparative analyses. Am. J. phys. Anthrop. 73: 289–303.

Endler, J. A., 1986. Natural selection in the wild. Princeton Univ. Press, Princeton, N.J., 337 pp.

Falconer, D. S., 1981. Introduction to quantitative genetics. Longman, Lond., 340 pp.

Faller-Fritsch, R. J. & R. H. Emson, 1985. Causes and patterns of mortality in *Littorina rudis* (Maton) in relation to intraspecific variation: A review. In P. G. Moore & R. Seed (eds), The ecology of rocky coasts. Hodder and Stoughton, Lond.: 157–177.

Fish, J. D. & L. Sharp, 1985. The ecology of the periwinkle, *Littorina neglecta* Bean. In P. G. Moore & R. Seed (eds), The ecology of rocky coasts. Hodder and Stoughton, Lond.: 143–156.

Fretter, V. & A. Graham, 1980. The prosobranch molluscs of Britain and Denmark. Part 5 – Marine Littorinacea. J. moll. Stud. suppl. 7.

Gould, S. J. & R. C. Lewontin, 1979. The spandrels of San Marco and the Panglossian paradigm: a critique of the adaptationist programme. Proc. r. Soc. Lond. B 205: 581–598.

Hannaford Ellis, C. J., 1979. Morphology of the oviparous rough winkle, *Littorina arcana* Hannaford Ellis 1978, with notes on the taxonomy of the *L. saxatilis* species-complex (Prosobranchia: Littorinidae). J. Conch. 30: 43–56.

Hannaford Ellis, C. J., 1980. British rough winkles: aspects of their anatomy, taxonomy and ecology. Ph. D. Thesis, University of Liverpool.

Hannaford Ellis, C. J., 1983. Patterns of reproduction in four *Littorina* species J. moll. Stud. 49: 98–106.

Hannaford Ellis, C. J., 1984. Ontogenetic change in shell colour patterns in *Littorina neglecta* Bean (1844). J. Conch. 31: 343–347.

Heller, J., 1975. The taxonomy of some British *Littorina* species, with notes on their reproduction (Mollusca: Prosobranchia). J. linn. Soc., Zool. 56: 131–151.

James, B. L., 1968. The characters and distribution of the subspecies and varieties of *Littorina saxatilis* (Olivi, 1792) in Britain. Cah. Biol. mar. 9: 143–165.

Janson, K., 1983. Selection and migration in two distinct phenotypes of *Littorina saxatilis* in Sweden. Oecologia (Berlin) 59: 58–61.

Janson, K., 1985. A morphologic and genetic analysis of *Littorina saxatilis* (Prosobranchia) from Venice, and on the problem of *saxatilis-rudis* nomenclature. Biol. J. linn. Soc. 24: 51–59.

Janson, K., 1986. Polymorphisms, causes and evolutionary consequences in a marine prosobranch species, *Littorina saxatilis*. Ph.D. thesis, University of Göteborg.

Janson, K. & P. Sundberg, 1983. Multivariate morphometric analysis of two varieties of *Littorina saxatilis* from the Swedish west coast. Mar. Biol. 74: 49–53.

Janson, K. & R. D. Ward, 1984. Microgeographic variation in allozyme and shell characters in *Littorina saxatilis* Olivi (Prosobranchia: Littorinidae). Biol. J. linn. Soc. 22: 289–307.

Janson, K. & R. D. Ward, 1985. The taxonomic status of *Littorina tenebrosa* Montagu as assessed by morphological and genetic analyses. J. Conch. 32: 9–15.

Jolicœur, P., 1963. The multivariate generalization of the allometry equation. Biometrics 19: 497–499.

Manly, F. J., 1986. Multivariate statistical methods: A primer. Chapman & Hall, Lond., 159 pp.

Naylor, R. & M. Begon, 1982. Variation within and between populations of *Littorina nigrolineata* Gray on Holy Island, Anglesey. J. Conch. 31: 17–30.

Neff, N. A. & L. F. Marcus, 1980. A survey of multivariate methods for systematics. Privately published, N.Y., 243 pp.

Pimentel, R. A., 1979. Morphometrics: the multivariate analysis of biological data. Kendall & Hunt, Dubuque, Iowa, 276 pp.

Raffaelli, D., 1979. The taxonomy of the *Littorina saxatilis* species-complex, with particular reference to the systematic status of *Littorina patula* Jeffrys. J. linn. Soc., Zool. 65: 219–232.

Raffaelli, D., 1982. Recent ecological research on some European species of *Littorina*. J. moll. Stud. 48: 342–354.

Reimchen, T. E., 1981. Microgeographical variation in *Littorina marie* Sacchi & Rastelli and a taxonomic consideration. J. Conch. 30: 341–350.

Reyment, R. A., R. E. Blackith & N. A. Campbell, 1984. Multivariate morphometrics. Academic Press, Lond., 233 pp.

Robertson, A. J. & K. H. Mann, 1982. Population dynamics and life history adaptations of *Littorina neglecta* Bean in an eelgrass meadow (*Zostera marina* L.) in Nova Scotia. J. exp. mar. Biol. Ecol. 63: 151–171.

Sneli, J. A. & P. van Marion, 1979. Nye strandsnegler i norsk fauna. Fauna 32: 4–8.

Sokal, R. R. & F. J. Rohlf, 1969. Biometry. Freeman, San Francisco, 776 pp.

Sundberg, P., 1988. Microgeographic variation in shell characters of *Littorina saxatilis* Olivi – a question mainly of size? Biol. J. linn. Soc. 35: 169–184.

Thorpe, R. S., 1983a. A biometric study of the effects of growth on the analysis of geographic variation: Tooth number in Green geckos (Reptilia: *Phelsuma*). J. Zool., Lond. 201: 13–26.

Thorpe, R. S., 1983b. A review of the numerical methods for recognizing and analysing racial differentiation. In J. Felsenstein (ed.), Numerical Taxonomy. Springer-Verlag, Berlin: 404–423.

*Hydrobiologia* **193**: 89–97, 1990.
*K. Johannesson, D. G. Raffaelli and C. J. Hannaford Ellis (eds), Progress in Littorinid and Muricid Biology.*
© 1990 *Kluwer Academic Publishers.*

# Genetic variation within *Littorina saxatilis* (Olivi) and *Littorina neglecta* Bean: Is *L. neglecta* a good species?

Kerstin Johannesson & Bo Johannesson
*Tjärnö Marine Biological Laboratory, Pl 2781, S-452 00 Strömstad, Sweden*

*Key words:* Prosobranchia, sibling species, genetic diversity, electrophoresis, ecotypes, population differentiation

## Abstract

Genetic variation was compared within- and between-samples of *Littorina saxatilis* and *L. neglecta* from five geographic regions of western Europe (Iceland, Norway, Sweden, Isle of Man and Anglesey). The variation at five highly polymorphic enzymes (*Aat-1*, *Pgm-1*, *Pgi*, *Mpi* and *Np*) were revealed in samples from eleven vertical transects extending upshore from the barnacle zone to the upper littoral fringe. Both morphological types, *L. saxatilis* and *L. neglecta*, were present in all geographic regions except in Sweden.

The results of the genetic analyses show that at four of the five loci between 83 and 95% of the between-sample variation was due to differentiation between geographic areas, while only 4% or less was attributable to differentiation between barnacle zone and high littoral fringe samples. An accompanying morphological study revealed that the barnacle zone snails were mostly of *L. neglecta* type, except in the Swedish locality (where although they were distinct from the upper shore snails they were not in accordance with the description of *L. neglecta*), and the littoral fringe snails were of *L. saxatilis* type. The conclusion is therefore that there is more gene flow between *L. neglecta* and *L. saxatilis* type snails within the same locality than there is between snails of similar morphological type, but from geographically separated shores. Although we have not examined material from the type locality of *L. neglecta*, we suggest it to be a junior synonym of *L. saxatilis*.

One locus, *Aat-1*, was, in contrast to the other polymorphic loci, more differentiated over the vertical transects (68% of the between sample variation was attributable to differences between barnacle and high littoral fringe samples) than over the different geographic areas (21%). However, two observations indicated selective rather than stochastic differentiation at *Aat-1*: (1) The same pattern was found independent whether or not *L. neglecta* was present. (2) The much smaller degree of differentiation at the other polymorphic loci indicated a gene flow which would prevent such a large differentiation at *Aat-1* solely by random genetic drift.

## Introduction

As remarked in an accompanying paper (B. & K. Johannesson, this volume) the taxonomic status of the barnacle-dwelling species *Littorina neglecta* Bean, 1844 has not been questioned (e.g. Fretter & Graham, 1980; Hannaford Ellis, 1980; Raffaelli, 1982; Fish & Sharp, 1985) since Heller

(1975) split the former *Littorina saxatilis* into four distinct species. *L. neglecta* was separated from *L. saxatilis* on the basis of shell form, size at maturity, shell ornamentation and colour, and morphology of the penis (Heller, 1975).

Although most taxonomic and ecological work on *L. neglecta* has been undertaken on Welsh populations, the species is reported to be widespread and has been recorded from both the west and east coasts of the British Isles, from Isles of Scilly to the Shetlands (Fish & Sharp, 1985), Norway (Sneli & Marion, 1979), Massachusetts, USA (Fretter & Graham, 1980), Belgium and France (Warmoes, pers. commun.).

Biochemical studies together with morphometric analyses have proven valuable for testing hypotheses about the specific status of entities within the *L. saxatilis* group. Thus, for example, Ward & Warwick (1980) and later Ward & Janson (1985) confirmed the specific status of *L. arcana* Hannaford Ellis, while Heller's (1975) *L. patula* Jeffrys, Smith's (1981) *L. rudis* (Maton) and *L. saxatilis*, and the brackish water from *L. tenebrosa* Montagu, were all demonstrated to be morphological varieties of *L. saxatilis* (Raffaelli, 1979; Janson & Ward, 1984; Janson & Ward, 1985).

Only a few enzyme loci have, however, been analysed in populations of *L. neglecta* by means of electrophoresis prior to this study. Both Heller (1975) studying general esterase patterns, and Wilkins & O'Regan (1980), who analysed allele frequency distributions at two enzymes (*Pgi* and *Pgm-1*), found small differences between individuals of *L. neglecta* and *L. saxatilis* and they interpreted these as indications of a reproductive barrier.

To evaluate whether or not *L. neglecta* is a separate species, we sampled a number of transects from the barnacle zone (the micro-habitat of *L. neglecta*) and up to the high littoral fringe (the typical habitat of *L. saxatilis*), within geographically distant localities of western Europe both from places where *L. neglecta* is reported to occur (Norway, Anglesey, Isle of Man and Belgium), and from areas where *L. neglecta* has, so far, not been recorded (Iceland and Sweden).

Morphological analyses of partly the same material (B. & K. Johannesson this volume), indicated that snails of *L. neglecta* phenotype were present at all study sites except the Swedish one. However, although there was an extensive variation at some sites over the intertidal gradient, the phenotypic variation among samples of barnacle-dwelling snails and among samples of littoral fringe snails was as large as the within-site variation, or even larger. Furthermore the within-site variation was continuous rather than discrete, and quantitative and qualitative characters did not appear associated with each other.

However, the morphological study alone was not sufficient to reject the possibility that two species were present in our material, since there could be two sibling species present with a high degree of morphological overlap (as is the case with *L. arcana* and *L. saxatilis*; Ward & Janson, 1985). If barnacle zone and littoral fringe snails were separate species, samples from the same shore levels of different geographical localities should be more genetically similar to each other than to neighbouring samples from another level on the same shore. Such a pattern of genetic variation would indicate gene flow within each morphological type of snail, but a reproductive barrier between *L. neglecta* and *L. saxatilis*. The present study was designed to test this hypothesis.

## Study sites

Snails from seven localities (Fig. 1), Akranes (A-) and Sandgerði in Iceland (S-), Herdla in Norway (H-), Ursholmen in Sweden (U-), Port St. Mary on the Isle of Man (P-), Cable Bay on Anglesey, UK (C-), and Oostende in Belgium (O-), were sampled. Samples from two parallel vertical transects (-1-, -2-) were analysed from Akranes, Sandgerði, Herdla, Ursholmen, and Port St. Mary (Fig. 2), and three samples were obtained from each transect with the lowest sample from the barnacle zone (-B), the intermediate sample from the low littoral fringe (-LF), and the uppermost sample from the high littoral fringe (-HF). In transect S2 the low fringe sample is missing. All

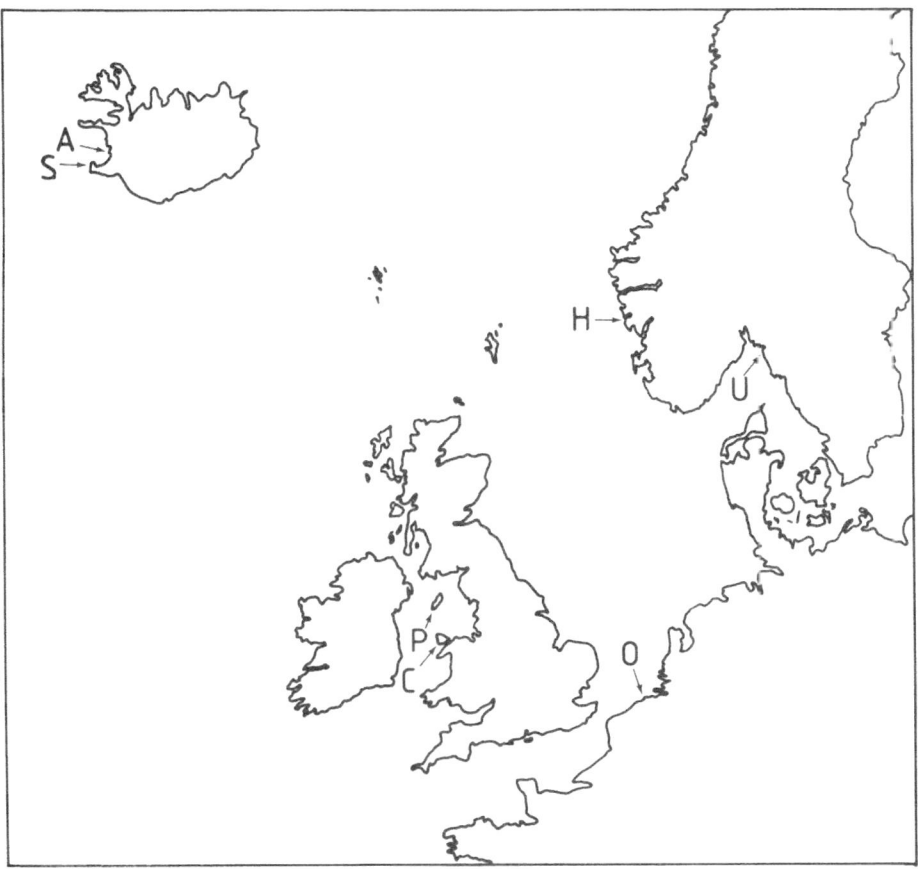

*Fig. 1.* Geographic areas from which *Littorina* of *saxatilis* and *neglecta* type were sampled for genetic analysis. The localities are abbreviated as follows: Akranes, Iceland (A), Sandgerði, Iceland (S), Herdla, Norway (H), Ursholmen, Sweden (U), Cable Bay, Anglesey, Wales (C), Port St. Mary, Isle of Man (P), and Oostende, Belgium (O).

these sites are of natural rocks and exposed to moderate to heavy wave action. The two samples from Oostende were both from the barnacle zone, but one was from a comparatively exposed vertical surface of a dike (O1B) whilst the other was from a more sheltered gently sloping breakwater (O2B) about 2 km from the dike. Six samples along two transects from Cable Bay were collected and sent to us by Eeva Furman (Menai Bridge), but unfortunately four of these included a high proportion of *L. arcana* and had to be excluded from the analyses. The remaining Welsh samples, one from the barnacle zone (CB) and one from the high littoral fringe (CHF) of a comparatively exposed rocky shore, could possibly include one or two juvenile or male *L. arcana*, but

the great majority are of *L. saxatilis* and *L. neglecta* types.

## Materials and methods

Each sample was collected within one square metre such that all size classes of snails (> 2 mm) were represented. No sorting or typing of the material was made prior to analysis, but either the whole sample or a random subsample was subsequently analysed. Fifteen of the 33 samples of this study were analysed morphometrically (B. & K. Johannesson, this volume) and this indicated that the barnacle zone samples from Sandgerði, Herdla, Port St. Mary and Cable Bay consisted

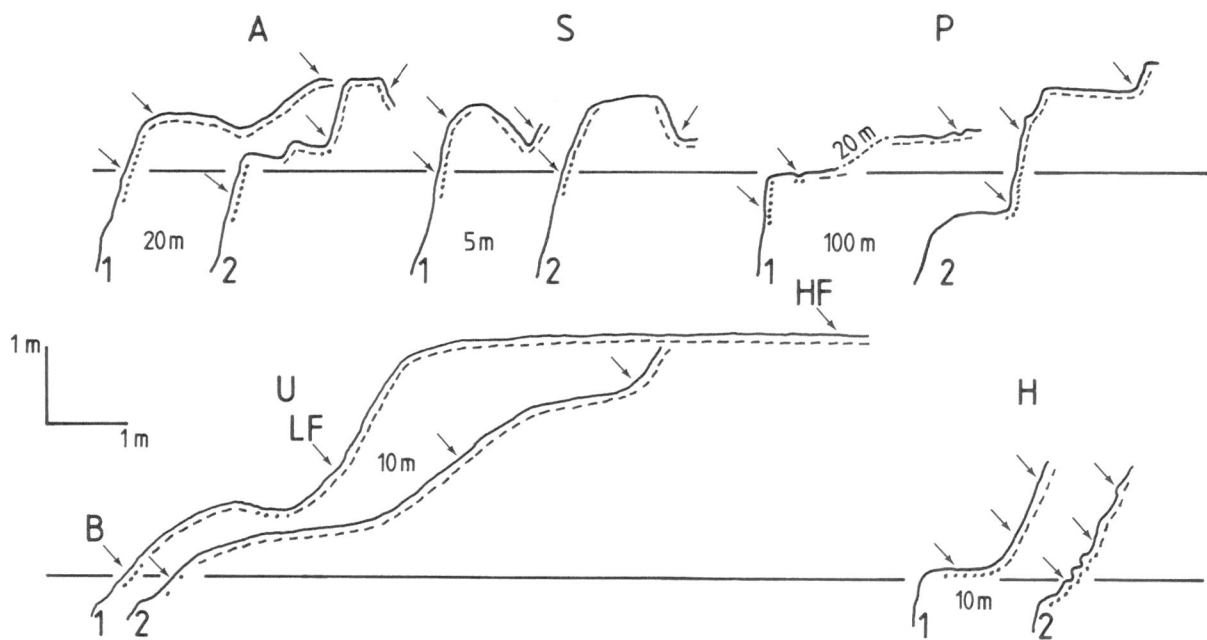

*Fig. 2.* Position of samples along the transects from Akranes (A), Sandgerđi (S), Port St. Mary (P), Ursholmen (U), and Herdla (H). In each transect a low sample from the barnacle zone (B), a high sample from the high littoral fringe (HF), and (except in S2) an intermediate sample from the low littoral fringe (LF) are included. The transects are all to the same scale and the distances between parallel transects (1 and 2) at each locality are indicated. The positions of the black lichen (broken line) and the barnacle zones (dotted line) are indicated. The horizontal line indicates mean high water of neap tides (MHWN).

of more or less *L. neglecta*-like snails, while the Ursholmen samples did not include snails of *L. neglecta* type. The Belgian samples were not analysed morphometrically but the majority of the snails from the dike in Oostende (O1B) were judged by us to be of *L. neglecta* type (material from this site has previously been examined by T. Warmoes who concluded that *L. neglecta* was present), while the breakwater sample (O2B) consisted of typical *L. saxatilis*.

Eleven enzyme loci (allele designations as in Ward & Warwick, 1980; Janson & Ward, 1984) were analysed in all samples and of these five (*Xdh, Idh-1, Mdh-1, Hdh* and *Est-1*) were monomorphic and five (*Aat-1, Pgm-1, Pgi, Mpi* and *Np*) were polymorphic. The monomorphic loci are not considered in the evaluation of the results. A further nine loci were analysed for some of the samples but were excluded from the study as they could not be resolved in all samples. Furthermore, one locus (*Est-2*) which was polymorphic, has a complex genetic basis (see e.g. Janson & Ward,

1984) and was excluded for that reason. On average 46 individuals (range 20 to 97) were analysed for the polymorphic loci. In two samples (A2HF and A1LF), however, only ten individuals were analysed for the enzyme *Np*. Due to technical problems *Pgm-1* failed to appear in the sample H1HF, and to be able to incorporate this sample in the genetic analyses we have copied the allele frequencies of the parallel sample (H2HF).

## Results

There was extensive genetic variation within and between samples in the five polymorphic enzymes (Table 1). The sources of this variation – the degree of genetic diversity within- and between-samples, and between habitats and geographic areas – were traced by a genetic diversity analysis (Nei, 1973; Chakraborty, 1980). In this analysis the low fringe samples, due to being intermediate in habitat type, and the Cable Bay and Oostende

Table 1. Genetic variation at five polymorphic loci of *Littorina* of *saxatilis* and *neglecta* type from seven localities in western Europe.

| Sample | Aat-1 | | Pgi | | | | Mpi | | | | Pgm-1 | | | | | Np | | | |
|---|---|---|---|---|---|---|---|---|---|---|---|---|---|---|---|---|---|---|---|
| | 120 | 100 | 110 | 100 | 90 | 80 | 140 | 120 | 100 | 75 | 115 | 105 | 100 | 85 | 175 | 140 | 100 | 65 | 35 |
| A1B | 0.314 | 0.686 | 0.021 | 0.670 | 0.309 | – | 0.008 | 0.539 | 0.320 | 0.164 | – | 0.305 | 0.506 | 0.188 | 0.011 | 0.330 | 0.489 | 0.170 | – |
| A1LF | 0.755 | 0.245 | 0.009 | 0.724 | 0.267 | – | – | 0.409 | 0.409 | 0.182 | – | 0.309 | 0.573 | 0.118 | 0.050 | 0.400 | 0.500 | 0.050 | – |
| A1HF | 0.917 | 0.083 | 0.006 | 0.718 | 0.276 | – | – | 0.230 | 0.620 | 0.150 | – | 0.300 | 0.541 | 0.153 | – | 0.467 | 0.450 | 0.083 | – |
| A2B | 0.203 | 0.797 | 0.019 | 0.699 | 0.282 | – | 0.006 | 0.590 | 0.244 | 0.160 | – | 0.301 | 0.583 | 0.090 | 0.050 | 0.150 | 0.575 | 0.200 | 0.025 |
| A2LF | 0.816 | 0.184 | – | 0.669 | 0.331 | – | – | 0.519 | 0.338 | 0.144 | – | 0.299 | 0.565 | 0.136 | 0.025 | 0.275 | 0.600 | 0.075 | 0.025 |
| A2HF | 0.991 | 0.009 | – | 0.709 | 0.273 | 0.018 | – | 0.412 | 0.353 | 0.235 | – | 0.692 | 0.269 | 0.038 | – | 0.200 | 0.600 | 0.200 | – |
| S1B | 0.095 | 0.905 | 0.012 | 0.786 | 0.202 | – | – | 0.059 | 0.485 | 0.456 | – | 0.476 | 0.298 | 0.226 | – | 0.321 | 0.583 | 0.095 | – |
| S1LF | 0.700 | 0.300 | – | 0.813 | 0.188 | – | – | 0.211 | 0.447 | 0.342 | – | 0.462 | 0.205 | 0.333 | – | 0.346 | 0.641 | 0.013 | – |
| S1HF | 0.853 | 0.147 | 0.024 | 0.786 | 0.190 | – | – | 0.038 | 0.450 | 0.513 | – | 0.525 | 0.250 | 0.225 | – | 0.333 | 0.583 | 0.083 | – |
| S2B | 0.303 | 0.697 | 0.028 | 0.717 | 0.255 | – | – | 0.163 | 0.490 | 0.346 | 0.006 | 0.527 | 0.203 | 0.270 | – | 0.324 | 0.546 | 0.120 | 0.009 |
| S2HF | 0.707 | 0.293 | 0.036 | 0.774 | 0.190 | – | 0.024 | 0.238 | 0.393 | 0.369 | – | 0.512 | 0.214 | 0.274 | 0.012 | 0.333 | 0.583 | 0.071 | – |
| H1B | 0.083 | 0.917 | 0.048 | 0.821 | 0.131 | – | – | 0.786 | 0.190 | – | – | 0.153 | 0.708 | 0.139 | – | – | 0.881 | 0.119 | – |
| H1LF | 0.538 | 0.462 | 0.038 | 0.700 | 0.263 | – | – | 0.663 | 0.338 | – | – | 0.324 | 0.459 | 0.216 | – | 0.025 | 0.875 | 0.100 | 0.012 |
| H1HF | 0.706 | 0.294 | 0.037 | 0.768 | 0.195 | – | – | 0.575 | 0.425 | – | – | 0.347 | 0.444 | 0.208 | – | – | 0.829 | 0.159 | 0.024 |
| H2B | 0.024 | 0.976 | 0.024 | 0.878 | 0.098 | – | 0.014 | 0.838 | 0.149 | – | – | 0.229 | 0.657 | 0.114 | – | – | 0.890 | 0.085 | 0.013 |
| H2LF | 0.319 | 0.681 | 0.013 | 0.838 | 0.150 | – | – | 0.713 | 0.275 | 0.013 | – | 0.311 | 0.527 | 0.162 | – | – | 0.788 | 0.200 | 0.013 |
| H2HF | 0.400 | 0.600 | 0.039 | 0.895 | 0.066 | – | – | 0.811 | 0.189 | – | – | 0.347 | 0.444 | 0.208 | – | – | 0.795 | 0.192 | 0.024 |
| U1B | 0.382 | 0.618 | – | 0.905 | 0.095 | – | – | 0.536 | 0.464 | – | – | 0.036 | 0.881 | 0.083 | – | – | 0.214 | 0.762 | – |
| U1LF | 0.952 | 0.048 | – | 0.845 | 0.155 | – | – | 0.571 | 0.429 | – | – | 0.012 | 0.952 | 0.036 | – | – | 0.262 | 0.738 | – |
| U1HF | 0.940 | 0.060 | – | 0.893 | 0.107 | – | – | 0.571 | 0.429 | – | – | 0.012 | 0.905 | 0.083 | – | – | 0.155 | 0.810 | 0.036 |
| U2B | 0.405 | 0.595 | – | 0.869 | 0.131 | – | – | 0.488 | 0.512 | – | – | 0.048 | 0.917 | 0.036 | – | 0.095 | 0.905 | – | – |
| U2LF | 0.905 | 0.095 | – | 0.821 | 0.179 | – | – | 0.583 | 0.417 | – | – | 0.036 | 0.929 | 0.036 | – | 0.190 | 0.798 | 0.012 | – |
| U2HF | 0.929 | 0.071 | – | 0.917 | 0.083 | – | – | 0.583 | 0.417 | – | – | 0.036 | 0.917 | 0.048 | – | 0.202 | 0.786 | 0.012 | – |
| P1B | 0.059 | 0.941 | – | 0.568 | 0.432 | – | – | 0.759 | 0.233 | 0.009 | – | 0.282 | 0.191 | 0.527 | – | 0.068 | 0.432 | 0.407 | 0.093 |
| P1LF | 0.154 | 0.846 | 0.019 | 0.620 | 0.343 | 0.019 | – | 0.792 | 0.208 | – | – | 0.297 | 0.378 | 0.324 | – | 0.028 | 0.454 | 0.463 | 0.056 |
| P1HF | 0.313 | 0.688 | 0.063 | 0.550 | 0.388 | – | – | 0.718 | 0.282 | – | – | 0.459 | 0.297 | 0.243 | – | – | 0.579 | 0.395 | 0.026 |
| P2B | 0.267 | 0.733 | 0.024 | 0.508 | 0.460 | 0.008 | – | 0.705 | 0.269 | 0.026 | – | 0.352 | 0.295 | 0.352 | – | 0.042 | 0.466 | 0.458 | 0.034 |
| P2LF | 0.372 | 0.628 | 0.013 | 0.513 | 0.475 | – | – | 0.738 | 0.238 | 0.025 | – | 0.385 | 0.436 | 0.179 | – | 0.051 | 0.346 | 0.590 | 0.013 |
| P2HF | 0.615 | 0.385 | 0.025 | 0.538 | 0.438 | – | – | 0.731 | 0.256 | 0.013 | – | 0.333 | 0.397 | 0.269 | – | 0.038 | 0.538 | 0.388 | 0.038 |
| CB | 0.049 | 0.951 | 0.060 | 0.762 | 0.179 | – | – | 0.650 | 0.325 | 0.025 | – | 0.551 | 0.308 | 0.141 | – | 0.128 | 0.513 | 0.244 | 0.115 |
| CHF | 0.513 | 0.488 | – | 0.825 | 0.175 | – | 0.013 | 0.625 | 0.350 | 0.013 | – | 0.346 | 0.462 | 0.192 | – | 0.025 | 0.363 | 0.513 | 0.100 |
| O1B (neg) | 0.021 | 0.979 | – | 0.724 | 0.276 | – | – | 0.303 | – | 0.697 | – | 0.140 | 0.550 | 0.310 | – | 0.016 | 0.667 | 0.313 | 0.005 |
| O2B (sax) | 0.017 | 0.983 | – | 0.552 | 0.448 | – | – | 0.414 | – | 0.586 | – | 0.034 | 0.603 | 0.362 | – | 0.069 | 0.707 | 0.224 | – |

A = Akranes, Iceland
S = Sandgerdi, Iceland
H = Herdla, Norway
U = Ursholmen, Sweden
P = Port St. Mary, Isle of Man
C = Cable Bay, Anglesey, UK

O – Oostende, Belgium
B = barnacle zone
LF = low littoral fringe
HF = high littoral fringe
1 = transect 1
2 = transect 2

samples, were omitted, the latter two because they did not include parallel transects.

The overall heterozygosity was in the range 0.4 to 0.6 (Table 2) and between 59 and 92% of this was differentiation within samples ($G_S$). The remainder – the variation between samples ($G_{ST}$) – derived from differentiation between barnacle zone and high littoral fringe habitats ($G_{HT}$) and from differentiation over geographic areas ($G_{AT}$). At *Pgm-1*, *Mpi*, *Pgi* and *Np*, between 83 and 95% of the variation between samples was attributable to genetic divergence between geographic areas ($G_{AT}/G_{ST}$), and only 4% or less was due to differentiation between shore levels (Table 2). The pattern of variation in these four loci indicates that gene flow between mid- and high-shore snails of the same shore is much larger than is gene flow between geographically distinct populations of barnacle zone or littoral fringe snails.

At *Aat-1*, however, most of the between-sample variation (68%) was attributable to differentiation between the barnacle zone and high littoral fringe samples ($G_{HT}/G_{ST}$); only 21% being due to geographic divergence (Table 2). A clinal variation was present in all transects with the frequency of the fast allele *Aat-1*[120] increasing rapidly upshores (Table 1). The same pattern was appearent also in the Swedish locality, yet no snails with *L. neglecta* type morphology were found. This indicates that the within-site differentiation at *Aat-1* is caused by diversifying selection in the two different micro-habitats rather than to a reproductive barrier between *neglecta* and *saxatilis* type snails.

Nei's genetic identity matrix for the polymorphic loci (*Aat-1* excluded) was converted into a dendrogram by the UPGMA procedure (Sokal & Sneath, 1963). The result is a nearly perfect clustering of samples in accordance with the geographical distribution, and differences between barnacle zone and littoral fringe samples within localities are small compared to geographic differences (Fig. 3).

Genotype distributions within each sample of the four loci, *Pgm-1*, *Mpi*, *Pgi* and *Np*, were in Hardy-Weinberg equilibrium. At *Aat-1*, however, four samples, all from the barnacle zone (S2B, A1B, A2B, and U2B), revealed significant deficiencies of heterozygotes ($P < 0.05$). This may be explained by either selection against heterozygotes within this habitat, or a downward migration of high shore snails, which have a different genotype of *Aat-1*.

Significant heterogeneity between samples within a transect was, as expected, found at *Aat-1* in all transects ($P < 0.005$, contingency chi-squared test, see Workman & Niswander, 1970). Whilst at the other polymorphic loci most within-transect differences between littoral fringe and barnacle zone samples were non-significant. Significant heterogeneities ($P < 0.05$) were, however, found at *Pgm-1* (transects: P1, A2 and H1), at *Pgi* (P1), at *Mpi* (S1 and A1), and at *Np* (C), that is,

*Table 2.* Gene diversity analysis of 20 samples of *Littorina* of *saxatilis* and *neglecta* type. The samples are from five different geographic areas, Port St. Mary, Sandgerði, Akranes, Herdla and Ursholmen. In each area two parallel transects were sampled, and within each transect one barnacle zone sample and one high littoral fringe sample are included in the analysis.

| Locus | Gene diversity | | | | | | | Coefficient of gene differentiation | | | | | |
|---|---|---|---|---|---|---|---|---|---|---|---|---|---|
| | $H_T$ | $H_S$ | $D_{ST}$ | $D_{HT}$ | $D_{SH}$ | $D_{AT}$ | $D_{SA}$ | $G_S$ | $G_{ST}$ | $G_{HT}$ | $G_{SH}$ | $G_{AT}$ | $G_{SA}$ |
| *Aat-1* | 0.499 | 0.294 | 0.205 | 0.137 | 0.068 | 0.044 | 0.161 | 0.589 | 0.411 | 0.275 | 0.136 | 0.088 | 0.323 |
| *Pgm-1* | 0.619 | 0.511 | 0.108 | 0.001 | 0.107 | 0.090 | 0.018 | 0.826 | 0.174 | 0.002 | 0.173 | 0.145 | 0.029 |
| *Mpi* | 0.589 | 0.484 | 0.105 | 0.003 | 0.102 | 0.088 | 0.017 | 0.822 | 0.178 | 0.005 | 0.173 | 0.149 | 0.029 |
| *Pgi* | 0.387 | 0.355 | 0.032 | 0.001 | 0.031 | 0.030 | 0.002 | 0.917 | 0.083 | 0.003 | 0.080 | 0.078 | 0.005 |
| *Np* | 0.603 | 0.454 | 0.149 | 0.000 | 0.149 | 0.142 | 0.007 | 0.753 | 0.247 | 0.000 | 0.247 | 0.235 | 0.012 |

| | | | |
|---|---|---|---|
| $H_T$ | total diversity (heterozygosity) | $D_{SH},\ G_{SH}$ | between samples of same habitat type |
| $H_S,\ G_S$ | within samples | $D_{AT},\ G_{AT}$ | between geographic areas |
| $D_{ST},\ G_{ST}$ | between samples | $D_{SA},\ G_{SA}$ | between samples within the same area |
| $D_{HT},\ G_{HT}$ | between habitats (high littoral fringe and barnacle zone) | | |

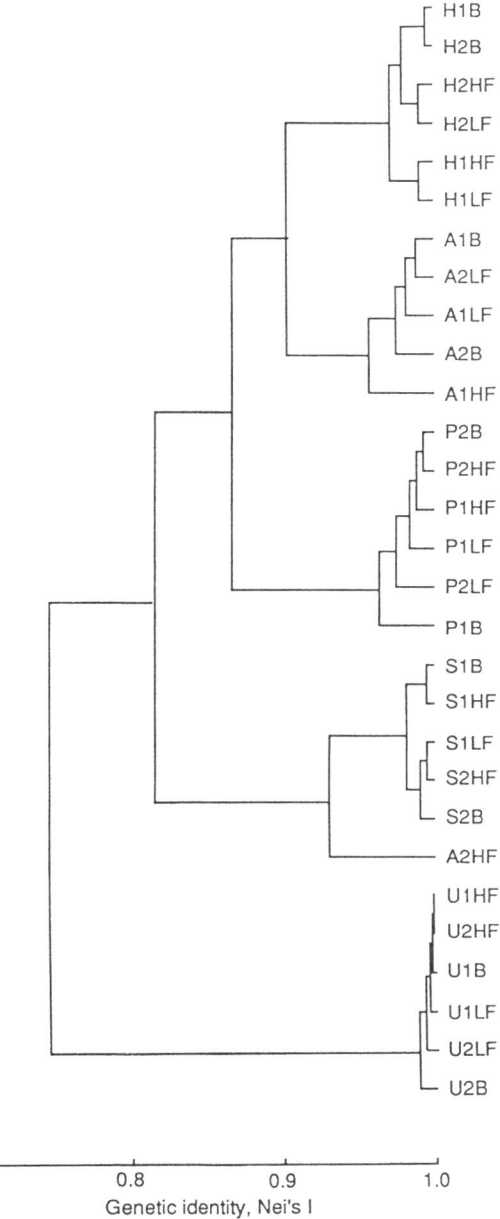

*Fig. 3.* Genetic relationships between 30 samples of *Littorina* of *saxatilis* and *neglecta* type from Akranes (A-), Sandgerdi (S-), Herdla (H-), Ursholmen (U-), and Port St. Mary (P-). Two parallel vertical transects (-1- and -2-) each including samples from the barnacle zone (-B), the low littoral fringe (-LF), and the high littoral fringe (-HF) were analysed from each geographic area. The genetic identity matrix is based on allele frequencies of the four polymorphic loci *Pgm-1*, *Pgi*, *Mpi* and *Np*, and converted into a dendrogram using an UPGMA clustering technique.

in 7 out of 44 possible cases. This suggests that there is not completely random mating between high- and mid-shore snails in, at least, some of the transects.

## Discussion

At four of five highly polymorphic loci we found a pattern of genetic variation which is not consistent with the hypothesis that there were two different species in our samples. Although the variation at the locus *Aat-1* when considered alone, may be interpreted as suggesting the presence of one barnacle zone and one littoral fringe species, a similar clinal variation in the Swedish population, which lacks *L. neglecta* phenotypes, indicates that this locus is selected over the microenvironmental gradient from high- to mid-shore (see also K. & B. Johannesson, 1989). Thus the differences between barnacle zone samples and littoral fringe samples are not caused by a reproductive barrier between the two types of snails.

Although we have support from the accompanying morphological analyses that our barnacle zone samples from Anglesey, the Isle of Man, Norway, and Iceland, included high numbers of snails of *L. neglecta* phenotype, according to the description of *L. neglecta* given by Heller (1975) and Fish & Sharp (1985), they did not form a discrete group but were rather at one end of a continuum of morphologies with snails from the littoral fringe at the other end (B. & K. Johannesson, this volume). This is to be expected if *L. neglecta* is a variety of *L. saxatilis* rather than a distinct species. In addition, the morphological study showed that there was a large between locality variation in the morphology of both the barnacle zone snails and the littoral fringe snails (B. & K. Johannesson, this volume). This is also to be expected as each particular site is likely to differ somewhat, despite the presence of a common component of the *L. neglecta* micro-habitat, i.e. the barnacles.

Our results show that the barnacle zone snails and the littoral fringe snails of our study are conspecific. From this we suggest *L. neglecta* to be a junior synonym of *L. saxatilis*. As we have not, however, examined samples from the type locality of *L. neglecta* (Scarborough, Yorkshire, UK) further biochemical analyses are required before the taxonomic status of *L. neglecta* can be unequivocally assessed.

In *L. saxatilis sensu stricto* genetic differences may occur over distances of 4 metres only (Janson & Ward, 1984), due to limited dispersal (in the range of a few metres, Janson, 1983), and direct development. Thus it seems likely that mating between snails of *L. neglecta* and *L. saxatilis* types is non-random as the two types are not sympatric on a micro-scale and there is likely to be a loss in fitness if they migrate into the habitat of each other. Our study supports the hypothesis of non-random mating as we found small but significant differences between mid- and high-shore samples at the presumably neutral loci (*Pgm-1*, *Np*, *Pgi* and *Mpi*) in some of the transects. Wilkins & O'Regan (1980) also found small differences between samples of *L. neglecta* and *L. saxatilis* at the locus *Pgm-1* at three sites. However, they suggested that this supported the specific status of *L. neglecta*. But although their samples of *L. neglecta* and *L. saxatilis* were from the same shore, it seems likely that they were not strictly sympatric, and thus Wilkins & O'Regan's conclusion may be premature.

A somewhat restricted gene flow and differential selection in the two types of micro-habitats may have established inherited differences in adaptive traits of *L. neglecta* and *L. saxatilis* phenotypes, although this needs to be tested by raising snails of the two types in similar environments.

If, as our results suggest, *L. neglecta* is to be included as an ecotype within *L. saxatilis*, this should not be too surprising. The morphological variability of *L. saxatilis sensu stricto* is by now a well documented fact and forms such as *L. tenebrosa*, E- and S-forms (*sensu* Janson, 1982) are as different from each other as each of them are to *L. neglecta*. Furthermore, intermediate forms are found between the different varieties where the environment changes (Janson & Sundberg, 1983; Janson & Ward, 1985).

Although there may be genetic differences in adaptive traits between different phenotypes of *L. saxatilis*, this study points to the fact that the gene flow between dissimilar ecotypes, e.g. *L. neglecta* and *L. saxatilis*, within the same shore is larger than gene flow between populations of similar phenotypes but from different geographic areas. This is an important distinction between ecotypes and sibling species. In the sibling species pair *L. saxatilis* and *L. arcana*, the differences in allozymes are very small (e.g. Ward, this volume), yet geographically distinct populations of the same species in general group together and are separate from populations of the other species (Knight *et al.*, 1987; Sundberg *et al.*, this volume) indicating that gene flow within species is larger than the rate of hybridisation between species at a sympatric site.

## Acknowledgements

A trip to Iceland was funded by the Nordic Council of Marine Biology, and we thank Prof. Agnar Ingolfsson for his hospitality during our stay at the Biology Institute of Reykjavik. We are also grateful to Eeva Furman and Thierry Warmoes who sent samples from Anglesey and Oostende, respectively. This study was also supported by a research grant to one of us (K.J.) from the Swedish Natural Science Research Council.

## References

Chakraborty, R., 1980. Gene diversity analysis in nested subdivided populations. Appendix 1 in R. Beckwitt. Genetic structure of *Pileolaria pseudomilitaris* (Polychaeta: Spirorbidae). Genetics 96: 711–726.

Fish, J. D. & L. Sharp, 1985. The ecology of the periwinkle, *Littorina neglecta* Bean. In P. G. Moore & R. Seed (eds), The ecology of rocky coasts. Hodder and Stoughton, London: 467 pp.

Fretter, V. & A. Graham, 1980. The prosobranch molluscs of Britain and Denmark. Part 5 – Marine Littorinacea. J. moll. Stud. suppl. 7.

Hannaford Ellis, C. J., 1980. British rough winkles: aspects of their anatomy, taxonomy and ecology. Ph.D. Thesis, University of Liverpool.

Heller, J., 1975. The taxonomy of some British *Littorina* species, with notes on their reproduction (Mollusca: Prosobranchia). J. linn. Soc., Zool. 56: 131–151.

Janson, K., 1982. Phenotypic differentiation in *Littorina saxatilis* Olivi (Mollusca, Prosobranchia) in a small area on the Swedish west coast. J. moll. Stud., 48: 167–173.

Janson, K., 1983. Selection and migration in two distinct phenotypes of *Littorina saxatilis* in Sweden. Oecologia (Berlin) 59: 58–61.

Janson, K. & P. Sundberg, 1983. Multivariate morphometric analysis of two varieties of *Littorina saxatilis* from the Swedish west coast. Mar. Biol. 74: 49–53.

Janson, K. & R. D. Ward, 1984. Microgeographic variation in allozyme and shell characters in *Littorina saxatilis* Olivi (Prosobranchia: Littorinidae). Biol. J. linn. Soc. 22: 289–307.

Janson, K. & R. D. Ward, 1985. The taxonomic status of *Littorina tenebrosa* Montagu as assessed by morphological and genetic analyses. J. Conch. 32: 9–15.

Johannesson, K. & B. Johannesson, 1989. Differences in allele frequencies of *Aat* between high- and mid-rocky shore populations of *Littorina saxatilis* (Olivi) suggest selection in this enzyme locus. Genet. Res., Camb. 57: 7–11.

Knight, A. J., R. N. Hughes & R. D. Ward, 1987. A striking example of the founder effect in the mollusc *Littorina saxatilis*. Biol. J. linn. Soc. 32: 417–426.

Nei, M., 1973. Analysis of gene diversity in subdivided populations. Proc. natn. Acad. Sci. USA 70: 3321–3323.

Raffaelli, D., 1979. The taxonomy of the *Littorina saxatilis* species-complex, with particular reference to the systematic status of *Littorina patula* Jeffrys. J. linn. Soc., Zool. 65: 219–232.

Raffaelli, D., 1982. Recent ecological research on some European species of *Littorina*. J. moll. Stud. 48: 342–354.

Smith, J. E., 1981. The natural history and taxonomy of shell variation in the periwinkles *Littorina saxatilis* and *Littorina rudis*. J. mar. biol. Ass. UK 61: 215–242.

Sneli, J. A. & P. van Marion, 1979. Nye strandsnegler i norsk fauna. Fauna 32: 4–8.

Sokal, R. R. & P. H. A. Sneath, 1963. Principles of numerical taxonomy. Freeman, San Francisco, 359 pp.

Ward, R. D. & K. Janson, 1985. A genetic analysis of sympatric subpopulations of the sibling species *Littorina saxatilis* (Olivi) and *Littorina arcana* Hannaford Ellis. J. moll. Stud. 51: 86–94.

Ward, R. D. & T. Warwick, 1980. Genetic differentiation in the molluscan species *Littorina rudis* and *Littorina arcana* (Prosobranchia: Littorinidae). Biol. J. linn. Soc. 14: 417–428.

Wilkins, N. P. & D. O'Regan, 1980. Generic variation in sympatric sibling species cf *Littorina*. Veliger 22: 355–359.

Workman, P. L. & J. D. Niswander, 1970. Population studies on southwestern Indian tribes. II. Local genetic differentiation in the Papago. Am. J. hum. Genet. 22: 24–49.

*Hydrobiologia* **193**: 99–108, 1990.
*K. Johannesson, D. G. Raffaelli and C. J. Hannaford Ellis (eds), Progress in Littorinid and Muricid Biology.*
© 1990 *Kluwer Academic Publishers.*

# Rapid colonization of Belgian breakwaters by the direct developer, *Littorina saxatilis* (Olivi) (Prosobranchia, Mollusca)

Kerstin Johannesson[1] & Thierry Warmoes[2]
[1] *Tjärnö Marine Biological Laboratory, Pl 2781, S-452 Strömstad, Sweden*; [2] *Laboratorium voor Algemene Dierkunde, Rijksuniversitair Centrum, Groenenborgerlaan 171, B-2020 Antwerpen, Belgium*

*Key words:* population genetics, founder effects, dispersal rate, *Littorina littorea*

## Abstract

The Belgian coast has no natural rocky sites but a number of man-made constructions are colonized by rocky shore organisms. The rough periwinkle, *Littorina saxatilis* (Olivi), lacks a planktonic larval stage but is found on most breakwaters along the Belgian coast, a few built as recently as 1986. This indicates a good potential of dispersal along this sandy shallow coast, nearly as good as for the planktonic developer *Littorina littorea* (L.) which is found on generally the same sites in Belgium. The breakwater populations of *L. saxatilis*, however, tend to be somewhat less variable (level of heterozygosity about 10% less) than non-Belgian *L. saxatilis* populations of natural sites. This suggests that the breakwater populations have passed through bottlenecks when founded, but probably restored population sizes fairly rapidly afterwards. No relationship is found between geographic and genetic distances between populations of *L. saxatilis*.

## Introduction

Little is known about spreading mechanisms for benthic invertebrate species which have a direct development and are sedentary (although not sessile) during their juvenile and adult stages. A number of more or less anecdotal observations or hypothetical arguments (e.g. Malone, 1965; Arnaud *et al.*, 1976; Gerlach, 1977; Highsmith, 1985; Scheltema, 1986) suggest that transport of juveniles, adults and egg masses may take place through rafting, attachment to the legs of birds, in ship ballast and other ways. Although it is generally appreciated that most benthic invertebrate species, especially those which lack a dispersal larval stage, may be passively transported, quantitative estimates of transport rates and possibili-

ties for a successful landing in the new site are almost completely lacking. The importance of passive transport in the biogeography of a direct developing species is thus hard to assess.

Because of reasons discussed elsewhere (Johannesson, 1988), the possibility of passive dispersal of adults and egg masses of benthic species can lead to the surprising conclusion that a direct developer may be more likely to establish populations at distant sites than a species with planktonic development.

The present study aims to estimate transport rates of the direct developer *Littorina saxatilis* (Olivi) over relatively short distances, in the range of hundreds of metres and up to a few tens of kilometres, and will allow an assessment of the dispersal potential of a direct developer, in comparison with the spread of species with plankto-trophic larvae over these fairly small distances. It is generally assumed that over distances in the range of kilometres, a species like *Littorina littorea* (L.) which has planktotrophic larvae should be much more effective in colonizing new substrata than a direct developer, such as *L. saxatilis*.

The Belgian coast offers an opportunity to study the relative colonization rates of these two congeneric species with different reproductive strategies. The coast is about 65 km in length and lacks completely natural rocky intertidal sites, but a number of artificial substrata are present, such as dikes, piers and breakwaters.

Populations of five species of *Littorina* are found in Belgium. *Littorina littorea* and *L. saxatilis* are both common all along the coast, while *L. neglecta* Bean (the specific status of which is however questioned by K. & B. Johannesson this volume), *L. obtusata* (L.) and *L. mariae* Sacchi & Rastelli are found in a few sites only. The occurrence of the different species of *Littorina* in Belgium has recently been described by Warmoes *et al.* (1988).

In this paper, we compare the distributions of *L. saxatilis* and *L. littorea* with special reference to the age of the substratum they occupy. In addition, we analyse the biochemical genetics of the direct developer, *L. saxatilis*, for any patterns of colonization and indications of founder effects.

## Description of *Littorina* sites of Belgium

The main substrata occupied by both species of snails are dikes and breakwaters. Dikes are stone barricades running parallel to the coastline; the first one was built in 1602 outside Oostende. Dikes run along a total of about 40 km of the coast, but only some 5 km at five sites are still in contact with the sea (Fig. 1). Breakwaters run perpendicular to the shore. They are about 400 m long, spaced out at intervals of 200 to 500 m, and mostly built of basalt stones, irregularly dumped in the lower part, but placed and sealed with asphalt in the upper courses. The first break-waters were built between 1815 and 1830, and the

oldest still in place date from around 1880. Some 150 breakwaters are found along 36 km of coastline (Fig. 1). Furthermore, docks, jetties, moles and wooden or concrete railings are associated with the ports of Nieuwpoort (built around 1900), Oostende (around 1440), Blankenberge (around 1860), and Zeebrugge (around 1910).

Only minor stretches of the sandy Belgian coast are left without breakwaters or dikes (Fig. 1). The longest distance is from the harbour of Dunkerque, in France, to the breakwaters of Koksijde, a distance of 18 km. A natural rocky coast is found to the east of Calais, some 60 km west of Koksijde. No natural rocky sites are present in the Netherlands, but here a large variety of man-made littoral substrata are also occupied by rocky shore organisms.

The sublittoral is sandy and gently sloping, reaching a depth of 10 m between 3 and 10 km from the coast of Belgium. A coastal current flows parallel to the shore from west to east, while tidal currents may flow either east to west or west to east. The currents may reach a speed of 5.5 km/h. Tidal amplitude of the area is between 2.8 and 4.3 m.

## Material and methods

Fifty-six sites from De Panne, in the west, to Het Zoute, in the east, were visited and the presence or absence of *L. saxatilis* and *L. littorea* was recorded. Densities of *L. saxatilis* were estimated roughly by counting the number of snails found within a certain period of active search by one of us (T.W.). Information about ages of the different breakwaters and dikes were obtained from the Ministry of Public Works, in Oostende.

Ten enzyme loci were screened in fresh samples of twenty or more individuals of *L. saxatilis* from 18 of the sites (Fig. 1, Table 1), applying standard methods of horizontal starch gel electrophoresis. In this study we use the same allele designations as in earlier studies by Ward & Warwick (1980) and Janson & Ward (1984).

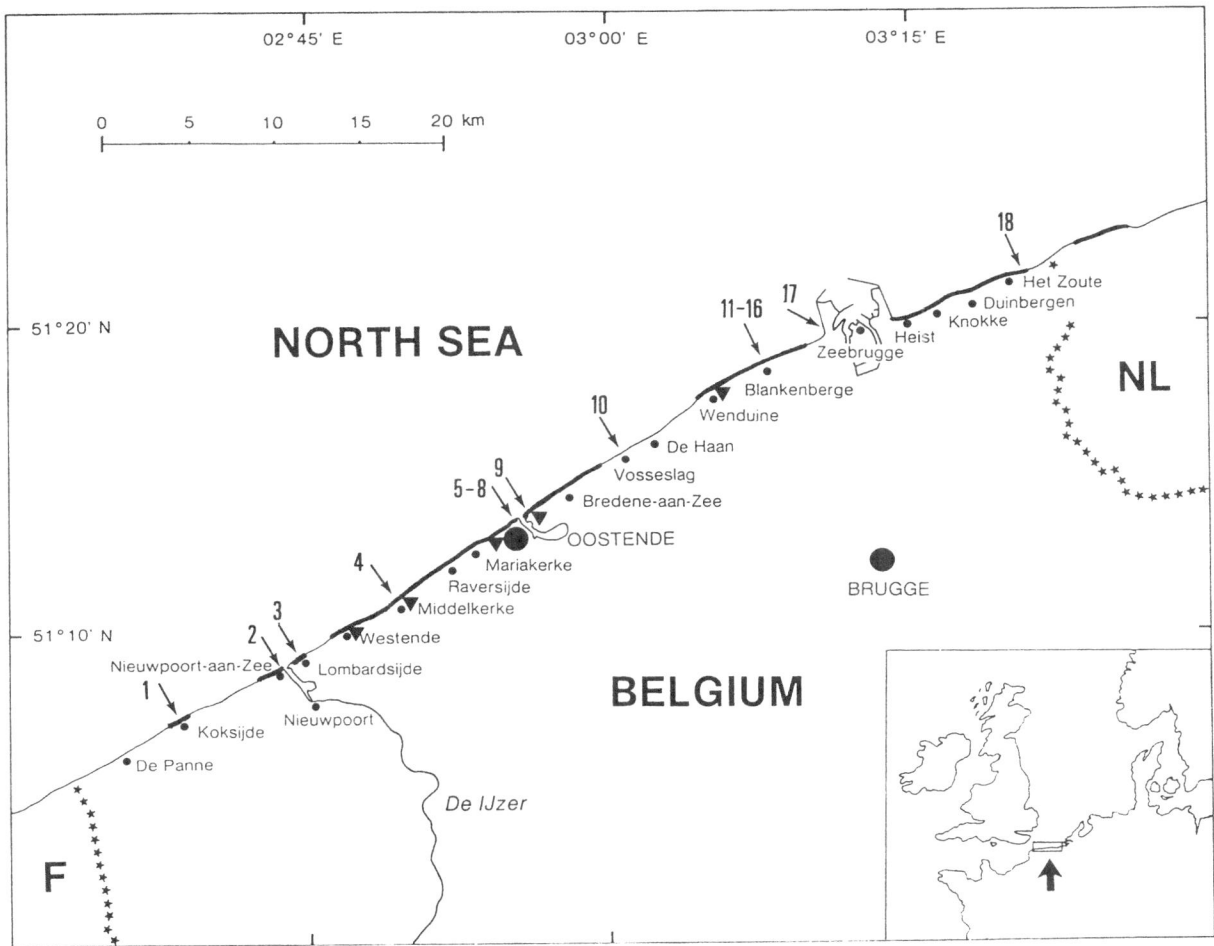

*Fig. 1.* Location of 18 Belgian sites from which samples of *Littorina saxatilis* were analysed genetically. Thick lines indicate breakwaters at interdistances of between 200 and 500 metres, and triangles the presence of dikes in contact with the sea.

## Results

*Littorina saxatilis* was found in 46 of the 56 sites visited, and *L. littorea* in 49 sites. No winkles of either species were found at five of the sites. Three of these sites were, however, in heavily polluted areas in the harbours of Oostende, Blankenberge and Zeebrugge. The two others were dikes in De Panne and Oostduinkerke (between Koksijde and Nieuwpoort-aan-zee). *Littorina saxatilis* was not found on three of the Westende breakwaters, one build in 1976, and two in 1987, as well as on the dike of Middelkerke, built in 1951, and on the eastern mole of Zeebrugge, finished in 1977.

*Littorina littorea* was found in all of these five sites but was missing in two of the sites where *L. saxatilis* occurred, namely on a breakwater in Lombardzijde built in 1953, and on a breakwater in Middelkerke built in 1966. The two Westende sites finished in March 1987, where *L. littorea* but not *L. saxatilis* were found, had a dense cover of young barnacles in the upper parts, and in the lower parts, several adult *L. littorea* were present together with large blue mussels.

Population densities of *L. saxatilis* varied somewhat between sites (Table 1) but were generally much lower than in natural populations in Europe where densities of hundreds or more per

*Table 1.* Type and age of the 18 sites sampled for genetic analysis. Population densities of *Littorina saxatilis* are indicated as low (less than 50 specimens found within an hour of active search), medium (between 50 and 100 found per hour) and high (between 100 and about 500 found per hour). All sites were populated by *L. littorea*. If a site has been rebuilt, the year of this event is indicated in brackets.

| Site | | Sample date | Type of substratum | Built | Density of *saxatilis* |
|---|---|---|---|---|---|
| 1 | Koksijde | May 87 | breakwater | 1958 | medium |
| 2 | Nieuwpoort | May 87 | breakwater | 1981 | low |
| 3 | Lombardsijde | May 87 | breakwater | 1953 | high |
| 4 | Middelkerke | May 87 | breakwater | 1969 | low |
| 5 | Oostende, west | April 88 | breakwater | ?(1949) | high |
| 6 | Oostende, west | April 88 | breakwater | 1962 | low |
| 7 | Oostende, west | Jan. 88 | breakwater | 1953(66) | low |
| 8 | Oostende, west | May 87 | dike, breakw. | ?(1949) | high |
| 9 | Oostende, east | Jan. 88 | dike | 1953 | high* |
| 10 | Vosseslag | May 87 | sandbags | 1978 | low |
| 11 | Blankenberge | April 88 | woodenrail. | 1871 | high |
| 12 | Blankenberge | April 88 | breakwater | 1986 | low |
| 13 | Blankenberge | April 88 | breakwater | 1890 | high |
| 14 | Blankenberge | April 88 | breakwater | 1890 | high |
| 15 | Blankenberge | April 88 | breakwater | 1890 | high |
| 16 | Blankenberge | April 88 | breakwater | 1986 | high |
| 17 | Zeebrugge | May 87 | harbour mole | 1983 | medium |
| 18 | Het Zoute | April 88 | breakwater | 1954(61) | high |

* The *L. saxatilis* sample includes specimens of *L. neglecta* type.

square metre will be reflected in a picking rate of about one thousand per hour (pers. obs.).

Five of the ten loci analysed were polymorphic (0.95 criterion) and in four of these (*Pgi*, *Aat-1*, *Np*, and *Mpi*) the frequency distribution of individual alleles differed between populations (Table 2), as indicated by significant heterogeneties over sites

(Table 3). This suggests that either gene flow is restricted between populations, or that the loci are not selectively neutral over the different sites, or both.

The level of intrapopulation variation, estimated as mean $H_L$ (expected) of all populations, is 0.216 (S.D. = 0.0152) for the Belgian popu-

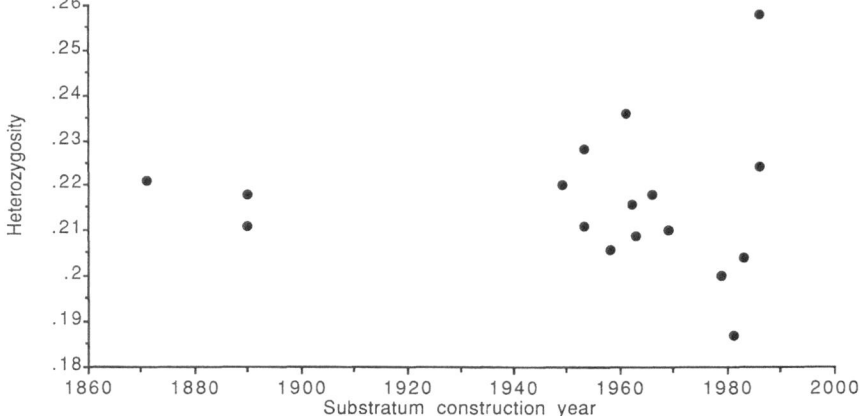

*Fig. 2.* Average population heterozygosity ($H_L$) of 9 loci (5 polymorphic and 4 monomorphic) plotted against the construction year of the substratum for 18 Belgian populations of *Littorina saxatilis*.

*Table 2.* Genetic variation in 18 populations of *Littorina saxatilis* from Belgium.

| Locus | allele | 1 | 2 | 3 | 4 | 5 | 6 | 7 | 8 | 9 | 10 | 11 | 12 | 13 | 14 | 15 | 16 | 17 | 18 |
|---|---|---|---|---|---|---|---|---|---|---|---|---|---|---|---|---|---|---|---|
| *Np* | 140 | 0.046 | – | 0.028 | 0.038 | 0.026 | – | 0.069 | – | 0.016 | 0.023 | 0.012 | – | – | – | – | 0.014 | – | – |
|  | 100 | 0.759 | 0.781 | 0.840 | 0.718 | 0.789 | 0.800 | 0.707 | 0.784 | 0.667 | 0.545 | 0.679 | 0.810 | 0.730 | 0.786 | 0.711 | 0.649 | 0.667 | 0.738 |
|  | 65 | 0.194 | 0.219 | 0.132 | 0.231 | 0.184 | 0.200 | 0.224 | 0.205 | 0.313 | 0.432 | 0.310 | 0.190 | 0.270 | 0.214 | 0.289 | 0.338 | 0.333 | 0.262 |
|  | 35 | – | – | – | 0.013 | – | – | – | 0.011 | 0.005 | – | – | – | – | – | – | – | – | – |
|  | *n* | 54 | 16 | 53 | 39 | 19 | 30 | 29 | 44 | 96 | 22 | 42 | 21 | 37 | 42 | 45 | 37 | 30 | 42 |
| *Aat-1* | 120 | 0.028 | – | 0.142 | – | 0.118 | 0.033 | 0.017 | 0.087 | 0.021 | – | – | 0.500 | – | 0.036 | 0.040 | 0.014 | – | – |
|  | 100 | 0.972 | 1.000 | 0.858 | 1.000 | 0.882 | 0.967 | 0.983 | 0.913 | 0.979 | 1.000 | 1.000 | 0.500 | 1.000 | 0.964 | 0.960 | 0.986 | 1.000 | 1.000 |
|  | *n* | 54 | 16 | 53 | 32 | 17 | 30 | 30 | 40 | 95 | 22 | 42 | 21 | 37 | 42 | 50 | 37 | 30 | 41 |
| *Pgm-1* | 105 | 0.121 | – | 0.085 | 0.133 | 0.026 | 0.083 | 0.034 | 0.058 | 0.140 | 0.045 | 0.071 | 0.048 | 0.095 | 0.071 | 0.057 | 0.122 | 0.052 | 0.048 |
|  | 100 | 0.576 | 0.688 | 0.594 | 0.583 | 0.684 | 0.550 | 0.603 | 0.558 | 0.550 | 0.523 | 0.619 | 0.548 | 0.581 | 0.512 | 0.670 | 0.432 | 0.690 | 0.583 |
|  | 85 | 0.303 | 0.313 | 0.311 | 0.267 | 0.289 | 0.367 | 0.362 | 0.384 | 0.310 | 0.432 | 0.310 | 0.405 | 0.324 | 0.417 | 0.273 | 0.446 | 0.259 | 0.369 |
|  | 75 | – | – | 0.009 | 0.017 | – | – | – | – | – | – | – | – | – | – | – | – | – | – |
|  | *n* | 33 | 16 | 53 | 30 | 19 | 30 | 29 | 43 | 50 | 22 | 42 | 21 | 37 | 42 | 44 | 37 | 29 | 42 |
| *Mpi* | 120 | 0.222 | 0.344 | 0.311 | 0.250 | 0.441 | 0.433 | 0.414 | 0.318 | 0.303 | 0.095 | 0.405 | 0.476 | 0.365 | 0.524 | 0.411 | 0.486 | 0.317 | 0.500 |
|  | 100 | 0.019 | – | 0.075 | – | 0.029 | 0.017 | – | – | – | 0.143 | 0.048 | 0.048 | 0.054 | 0.024 | 0.022 | – | 0.050 | 0.037 |
|  | 75 | 0.750 | 0.656 | 0.604 | 0.738 | 0.529 | 0.550 | 0.586 | 0.682 | 0.697 | 0.762 | 0.548 | 0.476 | 0.581 | 0.452 | 0.567 | 0.514 | 0.633 | 0.463 |
|  | 60 | – | – | 0.009 | 0.013 | – | – | – | – | – | – | – | – | – | – | – | – | – | – |
|  | *n* | 54 | 16 | 53 | 40 | 17 | 30 | 29 | 44 | 94 | 21 | 42 | 21 | 37 | 42 | 45 | 35 | 30 | 41 |
| *Pgi* | 110 | 0.010 | – | 0.009 | 0.012 | – | – | – | 0.011 | – | – | – | – | – | – | – | – | – | – |
|  | 100 | 0.657 | 0.656 | 0.660 | 0.583 | 0.647 | 0.419 | 0.552 | 0.716 | 0.724 | 0.773 | 0.512 | 0.690 | 0.676 | 0.702 | 0.554 | 0.681 | 0.661 | 0.690 |
|  | 90 | 0.333 | 0.344 | 0.330 | 0.405 | 0.353 | 0.581 | 0.448 | 0.273 | 0.276 | 0.227 | 0.488 | 0.310 | 0.324 | 0.298 | 0.446 | 0.319 | 0.339 | 0.310 |
|  | *n* | 51 | 16 | 53 | 42 | 17 | 31 | 29 | 44 | 96 | 22 | 42 | 21 | 37 | 42 | 46 | 36 | 28 | 42 |

The following loci were monomorphic; *Xdh*, *Idh-1*, *Hdh*, *Ldh* and *Est-1*.

lations. This value is not at first sight significantly different from the mean of 18 populations of *L. saxatilis* from Europe and America (data from Knight *et al.*, 1987 and from K. & B. Johannesson, this volume), where the mean of $H_L$ is 0.227 (S.D. = 0.0429) (student's t-test: $t = 1.03$, $df = 34$, and $P > 0.05$). Two non-Belgian populations are, however, much less variable than the others, namely the population from Venice and the American Stonnington population (Table 4). The Venice population is extremely isolated and probably introduced, and is known to be inbred (Janson, 1985; Knight *et al.*, 1987). The Stonnington population is from a breakwater and thus likely to be a discontinuous population founded in a similar way as the Belgian populations. Thus, to compare the heterozygosity

*Table 3.* Effective number of alleles ($n_e$), Wright's inbreeding coefficient ($F_{ST}$) analysed by a $\chi^2$ test (Snedecor & Irwin, 1933) where '*' indicates $P < 0.05$ and '**' $P < 0.005$, and the $\chi^2$, df and $P$ values of the genic contingency test by Workman & Niswander (1970) for 18 Belgian populations of *Littorina saxatilis*.

| Locus | $n_e$ | $F_{ST}$ | $\chi^2$ | df | $P$ |
|---|---|---|---|---|---|
| Np | 1.68 | 0.022* | 74.3 | 51 | <0.025 |
| Aat-1 | 1.12 | 0.161** | 221.6 | 17 | <0.005 |
| Pgm-1 | 2.18 | 0.016 | 53.4 | 51 | >0.1 |
| Mpi | 1.99 | 0.038** | 119.4 | 51 | <0.005 |
| Pgi | 1.85 | 0.029** | 52.1 | 34 | <0.025 |

*Table 4.* A comparison of the level of heterozygosity of 18 Belgian populations and 18 European and American populations of *Littorina saxatilis*. Mean locus heterozygosity is calculated over 9 loci (5 polymorphic and 4 monomorphic) analysed in all populations. Heterozygosity levels of non-Belgian populations is estimated from gene frequency data given by (*) Knight *et al.* (1987) and (**) K. & B. Johannesson (1989).

| Non-Belgian | $H_L$ | Belgian | $H_L$ |
|---|---|---|---|
| Scotland, N. Berwick* | 0.223 | 1 | 0.206 |
| Scotland, Isle of May* | 0.247 | 2 | 0.187 |
| England, Swanage* | 0.276 | 3 | 0.228 |
| England, Bude* | 0.255 | 4 | 0.210 |
| England, Watchet* | 0.205 | 5 | 0.220 |
| Channel Is., Jersey* | 0.262 | 6 | 0.216 |
| Wales, Bangor* | 0.217 | 7 | 0.218 |
| N. Ireland, Derry* | 0.209 | 8 | 0.209 |
| Eire, Doolin* | 0.274 | 9 | 0.211 |
| Italy, Venice* | 0.144 | 10 | 0.200 |
| USA, Maine* | 0.215 | 11 | 0.221 |
| USA, Stonnington* | 0.132 | 12 | 0.258 |
| Canada, Hudson's Bay* | 0.198 | 13 | 0.211 |
| Sweden, Ursholmen** | 0.192 | 14 | 0.211 |
| Norway, Herdla 1** | 0.245 | 15 | 0.218 |
| Isle of Man, St. Mary** | 0.281 | 16 | 0.224 |
| Iceland, Akranes** | 0.263 | 17 | 0.204 |
| Iceland, Sandgerdi** | 0.254 | 18 | 0.236 |
| Mean: | 0.227 | | 0.216 |

levels of the non-natural Belgian breakwater populations with those of natural populations it is sensible to exclude the Venice and Stonnington populations. If this is done, the mean $H_L$ of

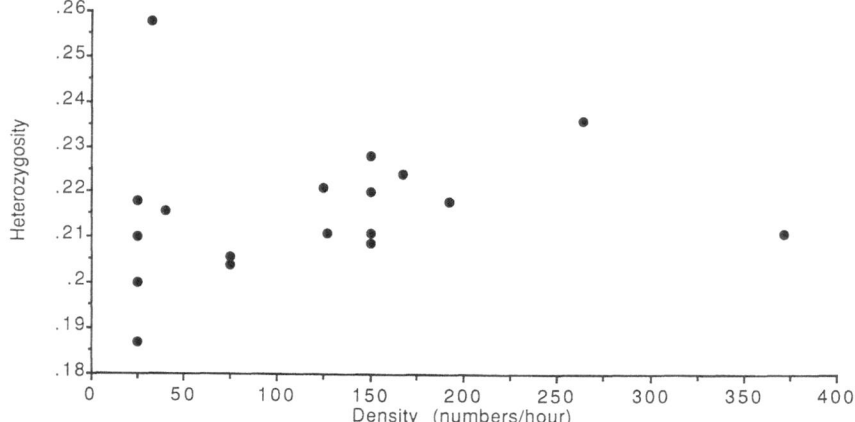

*Fig. 3.* Average population heterozygosity ($H_L$) of 9 loci plotted against the approximate density (numbers of snails found per hour of active search) of each of 18 Belgian populations of *Littorina saxatilis*.

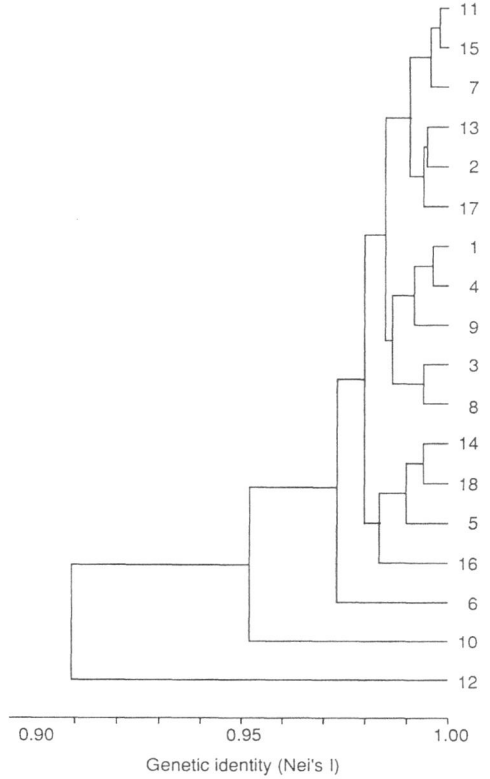

*Fig. 4.* A dendrogram of genetic identities between 18 Belgian populations of *Littorina saxatilis* grouped by an UPGMA analysis.

the Belgian group ($t = 2.93$, *df* = 32, $P < 0.01$). Although the difference is not large, the Belgian breakwater populations are on average 9.6% less heterozygous than the natural non-Belgian ones.

Heterozygosity levels varied somewhat between the Belgian populations, but no correlation was found between $H_L$ and habitat age ($R^2 = 0.0001$, $P > 0.05$, Fig. 2) or density of populations ($R^2 = 0.028$, $P > 0.05$, Fig. 3).

Genetic identities and distances between all possible pairs of populations (Nei's $I$ and $D$; Nei, 1972) calculated for the five polymorphic loci range from $I = 0.854$, $D = 0.158$ (populations 12 and 10) to $I = 0.998$, $D = 0.002$ (11 and 15), with an average of $I = 0.972$, $D = 0.028$. An UPGMA-cluster analysis of the genetic identities (Fig. 4) reveals no readily interpretable grouping of the populations. Yet, two populations seem to be somewhat separated from the others, namely 10 and 12. Population 12 differs mainly in the *Aat-1* locus, which is possibly explained by selection in this locus over intertidal gradients (K. & B. Johannesson, 1989). Population 10 is somewhat different in the *Mpi* locus.

As *Aat-1* may be selected, this locus was excluded from an analysis of the relationship between geographic and genetic distances between pairs of populations. However, there was still no relationship between these two parameters ($R^2 = 0.00002$, $P > 0.05$, Fig. 5).

natural (i.e. non-Belgian) populations 0.239 (S.D. = 0.0298) is significantly different to that of

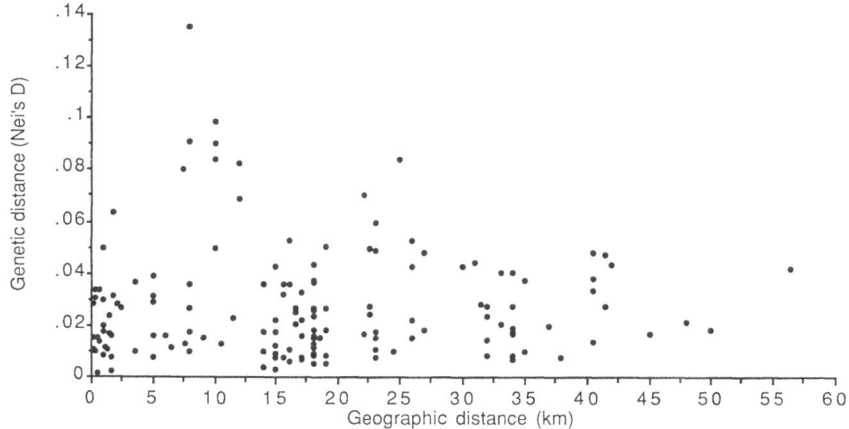

*Fig. 5.* Genetic distance (Nei's $D$) as a function of geographic distance between pairs of 18 Belgian populations of *Littorina saxatilis*. The comparison is based on 8 loci (*Aat-1* is excluded, see text).

## Discussion

It is generally assumed that a sedentary benthic invertebrate species with a direct development has a very restricted dispersal ability and will show tendencies of local inbreeding and genetic differentiation between populations (see reviews by Jablonski & Lutz, 1983 and Strathmann, 1986). Our study, however, indicates a fairly high colonization rate of the direct developer *L. saxatilis* to new (on an evolutionary time scale) sites along the Belgian coast which are discontinuously distributed and distant from natural sites.

An earlier study (Janson, 1987a) showed that *L. saxatilis* colonized recently emerged skerries (islands of a few square metres in size) in Sweden over centuries or decades. The time for colonization of an experimentally cleared skerry, 15 m from an island populated with snails, was about two years, and over not more than two generations after recolonization the population had attained the original density. Similar colonization rates of *L. saxatilis* were observed by K. & B. Johannesson in Iceland. A 5 km long rocky and gravel shore uplifted during volcanic activity in 1973 on Vestmannaeyja was searched for the presence of *L. saxatilis* in August 1986. About 4 km of shore had not been colonized although the substratum in many places was of the kind preferred by the species. However, two rocky outcrops of the new shore were inhabited by large numbers of *L. saxatilis*. One of these sites was separated from the old shore (which had a snail density of hundreds per metre square) by 200 m of sand and gravel where no snail or perennial algae were found due to the mobility of the substratum. The other site was about 1 km further away from the old shore and separated from the first site by another gravel beach of about 200 m width.

The observations from Sweden and Iceland indicate that dispersal over discontinuous habitats may be somewhat restricted, but over years or decades colonization may occur. Although the colonization of old breakwaters may be explained by the historical fact that in the past some 40 km of dikes were in contact with the sea, and

breakwaters built in contact with these dikes may have been colonized from dike populations, the establishment of populations on new breakwaters, which are most often discontinuous with the old dikes, can not be explained in this way. The rapid establishment of populations on two breakwaters built in 1986 (both populated by *L. saxatilis* in April 1988), supports the observations from Sweden and Iceland of a fairly high colonizing potential of *L. saxatilis*. Although it appears to be a difference in colonizing potential over short distances (see also Johannesson, 1988 about long distance colonizing potential) between *L. saxatilis* and the planktonic developer *L. littorea* which over a year populates new sites, for example, the two 1987 breakwaters in Westende, this difference is small in evolutionary terms. The reason why *L. saxatilis* was missing from three and *L. littorea* from two somewhat older sites (1953 to 1976) is not clear. Perhaps these sites were in some way unsuitable for *Littorina*, or snails may have been present in extremely low densities and therefore not found.

There seems to be, however, an important difference between the two species in the number of founding individuals. As argued by Johannesson (1988), a direct developer needs only a few founder individuals, perhaps only one fertilized female, to establish a new population, while in a planktonic developer a new population will arise only if a large number of larvae or adults reach the same site.

Founding a new population with a small number of individuals might lead to a heavy loss of genetic variation (a 'founder effect', Mayr, 1963). The enzyme analyses of the Belgian populations did, however, reveal a heterozygosity level only about 10% lower than that of populations from natural habitats. Although it is likely that most breakwater populations have passed through bottlenecks both when founded, and perhaps again when breakwaters have been rebuilt, a rapid expansion in population size after a bottleneck will reduce the genetic loss considerably (Chakraborty & Nei, 1977). As shown by Janson (1987a), the Swedish skerry populations lost only about 15% of their genetic variability when founded, and this may be explained by a rapid increase in size over the first few generations.

It is, however, somewhat surprising that a high level of variation is maintained in the Belgian populations despite the low density of snails. The low densities are possibly explained by the exacting physical environment and predation from waders, and other shorebirds rather than by a low reproductive potential. Predation from waders are high in the winter, a mean of six birds, mainly purple sandpiper (*Calidris maritima*), and turnstone (*Arenaria interpres*), foraging on each breakwater between January and April (Warmoes pers. obs.). Snail mortality has also been shown to be related to the availability of cracks and holes in the substratum (Raffaelli & Hughes, 1978). The Belgian breakwaters and dikes are fairly rich in small crevices and empty barnacles but the risk of smothering by sand may be high after rough weather. Possibly, the genetic variation within each population may be supported by some gene flow between breakwaters, and maintained as population size increases rapidly after periods of high selection pressure.

We found no relationship between genetic and geographic distances between the Belgian populations, although such a relationship was found along 300 km of Swedish shoreline (Janson, 1987b). Perhaps the shallow sandy Belgian coast and the fairly strong currents running parallel to the shoreline provides an environment where founders and migrants are as likely to end up on a distant as on a close breakwater, while in Sweden a rocky archipelago and generally weak coastal currents may act as barriers which makes long transports less likely than short ones.

Although we may conclude that *L. saxatilis* has a fairly good dispersal potential, we are still largely ignorant of the actual mechanism of dispersal. Some live juvenile *L. saxatilis* have been observed lying on the sandy shore in Heist, Belgium, (Warmoes *et al.*, 1988) and it is possible that individuals may be dislodged from a breakwater and rolled to another site along the smooth sandy bottom surface by waves and currents. While this may be a possible mechanism for the Belgian shoreline, rafting on pieces of macroalgae seems more likely in rocky areas, e.g. Sweden and Iceland. An interesting observation is that pre-hatched embryos may be accidentally released from the females brood-pouch, and that these embryos may survive until the crawling stage (Hannaford Ellis, pers. comm.). Although a passive transport of prehatched embryos may occur, it must be rather frequent in order to be important in the process of colonization of new breakwaters, otherwise the density of founders in a new site will not be high enough to establish a breeding population.

After a population bottleneck the genetic variation may be restored through mutations and through gene flow between populations. Mutation rate is generally low and thousands of generations are needed to reach equilibrium levels of variation (Chakraborty & Nei, 1977). High gene flow from populations of high variability would, on the other hand, increase genetic variation fairly rapidly. The lack of a relationship between habitat age and level of heterozygosity suggests that gene flow from natural sites is low – mutation rate as well as gene flow between inbred breakwater populations are not likely to raise the level of genetic variation considerably within populations. The lack of correlation between snail density and level of heterozygosity is somewhat unexpected as heterozygosity is expected to increase with effective population number (e.g. Ward, this volume). However, removing one single observation (population 12) gives $R = 0.487$ which is about significantly different from zero slope ($P = 0.0525$).

## Acknowledgements

Our thanks go to Mr. R. Simoen and Mr. J. Pauwels of the Ministry of Public Works in Oostende, and to Mr. R. Desaever, Port Authorities in Oostende. Furthermore to Dr. T. Backeljau, L. De Bruyn, D. Huisseune, and I. Dumon for field assistance, and to Prof. Dr. S. L. J. Hulselmans and Prof. Dr. W. N. Verheyen for laboratory facilities. This work was supported by a research grant to K.J. from the Swedish Natural Science Research Council.

# References

Arnaud, F., P. M. Arnaud, A. Intés & P. de Lœuff, 1976. Transport d'invertebres benthique entre l'Afrique du Sud et Sainte Hélene par les laminaires (Phaeophyceae). Bull. Mus. natn. Hist. nat., Paris 384: 49–55.

Chakraborty, R. & M. Nei, 1977. Bottleneck effects on average heterozygosity and genetic distance with the stepwise mutation model. Evolution, Lancaster, Pa. 31: 347–356.

Gerlach, S. A., 1977. Means of meiofauna dispersal. Microfauna Meeresboden 61: 89–103.

Highsmith, R. C., 1985. Floating and algal rafting as potential dispersal mechanisms in brooding invertebrates. Mar. Ecol. Prog. Ser. 25: 169–179.

Jablonski, D. & R. A. Lutz, 1983. Larval ecology of marine benthic invertebrates: paleobiological implications. Biol. Rev. 58: 21–89.

Janson, K., 1985. A morphologic and genetic analysis of Littorina saxatilis (Prosobranchia) from Venice, and on the problem of saxatilis-rudis nomenclature. Biol. J. linn. Soc. 24: 51–59.

Janson, K., 1987a. Genetic drift in small and recently founded populations of the marine snail Littorina saxatilis. Heredity 58: 31–37.

Janson, K., 1987b. Allozyme and shell variation in two marine snails (Littorina, Prosobranchia) with different dispersal abilities. Biol. J. linn. Soc. 30: 245–256.

Janson, K. & R. D. Ward, 1984. Microgeographic variation in allozyme and shell characters in Littorina saxatilis Olivi (Prosobranchia: Littorinidae). Biol. J. linn. Soc. 22: 289–307.

Johannesson, K., 1988. The paradox of Rockall: why is a brooding gastropod (Littorina saxatilis) more wide-spread than one which has a planktonic larval dispersal stage (L. littorea)? Mar. Biol. 99: 507–513.

Johannesson, K. & B. Johannesson, 1989. Clinal variation in Aat over vertical transects of rocky shores suggests selection in populations of Littorina saxatilis Olivi (Prosobranchia). Genet. Res., Camb. 54: 7–11.

Knight, A. J., R. N. Hughes & R. D. Ward, 1987. A striking example of the founder effect in the mollusc Littorina saxatilis. Biol. J. linn. Soc. 32: 417–426.

Malone, C. R., 1965. Killdeer (Charadrius vociferus Linnaeus) as a means of dispersal for aquatic gastropods. Ecology 46: 551–552.

Mayr, E., 1963. Animal Species and Evolution. Harvard Univ. Press, Camb., Mass., 797 pp.

Nei, M., 1972. Genetic distance between populations. Am. Nat. 106: 283–292.

Raffaelli, D. G. & R. N. Hughes, 1978. The effects of crevice size and availability on populations of Littorina rudis and Littorina neritoides. J. anim. Ecol. 47: 71–83.

Scheltema, R. S., 1986. On dispersal and planktonic larvae of benthic invertebrates: an eclectic overview and summary of problems. Bull. mar. Sci. 39: 290–322.

Snedecor, G. & M. R. Irwin, 1933. On the chi-square test for homogeneity. Iowa St. J. Sci. 8: 75–81.

Strathmann, R. R., 1986. What controls the type of larval development? Summary statement for the evolution session. Bull. mar. Sci. 39: 616–622.

Ward, R. D. & T. Warwick, 1980. Genetic differentiation in the molluscan species Littorina rudis and Littorina arcana (Prosobranchia: Littorinidae). Biol. J. linn. Soc. 14: 417–428.

Warmoes, T., T. Backeljau & L. De Bruyn, 1988 (1989). The littorinid fauna of the Belgian coast (Mollusca, Gastropoda). Bull. Inst. r. Sci. nat. Belg. Biologie 58.

Workman, P. L. & J. D. Niswander, 1970. Population studies on southwestern Indian tribes. II. Local genetic differentiation in the Papago. Am. J. hum. Genet. 22: 24–49.

*Hydrobiologia* **193**: 109–116, 1990.
*K. Johannesson, D. G. Raffaelli and C. J. Hannaford Ellis (eds), Progress in Littorinid and Muricid Biology.*
© 1990 *Kluwer Academic Publishers.*

# Hybridisation in the *Littorina saxatilis* species complex (Prosobranchia : Mollusca)

T. Warwick[1], A.J. Knight[2,]*  & R.D. Ward[2]
[1] *Department of Zoology, University of Edinburgh, West Mains Road, Edinburgh EH9 3JT, East Lothian Scotland, UK*; [2] *Environmental Biology Unit, Department of Human Sciences, University of Technology, Loughborough LE11 0NU, Leics, England, UK (* author for correspondence and reprint requests)*

*Key words:* Littorina, interspecific hybrids

## Abstract

This paper details attempts to breed members of the *L. saxatilis* species complex in the laboratory and to construct interspecific hybrids between them. Success in reciprocal crosses between animals from the type localities of the 'species' of *Littorina saxatilis* (Olivi) and *Littorina rudis* (Maton) indicates that these two taxa are synonymous. Six of the twelve possible reciprocal hybridisation crosses between the four species (*L. saxatilis*, *L. arcana* Hannaford Ellis, *L. nigrolineata* Gray and *L. neglecta* Bean) have been attempted, with only one proving successful, that between male *L. saxatilis* and female *L. arcana*. This hybrid cross produces viable offspring, although at a lower frequency than either of the parental crosses in the laboratory, while the reciprocal cross has, as yet, proved unsuccessful. Limited work on the F1 hybrids shows them to be inter-fertile and also capable of backcrossing with male *L. saxatilis*, but not with female *L. saxatilis* or *L. arcana*. Details are also given of attempts to find natural hybrids in sympatric populations of the species of the complex, using gel electrophoresis.

## Introduction

The genus *Littorina* (Prosobranchia: Gastropoda) constitutes a most important part of the intertidal marine fauna, yet the species composition of the genus is still a source of controversy and debate. These problems are perhaps most marked in the *Littorina saxatilis* complex where a number of different species are recognised by different researchers.

Recent reviews of the taxonomy of the group (S. M. Smith, 1982; Raffaelli, 1982) list four saxatilid species: *Littorina saxatilis*, *L. neglecta*, *L. arcana* and *L. nigrolineata*. The first two of

these species are ovoviviparous (produce live young or 'crawlaways'), the latter two are oviparous (lay egg masses from which live young hatch). This classification is not universally recognised, with some authors preferring to regard *L. arcana* as a reproductive morph of *L. saxatilis* (= *L. rudis*) (e.g. Caugant & Bergerard, 1980; Hughes & Roberts, 1981) and one author suggesting that *L. rudis* and *L. saxatilis* are two separate species from boulder shores and cliffs respectively (J.E. Smith, 1981). It has also been suggested that a further variety of *Littorina saxatilis*, *tenebrosa* is a distinct species (S.M. Smith, 1982), although recent biochemical work casts doubt on this interpretation (Janson & Ward, 1985).

In earlier work on the electrophoretic taxonomy of the saxatilids (Ward & Janson, 1985; Ward & Warwick, 1980; Ward, this volume; Knight & Ward, unpubl.), data indicated that three of the four members of the species complex were true species, with the sibling species of *Littorina saxatilis* and *L. arcana* being particularly closely related, and *L. nigrolineata* being genetically more distinct. The taxonomic status of the remaining species, *L. neglecta*, remains unclear (see K. Johannesson & B. Johannesson and B. Johannesson & K. Johannesson, both this volume, for recent views).

Any work involved with speciation must pay careful attention to the classical definition of a species as being a reproductively isolated entity. The sporadic success in constructing fertile interspecific hybrids between *L. saxatilis* and *L. arcana* in the laboratory is of interest in this respect, yielding information on the frequency and morphology of such hybrids, and so helping to facilitate the identification of natural hybrids in wild populations. Here we discuss data from laboratory hybridisation experiments involving animals of the *saxatilis* complex, and, also using electrophoretic data, examine whether hybrids are found in natural sympatric populations of *L. arcana* and *L. saxatilis*.

## Methods

The techniques for breeding and maintenance of *Littorina saxatilis* and other members of the group have been previously described (Warwick, 1983). The electrophoretic methods used have also been previously described (Ward & Warwick, 1980; Janson & Ward, 1984).

## Results and discussion

### i) Breeding work in the Littorina saxatilis species complex

A series of inter- and intraspecific crosses between member species of the *L. saxatilis* complex from British populations was carried out between late 1985 and 1987, and Table 1 summarises the results. In order to facilitate comparisons, only experiments established when potential parents are likely to be in breeding condition are considered (Hannaford Ellis, 1983; Warwick, unpubl.). Only crosses in which both parents survived to the end of the experimental period are included. Virginity of females used in the experiments is assured either by rearing juveniles typed as females from infancy or isolating mature females for a period of months. In many of the crosses, both inter and intraspecific progeny and parents were checked electrophoretically. This procedure had two functions: firstly to confirm the parentage of the offspring and secondly to

*Table 1.* Summary of breeding experiments in the *Littorina saxatilis* species complex. *N* is number of crosses attempted, figure in % is percentage of successful crosses.

| Male parent | Female parent | | | | | | | |
|---|---|---|---|---|---|---|---|---|
| | *L. arcana* | | *L. nigrolineata* | | *L. neglecta* | | *L. saxatilis* | |
| | *N* | % | *N* | % | *N* | % | *N* | % |
| *arcana* | 41 | 41% | 20 | 0% | 6 | 0% | 36 | 0% |
| *nigrolineata* | 39 | 0% | 16 | 13% | – | – | – | – |
| *neglecta* | – | – | – | – | 3 | 0%* | – | – |
| *saxatilis* | 114 | 20% | – | – | 12 | 0% | 64 | 70% |

* This cross has been successful in previous experiments (Warwick, 1983).

provide information on the Mendelian inheritance and linkage grouping of the enzyme loci screened. In all crosses examined, 98–100% of progeny examined had genotypes consistent with the genotypes of the putative parents, with the remainder being contaminants (Ward *et al.*, 1986).

Examination of the intraspecific crosses shows that the most successful involve *L. saxatilis*. This is perhaps not unexpected, as *L. saxatilis* is both the most abundant and ecologically diverse species in the group and, perhaps more importantly, is the only species of the complex to be found naturally feeding on the algae-covered topshore pebbles used as the food source in all these experiments. The apparent gestation period in this species, involving as it does both the production of eggs and their subsequent hatching and development in the brood pouch, averages 110 days in the laboratory (range 72–162), measured from the date of establishing the pair, while *L. arcana* crosses took an average period of 75 days to produce a spawn mass (range 41–109) and these typically took 20–30 days further to hatch. The third species, *L. nigrolineata*, shows lower fertility under these conditions, perhaps due to the larger maturation size of the species, so that some of the virgin animals used in the crosses had possibly not developed to sexual maturity at the beginning of the experiment. Alternatively, this midshore species may suffer some dietary deficiency under these conditions, or it may simply respond less well to the unnatural environment. No successful crosses were obtained using *L. neglecta* in this period of the experimental programme, possibly due to its low survival rates in laboratory conditions during the summer months. In previous experimental sets, some crosses of this species have produced live young (Warwick, 1983; Warwick, unpubl.).

Six of the twelve possible reciprocal interspecific crosses have so far been attempted, but only one has proved successful to date, that between male *L. saxatilis* and female *L. arcana* (although at a lower frequency than either of the comparable intraspecific crosses). The cross between the two species showed apparent reciprocal hybridisation failure, as the cross between male *L. arcana* × female *L. saxatilis* has as yet proved unsuccessful. The spawn masses produced in the hybrid cross are frequently fertile and in a number of cases electrophoretic analysis of the cross parents and their offspring was undertaken in successful attempts to eliminate the possibility that the *L. saxatilis* was not the true male parent.

The reason for the lack of success in the reciprocal cross is unknown, but it does not appear to be due to the absence of mating in the pair, as such behaviour was observed. Mating attempts were also noted in two of the other unsuccessful interspecific crosses (*L. nigrolineata* male × *L. arcana* female and the reciprocal cross), and as expected from previous work in natural environments (Raffaelli, 1977) and in the laboratory (Saur, this volume), it does not seem that these animals have an effective physical barrier to interspecies mating.

A recent study of the Venetian populations of *L. saxatilis*, the type locality of that species (Olivi, 1792), indicated that the European populations of the Atlantic described under the name of *L. rudis*

*Table 2.* Comparative success rates of three classes of the interspecific cross *L. saxatilis* (male) × *L. arcana* (female).

| Cross type | No. of crosses attempted | No. of crosses successful | Success rate (%) |
|---|---|---|---|
| allopatric population *saxatilis* male × allopatric population *arcana* female | 33 | 10 | 30.3 |
| allopatric population *saxatilis* male × sympatric population *arcana* female | 42 | 10 | 23.8 |
| sympatric population *saxatilis* male × *arcana* female from same population | 57 | 8 | 14.0 |

(Maton, 1797) are conspecific, but it was suggested that cross-breeding experiments should be carried out to confirm this interpretation (Janson, 1985). Such experiments have now been performed and show successful hybridisation in both directions between males and females of British and Venetian populations, including a series of reciprocal crosses involving animals from the type localities of the two taxa (Venice lagoon × Cargeen, North Devon). We therefore suggest that Janson's interpretation is correct, and that *L. rudis* is a junior synonym of *Littorina saxatilis*.

The existence of *L. saxatilis* and *L. arcana* in both sympatry and allopatry may lead to variation in the potential for hybridisation in natural colonies. Some factor or factors causing reduced success of the interspecific cross (compared to the intraspecific crosses), may be more marked in populations where the two species naturally occur together, since the function of this hypothetical mechanism is to prevent hybridisation between the species, and so prevent the breakdown of integrated genotypes (Ayala, 1975; Mayr, 1963). We might therefore predict greater mating success in crosses between animals taken from allopatric populations of the two species than in crosses where the prospective parents are derived from the same sympatric population. A number of intermediate cases exist, where only one parent is taken from an allopatric population, or where both parents are from different sympatric populations, and in such cases we might expect an intermediate level of mating success. Data from three classes of mating for the cross of male *L. saxatilis* × female *L. arcana* are considered in Table 2: no significant difference in mating success between the experimental types of cross is observed ($\chi^2 = 3.56$, df = 2, $P = 0.167$), although the data for the percentage of mating success of the three classes examined are consistent with these prior expectations. A comparison of mating success between allopatric and fully sympatric populations of the species, the most extreme cases, reveals large but non significant differences ($\chi^2 = 3.46$, df = 1, $P = 0.063$).

The F1 interspecific hybrids have the capacity for both hatching and development to sexual

*Table 3.* Summary of breeding experiments between *Littorina saxatilis*. *L. arcana* and the *L. saxatilis* (male) × *L. arcana* (female) F1 hybrid. *N* is number of crosses attempted, figure in percent is % of successful crosses.

| Male parent | Female parent | | | | | |
|---|---|---|---|---|---|---|
| | *L. arcana* | | F1 hybrid | | *L. saxatilis* | |
| | *N* | % | *N* | % | *N* | % |
| *arcana* | 41 | 41% | – | – | 36 | 0% |
| Hybrid | 7 | 0% | 9 | 33% | 5 | 0% |
| *saxatilis* | 114 | 20% | 19 | 42% | 64 | 70% |

maturity, for both sexes (Table 3). The female hybrids have *arcana*-like sexual organs and, like *arcana*, reproduce by oviparity. Male and female hybrids will mate successfully with each other, again to produce spawn masses, and the backcross of hybrid (female) × *saxatilis* (male) has produced spawn masses in 8 of 19 crosses. The corresponding backcross (hybrid (female) × *arcana* (male)) has so far proved unsuccessful as have attempts to produce successful crosses between either *saxatilis* or *arcana* females and the hybrid males.

Comparison of a series of crosses involving the oviparous females of *L. arcana* and the F1 hybrid × male *L. saxatilis* shows a trend to an increasing number of small, multiple egg masses when compared to the intraspecific *L. arcana* cross (Table 4). The higher frequency of such repeated brooding may be related to the increasing proportion of the *L. saxatilis* genotype in the egg mass – in the F1 backcross (F1 hybrid female × *L. saxatilis* male) 75% of the genome of the progeny would be expected to come from the ovoviviparous lineage.

In these experiments an unsuccessful cross is one that fails to produce a spawn mass or small live young within three to four months of setting up the cross for the oviparous species (*L. arcana* and *L. nigrolineata*), and five to six months for the ovoviviparous species (*L. saxatilis* and *L. neglecta*). Data from a small number of long-term crosses (up to 18 months in duration) in *L. saxatilis* and *L. arcana* show that pairs that fail

*Table 4.* Number of spawn masses laid in oviparous crosses. *N* is number of crosses. (Some crosses included in Tables 1 and 3 cannot be included here, as the number of spawn masses was not recorded.) $\bar{x}$ is the average number of spawn masses per successful cross.

| Cross type | No. of spawn masses laid | | | | | | |
|---|---|---|---|---|---|---|---|
| | 0 | 1 | 2 | 3 | 4 | *N* | $\bar{x}$ |
| L. arcana × L. arcana | 17 | 14 | 1 | 0 | 0 | 32 | 1.06 |
| L. saxatilis × L. arcana | 79 | 18 | 4 | 0 | 0 | 101 | 1.18 |
| L. saxatilis × F1 hybrid | 11 | 4 | 1 | 2 | 1 | 19 | 1.78 |

to produce young within a few months generally prove infertile over longer periods. This definition therefore includes as 'successes' spawn masses laid by the oviparous species in the group, or by oviparous hybrids, which fail to hatch. Such cases of incomplete development occur frequently, but do not seem to be indicative of the 'dumping' of unfertilized egg masses, since such behaviour is rarely seen in cultures of virgin females kept for long periods. Many of these spawn masses begin to develop, but die in the early stages of nucleation, implying that the culture conditions of extreme shelter, relative warmth and high algal growth under which the parents grew to reproductive maturity are sub-optimal for the maturation of their spawn. In natural populations of oviparous species, spawn masses are very seldom encountered on sunlit surfaces, even at high population densities, and are found mainly in cryptic habitats such as under boulders (*L. nigrolineata*) or in deep crevices (*L. arcana*) (Warwick & Knight, pers. obs.).

*ii) The search for natural hybrids*

From the laboratory studies, it appears that there is the possibility of the exchange of genetic mate-rial between *Littorina saxatilis* and *Littorina arcana* in natural sympatric populations, while the as yet incomplete series of reciprocal crosses performed failed to show hybridisation in other pairs of species within the complex. We therefore concentrate our efforts here on attempts to estimate the frequency of interspecific hybrids between the sibling species of *L. saxatilis* and *L. arcana* in wild populations. What then are the characteristics of these natural hybrids that will enable them to be distinguished from the species from which they are derived? In the laboratory the F1 hybrids produced (from male *L. saxatilis* and female *L. arcana*) show the primary sexual characteristics of the *arcana* parent, in that females have a pallial oviduct and a jelly gland. Morphological characters such as shell colour and shape may resemble either parent to the exclusion of the other, but such relationships are often seen to break down in the F2 generation to reveal the hidden variation. The more usual case is that the hybrids resemble both parents in part, and in populations where the two species differ greatly in shell morphology, it may be possible to hypothesise the existence of natural hybrids, and even assign individuals to a 'hybrid' group on such a basis. Such identifications are however likely to be uncertain, since such oviparous animals could be merely rare morphs of *L. arcana*, and a more objective method should enable a less tentative classification to be made.

We know from earlier work that the enzyme polymorphisms detectable by electrophoretic methods in the parents of an experimental cross are expressed co-dominantly in the offspring of the cross, and that this is also true of the F1 hybrids of the interspecific cross between *L. saxatilis* and *L. arcana* (Ward *et al.*, 1986; Knight & Ward, 1986 and unpubl. data). Unfortunately we have been unable to discover a universally diagnostic locus, one that will permit unequivocal discrimination between animals of the two species throughout their range, although there may be extensive local differentiation for individual gene loci in sympatric populations (Knight *et al.*, 1987). It is therefore difficult to identify natural hybrids on the basis of allozyme data from single enzyme loci. Further difficulties exist, as no universally diagnostic anatomical or morphological character has been described which permits unequivocal separation of the two species, although mature females can readily be separated by differences in the reproductive tract. However, remembering that the laboratory studies showed that any female hybrids present are likely to be *arcana*-like in reproductive morphology, and are thus likely to be included as contaminants among females identified as *L. arcana*, then if an animal is found, in a sympatric population of the two species, which is phenotypically identifiable as *L. arcana* (and thus will have to be female), but which has a strong infusion of alleles unique at this location to *L. saxatilis*, then the strong probability exists that it is, in fact, an F1 hybrid. In the absence of diagnostic loci for the species, the identification of F2 or other hybrids will be extremely difficult or impossible.

Ten pairs of sympatric populations selected (from data collected for a wider study) by virtue of the two species having widely divergent allele frequencies, were examined for evidence of interspecific hybridisation. In most of these populations, the species differed significantly in allele frequencies at five or more enzyme loci. Data from one of these populations, from Bracelet Bay, Mumbles Head, South Wales (O.S. ref SS629873) are presented by way of an example in Table 5. For the two loci (*Np*, *Aat-1*) in the table showing the highest significant difference in allele frequencies between species, the most common allele in *L. saxatilis* is completely absent from the *L. arcana* sample. If the *Aat-1* locus is considered first, any hybrid in the *L. arcana* sample will have either the 120/100 or 100/100 genotype. The latter genotype would not be distinct from the *L. arcana* 100/100 genotype, and therefore at this locus there is a probability of 0.343 of hybrids in the *L. arcana* sample not being detected. At the *Np* locus, there is a corresponding probability of 0.323 of such hybrids not being detected. Combining these two estimates (which is permissible since we have data showing that these two loci are not tightly linked and not in linkage disequilibrium) yields a probability of 0.111 (cal-

*Table 5.* Comparison of selected enzyme loci in a strictly sympatric sample of *L. saxatilis* and *L. arcana* from a site at Bracelet Bay on Mumbles Head, South Wales. $N$ is number of animals examined.

| Locus | | *L. saxatilis* | *L. arcana* | Rare genotypes in *L. arcana* |
|---|---|---|---|---|
| *Mpi* | 120 | 0.485 | 0.940 | 3 120/100 |
| | 100 | 0.279 | 0.060 | heterozygotes in |
| | 75 | 0.235 | – | 25 animals |
| | $N$ | 34 | 25 | |
| *Pgi* | 115 | 0.015 | – | 3 100/90 |
| | 100 | 0.603 | 0.944 | heterozygotes in |
| | 90 | 0.382 | 0.056 | 27 animals |
| | $N$ | 34 | 27 | |
| *Pgm-1* | 105 | 0.120 | – | 1 100/85 |
| | 100 | 0.660 | 0.976 | heterozygote in |
| | 85 | 0.220 | 0.024 | 21 animals |
| | $N$ | 25 | 21 | |
| *Aat-1* | 120 | 0.657 | – | – |
| | 100 | 0.343 | 1.000 | |
| | $N$ | 32 | 27 | |
| *Np* | 140 | – | 0.111 | – |
| | 100 | 0.323 | 0.889 | |
| | 65 | 0.677 | – | |
| | $N$ | 34 | 27 | |
| *Hdh* | 145 | – | 0.320 | – |
| | 100 | 0.985 | 0.680 | |
| | 60 | 0.015 | – | |
| | $N$ | 34 | 25 | |

culated from $0.343 \times 0.323$). In other words, using these two loci alone, the probability of detecting an F1 hybrid among the *L. arcana* sample, had one been present, is approximately 0.9. An element of bias has been introduced by using enzymes showing the greatest difference in alleles between the two species to estimate the rate of hybridisation, since if hybrids were present in the sample this difference would be reduced, and we therefore look at three other enzymes showing highly significant differences between the two species, to see if there is any concentration of rare alleles in individuals of *L. arcana* that might indicate that they are hybrids, deriving these alleles from their *L. saxatilis* parent. Of the seven rare heterozygotes from the three loci (*Mpi*, *Pgi* and *Pgm-1*) in the *L. arcana* sample, one animal is heterozygous for two separate enzyme loci, but we can be certain that it is not an F1 hybrid, since it has the 145/145 genotype for *Hdh*, and a *L. saxatilis* parent could not have contributed this allele to its progeny, since in this population it is restricted to *L. arcana*.

No animals thought to be hybrids were identified in a total of 223 female *L. arcana* animals screened in the ten pairs of sympatric populations, although the average probability of recognising a hybrid had it been present in any population was higher than 0.65. Similarly no potential hybrids were detected in the sympatric *L. saxatilis*, although the chance of detecting them in these populations is small. Natural hybrids would appear to be rare in these populations and if we can assume that hybridisation is not more frequent in those sympatric populations where the two species are genetically more similar than the selected populations we have examined here, then the frequency of natural hybrids on British shores between *L. saxatilis* and *L. arcana* seems unlikely to exceed 1–2%. This is perhaps an unexpectedly low figure, given the laboratory hybridisation results, and given also that studies in some other marine organisms have shown extensive hybridisation between species less closely related than these two saxatilids. An example concerns the mussel species *Mytilus edulis* L. and *M. galloprovincialis* Lmk. which have a genetic identity of 0.85 (Skibinski *et al.*, 1980) compared to a figure of 0.95 for *L. saxatilis* and *L. arcana* (Ward & Warwick, 1980), yet hybridise relatively readily (Skibinski *et al.*, 1983). *Mytilus* has external fertilization, and thus the capacity for hybridisation depends on the extent of temporal overlap in spawning period and the biochemical affinities of the gametes of the species involved. In natural populations of *Littorina*, behavioural or microgeographic isolating mechanisms may act to reduce the incidence of interspecific copulation between reproductively competent adults. Under laboratory conditions, where only a single potential mate is provided, and the pair left together for long periods, these mechanisms may break down. It seems unlikely that temporal isolating mechanisms have a significant role in preventing hybridi-

sation in natural populations, as *L. saxatilis* populations are reproductively active throughout most of the year. Alternatively, the exclusive use of sexually mature females in the electrophoretic work may have resulted in discrimination against hybrids present in the population, if these show reduced viability or growth rates.

## Acknowledgements

RDW gratefully acknowledges receipt of a NERC research grant. We wish to thank Kerstin Johannesson and an unknown referee for providing useful criticism of an earlier version of this manuscript, and Mike Lord, Julie Knight, John Perry and others for collecting samples.

## References

Ayala, F. J., 1975. Genetic differentiation during speciation process. In T. Dobzhansky, M. Hecht & W. C. Steer (eds), Evolutionary Biology, volume 8. Plenum Press, New York.

Caugant, D. & J. Bergerard, 1980. The sexual cycle and reproductive modality in *Littorina saxatilis* Olivi (Mollusca: Gastropoda). Veliger 23: 107–111.

Hannaford Ellis, C. J., 1983. Patterns of reproduction in four *Littorina* species. J. moll. Stud. 49: 98–106.

Hughes, R. N. & D. J. Roberts, 1981. Comparative demography of *Littorina rudis*, *L. nigrolineata* and *L. neritoides* on three contrasted shores in North Wales. J. anim. Ecol. 50: 251–268.

Janson, K., 1985. A morphological and genetic analysis of *Littorina saxatilis* (Prosobranchia) from Venice, and on the problem of *saxatilis-rudis* nomenclature. Biol. J. linn. Soc. 24: 51–59.

Janson, K. & R. D. Ward, 1984. Microgeographic variation in allozyme and shell characters in *Littorina saxatilis* Olivi (Prosobranchia: Littorinidae). Biol. J. linn. Soc. 22: 289–307.

Janson, K. & R. D. Ward, 1985. The taxonomic status of *Littorina tenebrosa* Montagu as assessed by morphological and genetic analyses. J. Conch. 32: 9–15.

Knight, A. J. & R. D. Ward, 1986. Purine nucleoside phosphorylase polymorphism in the genus *Littorina* (Prosobranchia: Mollusca) Biochem. Genet. 24: 405–413.

Knight, A. J., R. N. Hughes & R. D. Ward, 1987. A striking example of the founder effect in the mollusc *Littorina saxatilis*. Biol. J. linn. Soc. 32: 417–426.

Raffaelli, D. G., 1977. Observations on the copulatory behavior of *Littorina rudis* (Maton) and *Littorina nigrolineata* Gray. Veliger 20: 75–77.

116

Raffaelli, D. G., 1982. Recent ecological research on some European species of *Littorina*. J. moll. Stud. 48: 342–354.

Maton, W. G., 1797. Observations relative Chiefly to the Natural History, Picturesque Scenery and Antiquities of the Western Counties of England made in the Years 1794 and 1796. Eaton, Salisbury.

Mayr, E., 1963. Animal Species and Evolution. Belknap Press of Harvard University Press, Cambridge, Mass.

Olivi, A. G., 1792. Zoologica Adriatica. Bassano. pp. 334 pl 9.

Skibinski, D. O. F., J. A. Beardmore & T. F. Cross, 1983. Aspects of the population genetics of *Mytilus* (Mytilidae: Mollusca) in the British Isles. Biol. J. linn. Soc. 19: 137–183.

Skibinski, D. O. F., T. D. Cross & M. Ahmad, 1980. Electrophoretic investigation of systematic relationships in the marine mussels *Modiolus modiolus* L., *Mytilus edulis* L., and *Mytilus galloprovincialis* Lmk. (Mytilidae; Mollusca). Biol. J. linn. Soc. 13: 65–73.

Smith, J. E., 1981. The natural history and taxonomy of shell variation in the periwinkles *Littorina saxatilis* and *Littorina rudis*. J. mar. biol. Ass. U.K. 61: 215–241.

Smith, J. M., 1982. A review of the genus *Littorina* in British and Atlantic waters (Gastropoda: Prosobranchia). Malacòlogia 22: 535–539.

Ward, R. D. & K. Janson, 1985. A genetic analysis of sympatric subpopulations of the sibling species *Littorina saxatilis* (Olivi) and *Littorina arcana* Hannaford Ellis. J. moll. Stud. 51: 86–94.

Ward, R. D. & T. Warwick, 1980. Genetic differentiation in the molluscan species *Littorina rudis* and *Littorina arcana* (Prosobranchia: Littorinidae). Biol. J. linn. Soc. 14: 417–428.

Ward, R. D., T. Warwick & A. J. Knight, 1986. Genetic analysis of ten polymorphic enzyme loci in *Littorina saxatilis* (Prosobranchia: Mollusca). Heredity 57: 233–241.

Warwick, T., 1983. A method of maintaining and breeding members of the *Littorina saxatilis* (Olivi) species complex. J. moll. Stud. 48: 368–370.

*Hydrobiologia* **193**: 117–138, 1990.
*K. Johannesson, D. G. Raffaelli and C. J. Hannaford Ellis (eds), Progress in Littorinid and Muricid Biology.*
© 1990 *Kluwer Academic Publishers.*

# Scraping a living: a review of littorinid grazing

T.A. Norton[1], S.J. Hawkins[1], N.L. Manley[1], G.A. Williams[1] & D.C. Watson[2]
[1] *Port Erin Marine Laboratory, Liverpool University, Port Erin, Isle of Man, UK (address for reprint requests)*; [2] *Department of Botany, University of Glasgow, Glasgow, Scotland, UK*

*Key words:* feeding behaviour, algal defences, intertidal ecology

## Abstract

Littorinid snails are predominantly herbivorous and the versatility of their radulae enables them to feed on a variety of macroscopic and microscopic plants in a diversity of habitats. Some are selective feeders preferring some species of algae to others, and rejecting some even after a prolonged period of starvation. Different species of snail exhibit different preferences. The factors affecting the attractiveness and edibility of food plants are discussed and food value considered.

Foraging behaviour of littorinids is briefly reviewed in relation to the influence of chemical cues from the algae. Littorinids appear to be able to select or reject algae without having ingested them, having perceived the plants from a distance, moving towards favoured foods (or habitat-providing plants) and away from those that it rejects. The nature of the chemical cues emitted by the algae is discussed. Temporal patterns of foraging activity show some evidence of an endogenous component which can be overridden by responses to environmental conditions. These patterns place restraints on energy intake.

The structural and chemical defences used by algae against littorinid grazing are considered. The importance of polyphenolic compounds is evaluated. The effects of grazing as a selective agency and a factor influencing algal populations are discussed. There is some evidence that life history patterns are a response to grazing. The influence of external physical factors, such as salinity on grazing pressure is demonstrated.

Finally, the impact of littorinid snails on intertidal communities is assessed in relation to their abundance and biogeographical distribution. The relative importance of littorinids is contrasted on shores possessing or lacking limpets.

## Introduction

Littorinid snails are predominantly herbivorous. Some feed largely on epilithic microalgae and the germlings of seaweeds: e.g. *Littorina mariae* Sacchi & Rastelli (Watson & Norton, 1987; Williams, 1987), *Littorina keenae* Rosewater (= *L. planaxis*, Dahl, 1964; Foster, 1964), *Nodilittorina cincta* (Quoy & Gaimard), and

*Nodilittorina antipodum* (Phillipi) (= *L. unifasciata*) (Luckens, 1974) or microflora on mud in mangroves: e.g. *Bembicium auratum* Quoy & Gaimard (Branch, 1979; Branch & Branch, 1981a). High on rocky shores lichens can form an important component of the diet (e.g. *N. antipodum* (= *L. unifasciata*) Branch & Branch, 1981b). Many browse on the surface of seaweed, seagrasses or mangroves consuming epiphytic algae, fungi or micro organisms such as diatoms, protozoa, cyanobacteria and bacteria: e.g. *L. obtusata* L. (Reimchen, 1974) *L. neglecta* Bean (Robertson & Mann, 1982) *L. mariae* (Williams, 1987), *Littoraria irrorata* Say (Bebout, 1986) and *Littoraria angulifera* (Lamark) (Kohlmeyer & Bebout, 1986). Some can also consume adult macroalgae e.g. *Littorina littorea* L. (Watson & Norton, 1985a), *Littorina obtusata* (Watson & Norton, 1987; Williams, 1987), *Littorina scutulata* Gould (Dahl, 1964), *Littorina saxatilis* (Olivi) and *Littorina nigrolineata* (Gray) (Sacchi *et al.*, 1981). Both *L. scutulata* and *L. littorea* are versatile opportunists able to feed with equal facility on macroalgae and microphagously on the surface of rock, mud, sand or gravel (Ankel, 1937; Frid & James, 1988; Newell, 1958a; Watson & Norton, 1985a; Brenchley, 1987; Voltolina & Sacchi, this volume).

Where macroalgal food is scarce, as on the upper shore, *L. littorea* and *L. sitkana* Philippi can rely heavily upon drift algae newly washed up on the shore (Voltolina & Sacchi, this volume; Watson & Norton, 1985a). The former species is repelled by seaweed that is not fresh (Norton, unpublished), but *L. keenae*, *L. scutulata* and *L. saxatilis* have been observed scavenging on decaying pieces of algae (Dahl, 1964; Fretter & Graham, 1962 p. 536).

The present paper reviews the feeding mechanisms and behaviour of littorinid snails and examines the consequences of their grazing for algal populations and community structure. Competition between grazing gastropods has been covered in several recent reviews (e.g. Underwood, 1979; Hawkins & Hartnoll, 1983; Branch, 1984). Therefore we have not considered it here. Of necessity we will concentrate on the species from northern temperate waters that have been most intensively studied and with which we are most familiar. Nomenclature follows Reid (this volume). To avoid confusion, the names used by original authors have been given where appropriate.

## The feeding process

### Feeding mechanisms

Not all algae are potential food for all herbivorous molluscs. What can and cannot be consumed will be determined ultimately by the capabilities of the feeding apparatus. The taenioglossan radula of *Littorina* is a highly versatile structure able to utilise efficiently a wide variety of food resources. The function and importance of each tooth changes according to the nature of the food. Fundamentally the taenioglossan radula is able to break off or scrape both unicellular algae or seaweed tissue. The basal membrane and the individual teeth attain a high degree of mobility, an attribute fundamental to littorinid feeding. Comparison of the littorinid feeding apparatus with the less specialized and possibly more primitive rhipidoglossan radula suggests that two distinct factors are instrumental in the increased versatility of the taenioglossans:
1. A reduction in the number of marginal teeth.
2. A loss of ancillary muscles used to adjust radular tension and position, shifting the emphasis from positional adjustments of the radula to the force with which the radula is applied to the substratum (Steneck & Watling, 1982).

The structure and mode of action of the radula of *Littorina littorea* has been described and compared with that of other prosobranchs by Fretter & Graham (1962) and more recently by Hawkins *et al.* (1989). Many of the distinctive features of the teeth of the radulae of the sibling species *L. obtusata* and *L. mariae* hint at a degree of divergence in feeding behaviour (Watson & Norton, 1987). These relatively minor differences in the structure and functioning of the feeding apparatus may impose dietary restrictions and could reflect partitioning of food resources between co-existing species (see also Steneck & Watling, 1982). Raffaelli (1985) and Hawkins *et al.* (1989) warn against predictions of diet from radula morphology – markedly different radula may be used to feed on surprisingly similar foods.

Successive grazing strokes of *L. littorea* leave a

zig-zag pattern on the substratum caused by the characteristic side-to-side movement of the head during foraging. The angle between successive sections of the trace decreases as food abundance increases so that abundant food sources are utilised more efficiently (Watson, 1983). During each grazing stroke 13–14 rows of teeth are applied to the substratum in succession. The rotation and folding of the marginal and intermediate teeth produces an elaborate pattern. In most respects the grazing trace of *L. littorea* and of *L. obtusata* are similar but the broad, blunt cusps of the latter leave much wider scratch marks. Algal germlings are, however, grazed in a similar manner by both species. Entire tiny thalli are scraped or even 'bulldozed' from the substratum and brushed into the mouth (personal observation). The denticles on the radula are about as hard as a human finger nail (2–2.5 on Moh's scale of hardness compared with 4.5–5 for the radula of *Patella vulgata* L. (Watson, 1983; Hawkins *et al.*, 1989). Forceful scraping of the rock surface causes considerable wear on the teeth.

Snails that browse directly on the surface of rock may also inadvertently ingest detritus, small live animals, egg capsules, sand grains and fragments of rock (Dahl, 1964; Hylleberg & Christensen, 1978; Brenchley, 1982). An analysis of faecal pellets (McLean, 1967) indicated that *Nodilittorina meleagris* (Potiez & Michaud) can pass 0.2 g. of rock per year, *L. ziczac* 0.6 g. and *Nodilittorina tuberculata* (Menke) 1.2 g.. North (1954) has estimated that browsing by populations of *L. keenae* and *L. scutulata* at a combined density of 833 snails m$^{-2}$ can deepen the sandstone floor of the pool in which they live by just over 0.6 mm per year.

*Littorina littorea* feeds on microalgae by scraping the surface with the rachidian and lateral teeth. A single stroke of the radula removes only about 10% of the area of the surface film, even from a very smooth surface such as glass, because of the spacing of the teeth (Harding & Norton, unpublished). However, the same area may be scraped several times by the same snail, or other trail following snails, thus removing progressively more of the biofilm. Loosened material is captured by the inward folding laterals and is carried into the buccal cavity. The rachidian tooth assumes greater importance when feeding on the foliose thalli of algae such as *Ulva lactuca* L. and *Enteromorpha intestinalis* (L.) Link which *L. littorea* is able to tear effectively.

Grazing by *Littorina obtusata* usually begins at the centre of the frond and often leaves a characteristic 'rib' of ungrazed material along the thallus edge. The blunt teeth permit rapid excavation of the softer internal tissue of the fucoid thalli. Although the tougher outer epithelial layer must first be breached to allow access to the inner tissues, thereafter it is rarely consumed (per. obs.). *L. mariae* browses over the surface of seaweed and takes in very little epithelial tissue (Williams, 1987).

*Feeding preferences*

On many shores the choice of potential food items is very wide. Yet, despite the broad capabilities of the feeding apparatus, littorinids are highly selective grazers, strongly preferring some species of algae to others. Different species of co-occurring snail, however, do not have the same preferences (Table 1).

The food preferences of *L. littorea*, are best known having been examined by workers on both sides of the Atlantic. Attractiveness was examined by Lubchenco (1978), unfortunately without obviating mucus trail following (Dinter & Manos, 1972) which can cause problems with choice experiments. Bertness *et al.* (1983) and E.M. Bell & D.P. Cheney (unpublished) studied edibility of a variety of seaweeds. Watson & Norton (1985a, 1987) studied both the attractiveness and edibility of many algae not only to *L. littorea*, but also to *L. mariae* and *L. obtusata*. To test edibility Watson & Norton *loc. sit.* compared consumption with a standard preferred species (*Ulva lactuca*) to compensate for differences in the appetite of the snails (Table 1). The results achieved by all these authors are broadly similar: e.g. *L. littorea* prefers ephemeral greens and the ephemeral red *Porphyra* to the various fucoids,

*Table 1.* The attractiveness and edibility (Palatability Index) of seaweeds to littorine snails in laboratory experiments (compiled from Watson & Norton, 1985a and Norton unpublished). Significance of difference from consumption of *Ulva*, Wilcoxon rank Tests. * $0.01 < P < 0.05$, ** $0.001 < P < 0.01$, *** $P < 0.001$.

| | Attractiveness | | | Palatability Index | | |
|---|---|---|---|---|---|---|
| | (Preference rankings determined $\chi^2$ analysis) | | | (Amount consumed/amount of *Ulva* consumed) | | |
| | *L. littorea* | *L. obtusata* | *L. mariae* | *L. littorea* | *L. obtusata* | *L. mariae* |
| Chlorophyta | | | | | | |
| Cladophora rupestris | II | IV | V | 0 | 0 | 0 |
| Enteromorpha intestinalis | I | – | – | 0.54* | – | – |
| Ulva lactuca | I | III | IV | 1.00 | 1.00 | 1.00 |
| Phaeophyta | | | | | | |
| Ascophyllum nodosum | II | II | I | 0 | 2.05 | 0.94 |
| Fucus ceranoides | – | – | – | 0.21*** | 6.20*** | – |
| F. serratus | II | II | I | 0 | 3.63* | 3.10** |
| F. spiralis | – | I | I | 0.17*** | 9.67** | 5.36** |
| F. vesiculosus | – | I | I | – | 3.72** | 4.04 |
| Pelvetia canaliculata | II | II | II | 0.04*** | 3.51** | 2.51* |
| Rhodophyta | | | | | | |
| Corallina officinallis | III | – | – | 0 | 0 | 0 |
| Laurencia pinnatifida | II | – | – | 0.11** | – | – |
| Mastocarpus stellatus | II | – | – | 0 | – | – |
| Polysiphonia lanosa | II | III | III | 0 | 0 | 0 |
| Porphyra umbilicalis | I | – | – | 0.19** | – | – |

which in turn are preferred to perennial reds such as *Mastocarpus*, *Chondrus*, and *Corallina*. In laboratory experiments with *L. littorea*, attractiveness and edibility rankings for mature seaweeds are broadly similar (Table 1). *L. littorea* consistently showed a strong preference for the ephemeral green algae over the canopy forming fucoid species *Ascophyllum nodosum* (L.). Le Jolis. and the calcareous alga *Corallina officinalis* L. which were rejected even after prolonged starvation. Although starvation of the snails does not influence the apparent edibility of different algae, it may increase their relative attractiveness and can prompt an elevated feeding rate when food becomes available (Table 2). Attractiveness may also depend on the use of plants as a habitat or as a cue to suitable habitats. Habituation to a non-preferred food plant prompts less discriminate feeding (Watson & Norton, 1985a).

Most research of littorinid feeding behaviour has concentrated on the selection and consumption of adult seaweeds to the exclusion of the more vulnerable juvenile stages. There is evidence that small plants ($< 3$ cm) are more susceptible to grazing by *Littorina littorea* than larger plants (Lubchenco, 1983), allowing fast-growing plants to rapidly escape from danger. The tiny germlings are even more at risk and in choice experiments *L. littorea* selects juvenile *Fucus* in preference to adult plants (Watson & Norton, 1985a). Similarly, juvenile plants of *Chondrus crispus* Stackh. ($< 2$ mo. old, $< 2$ cm tall) are readily consumed in

*Table 2.* The effects of 50 days starvation on the feeding of *Littorina littorea*. The figures represent the post starvation results as a percentage of the pre starvation values. Relative attractiveness is expressed as attractiveness of *Pelvetia* relative to *Enteromorpha*. Significance of differences, Chi-square analysis, * $0.01 < P < 0.05$.

| | Relative attractiveness | Feeding rate | Edibility |
|---|---|---|---|
| Enteromorpha intestinalis | 88 | 256* | 102 |
| Pelvetia canaliculata | 112* | 590* | 175 |

laboratory experiments whereas adult plants are not eaten (D.P. Cheney unpublished).

Grazing on microalgae has been generally presumed to be unselective. Indeed the size range of diatoms ingested by *L. littorea* has been found to be similar to that of those on the rock surface, although smaller snails ingest a higher proportion of smaller diatoms than do larger snails (Hylleberg & Christensen, 1978). Some inadvertent selection may take place. *L. scutulata* ingests more loosely adherent diatoms than those that are firmly anchored, and takes more chain-forming diatoms. These are accessible on the top of the microalgal layer and if any part of the chain is contacted by the radula the entire chain is usually ingested (Castenholz, 1961; Nicotri, 1977)). Snail grazing may dramatically change the composition of the microflora on artificial substrata fixed on the shore (Nicotri, 1977; Hunter & Russell-Hunter, 1983). *L. littorea* for example causes a tenfold increase in the relative abundance of five taxa of diatoms (Hunter & Russell-Hunter, *loc. cit.*). The authors assumed that selection was inadvertent and probably resulted from the snails inability to remove less susceptible adherent diatoms.

More surprisingly, in the laboratory *L. littorea* clearly chooses to consume lawns of germlings of *Ulva lactuca* in preference to juvenile plantlets of several other species, and rejects those of *Ascophyllum* (Table 3). *L. littorea* can also distinguish between small areas of cellulose impregnated with water soluble algal extracts (Imrie *et al.*, this volume). In nature the snails may be unable to differentiate such tiny plants in well mixed stands, but could clearly select between the small patches often found on the shore.

*Food value*

Empirical and theoretical studies of feeding preferences assume that animals are subject to a strong selective pressure to eat those foods which yield maximum 'value' per unit metabolic cost (Townsend & Hughes, 1981). The 'value' of a given food item represents an amalgam of the quality and quantity of nutrients and energy it contains and the capacity of the animal to obtain them. The costs include those involved in finding, handling, ingesting and digesting the food.

It seems likely that highly-ranked plants offer more nutrition to the snail than other seaweeds thus compensating for the energy expended in searching for them (Lubchenco & Gaines, 1981). If so, this cannot be simply a matter of the energy that the plants contain, for different species of snail rank the same seaweeds differently in edibility trials (Table 1). The benefit that an animal derives from its food may have as much to do with its digestibility as with its nutritional content. Habitat requirements will also influence the behaviour of littorinids such as *L. obtusata* which live on their food source.

It might be assumed that the efficiency of digestion in littorinids is low, for their faeces have sometimes been found to contain recognisable fragments of macroalgae (e.g. in *L. keenae*, Dahl, 1964), seaweed fragments able to release swarmers (e.g. *Nodolittorina peruviana* (Lamark), Santelices & Ugarte, 1987) and live diatoms (*L. littorea*, Hylleberg & Christensen, 1978, *L. scutulata* Nicotri, 1977), but incomplete digestion is often a consequence of a surfeit of food. The assimilation efficiency for *L. keenae* has been calculated to be 36% of ingested material in dry weight, and for *L. irrorata* it is estimated to be 45% efficiency in terms of energy (Odum & Smalley, 1959 and Grahame, 1973 calculated from North 1954). More critical estimates (in terms of energy) have given assimilation efficiences as high as 73% for *L. obtusata* (Wright,

*Table 3.* The relative edibility to *Littorina littorea* of seaweed germlings when offered in paired choice experiments. The preferred species is underlined. N.S. = not significant, * 0.01 < P < 0.05, ** 0.001 < P < 0.01, *** P < 0.001, Wilcoxon signed-rank test.

| | | |
|---|---|---|
| *Enteromorpha intestinalis* | *Ulva lactuca* | N.S. |
| *Ascophyllum nodosum* | *Ulva lactuca* | *** |
| *Fucus serratus* | *Ulva lactuca* | * |
| *Mastocarpus stellatus* | *Ulva lactuca* | *** |
| *Ascophyllum nodosum* | *Fucus serratus* | ** |
| *Ascophyllum nodosum* | *Pelvetia canaliculata* | ** |

1977; Wright & Hartnoll, 1981), and 82% for *L. littorea* fed on *Ulva lactuca* (Grahame, 1973). These figures were calculated without allowance for the energy diverted into mucus production. They also represent exceptionally high figures for a herbivore (see Hawkins & Hartnoll, 1983 for review of energetics of marine invertebrate grazers) and should be viewed with caution.

Different foodstuffs may be differentially digestible, e.g. *L. scutulata* seems able to digest completely several species of diatom, but for other diatoms (even some preferred species) the mean digestibility was only 37–55% (Nicotri, 1977). Even relatively indigestible substances such as cellulose can be assimilated. *Littorina littorea* fed on its preferred food of *Ulva lactuca* digested almost 73% of the ingested cellulose, and *L. obtusata* can assimilate around 35% of the cellulose contained in its preferred food, *Fucus serratus* L. (Watson, Brett & Norton, unpublished). Food value is perhaps best assessed in terms of how well the animal thrives on a particular diet. *Littorina littorea* can survive for several months if maintained on a unialgal diet, but thrives best if offered a mixed diet (D.C. Cheney pers. commun., Imrie *et al.*, this volume).

## Foraging behaviour

### Finding appropriate food

Not only must a foraging littorinid find food but it must remain within its normal zone. In consequence, much attention has been devoted to the cues involved in the behavioural maintenance of distribution patterns (see Underwood, 1979; Underwood & Chapman, 1985, for reviews) and directional recovery following natural or experimental displacement (Gendron, 1977; McQuaid, 1981; Petraitis, 1982; Chapman, 1986). Littorinids make relatively limited foraging excursions sometimes returning to their original location (eg. *L. littorea* 1.5 m. Newell, 1958a; 2 m. Thamdrup 1935). Individuals of high shore species such as *Littorina saxatilis* or *Littorina acutispira* Smith that fail to find sheltering crevices

after foraging suffer an increased risk of mortality (Raffaelli & Hughes, 1978; Underwood & McFadyen, 1981). Remaining within a circumscribed zone limits the distance over which food can be encountered. Within the normal distribution range foraging may be largely at random (Haseman, 1911; Petraitis, 1982; Williams, 1987) or may follow a more directed search path (Dahl, 1964). Underwood & Chapman (1985) point out that behaviour patterns are usually the response to a suite of interacting cues and they suggest techniques for statistical analysis of directional data.

Littorinids are not only selective feeders, they also often reject algae without leaving any sign of having physically sampled them (Watson & Norton, 1985a). Visual cues are unlikely to be involved; although it has been suggested that in air littorinids can form sharp retinal images (Newell, 1965). Their acuity is likely to be limited (Land, 1968, see Underwood, 1979 for comments) particularly in water. When submerged the snails are more likely to be heavily reliant on olfactory and gustatory stimulii. Thus the snails are almost certainly responding to exudates from the algae which may enable them to select or reject food items from a distance, thereby minimizing foraging movements.

In laboratory experiments most individuals of *L. littorea*, *L. mariae* and *L. obtusata* tested respond rapidly to exudates from some marine algae by reorientating themselves and moving off in a consistent direction, sometimes at an accelerated rate of crawling (Manley & Norton, unpublished). They clearly differentiate between the exudates from various seaweeds and respond by crawling towards some and away from others (Manley & Norton, unpublished). Moreover, different species of snails may respond differently to the same alga. They appear to be attracted towards their favoured food plants or to their preferred algal habitat (Table 4). Particularly striking are the contrasting responses of different species of snail to exudates of *Ascophyllum* (Table 5): *L. littorea* is repelled as effectively by the exudates as by the plant itself; whereas *L. obtusata* is strongly attracted towards the

*Table 4.* A summary of the habitat and food preferences of three species of littorinid snails from sheltered shores in relation to their responses to algal exudates. + indicates a positive response, − indicates repulsion or rejection. and 0, indifference. Data from Watson & Norton (1985a, 1987) and G. A. Williams (unpublished).

| Snails | Algae | Snail habitat | Attractiveness | Edibility |
|---|---|---|---|---|
| | | On Rock Surfaces | | |
| *L. littorea* | *Fucus serratus* | | 0 | 0 |
| | *Ascophyllum* | | − | − |
| | *Ulva lactuca* | | + + + | + + + |
| | | On Seaweeds | | |
| *L. mariae* | *Fucus serratus* | Preferred host | + | + + |
| | *Ascophyllum* | Occasionally associated | 0 | + |
| | *Ulva lactuca* | Rarely associated | + + + | + |
| | | On Seaweeds | | |
| *L. obtusata* | *Fucus serratus* | Occasionally associated | + + + | + + |
| | *Ascophyllum* | Preferred host | + + + | + + |
| | *Ulva lactuca* | Rarely associated | + + | + |

exudates and *L. mariae* is indifferent to them.

Under ideal conditions in the laboratory littorinid snails can perceive exudates when at least a metre away from the source (Van Dongen, 1956). It would however, be much more difficult to detect the concentration gradient of an exudate amidst the turbulent water motion of the shore. In culture even aerating the water may confuse an animal's directional response (Frings & Frings, 1965). On the shore the multitudinous exudates emanating from various algae may further confuse the herbivore. *Aplysia jubiana* (Quoy & Gaimard), for example, is like *L. littorea* attracted by *Ulva lactuca*, and in the presence of exudate from *Ulva*, other algae, normally rejected, are eaten in error (Frings & Frings, 1965). It is perhaps significant that the algae that are most attractive to at least some littorinids are often present on the shore in

huge quantities. The magnitude of the chemical signs that emanate from large monospecific stands of such algae may be sufficient to swamp any interference from subsidiary members of the flora.

It is important that the animal should find food and shelter without excessive searching, and hence energetic costs. The oxygen consumption of *L. littorea* when crawling has been estimated to be 15 times higher than when at rest (Newell, 1970). Furthermore, the production of the enormous quantities of mucus required for locomotion constitutes a large additional drain on the snail's resources (Calow, 1974). In addition a foraging snail runs greater risks of drying out or predation. In taking a more direct route towards the apparent source of a preferred food the snail risks disregarding lower-ranked but edible algae

*Table 5.* Directional responses of littorinid snails from the Firth of Clyde to exudates of *Ascophyllum nodosum*. NS = not significant, * $0.01 < P < 0.05$, *** $P < 0.001$. (Watson, 1983).

| Snails | N | Towards | Away | Significance of difference (Wilcoxon Signed-Rank Test) |
|---|---|---|---|---|
| *L. littorea* | 57 | 16 | 41 | *** |
| Control | 60 | 39 | 21 | * |
| *L. mariae* | 51 | 30 | 21 | NS |
| Control | 60 | 36 | 24 | NS |
| *L. obtusata* | 56 | 50 | 6 | *** |
| Control | 54 | 32 | 32 | NS |

en route. This would only be a sensible procedure where the most desirable food plants are relatively abundant and are large in relation to the size of the snail (Lubchenco & Gaines, 1981). In such cases there would be a high probability of finding a suitable plant close by, and once encountered, the plant could provide sustenance for a lengthy period. Even intact and healthy seaweeds excrete a wide variety of substances into the surrounding water and herbivorous molluscs are known to use gradients of some of them to detect food and other targets in the environment (Croll, 1983). Our own unpublished experiments indicate that the exudates to which littorinids respond are water soluble, and that sufficient is excreted by a seaweed within 60 minutes to 'activate' 25 l of seawater. Exudates from *Ulva lactuca* that stimulated feeding behaviour in *Aplysia jubiana* were also water soluble and could be obtained by soaking fresh seaweed for only 10 minutes (Frings & Frings, 1965). These exudates were both non-volatile and heat-stable and still elicited a feeding response when diluted to less than one part in 15 000 000. In the case of *Aplysia dactylomela* Rang the phagostimulants from *Ulva* appear to include starch, glucose and glutamic acid (Carefoot, 1980). For the freshwater snail *Biomphalaria glabrata* (Say) some sugars, particularly maltose, as well as certain amino acids and carboxylic acids proved to be both powerful attractants and phagostimulants (Thomas, 1986; Thomas *et al.*, 1983, 1986; Uhazy *et al.*, 1978).

*Temporal patterns*

Many shore animals show rhythms of behavioural activity in response to the ebb and flow of the tides and diurnal changes (see Naylor, 1985 for review). These rhythms usually have an endogenous component due to internal biological clocks, and an exogenous component in direct response to environmental changes which also help to reset the clock. Rhythms in foraging activity of intertidal gastropods have been reviewed by Hawkins & Hartnoll (1983) and more recently by Little (1989). It is difficult to generalize about gastropod foraging rhythms as even in the same species activity patterns will vary from place to place, with season, shore level, microhabitat and stage of the spring/neap cycle (e.g. in *Patella vulgata*, Little *et al.*, 1988).

In *Littorina nigrolineata* an endogenous rhythm continues when constantly illuminated and submerged in the laboratory (Fig. 1a). Peak activity was shown to occur at expected high water, maxima being greater in expected daylight. There is an additional smaller burst of activity at the time of expected exposure by the ebbing tide (Fig. 1b). There is evidence that this rhythm is modulated on a semilunar and seasonal basis. *Littorina irrorata* has been shown to have a circatidal rhythm; migrating vertically along *Spartina* blades with the tidal cycle. This behaviour is suggested to be a response to osmotic stresses and predation pressure experienced if the littorinid remained on its mud-flat feeding ground during periods of immersion (Warren, 1985). Other species have been recorded as showing no particular rhythm (*Melaraphe neritoides*, Raffaelli & Hughes, 1978; *Littorina aspersa*, Garrity, 1984). It seems that in many littorinids any underlying endogenous rhythm is easily overridden or modified by direct responses to the environment. Many species (e.g. *L. littorea*; *L. obtusata* and *L. mariae*) will feed both when submerged and when emersed provided conditions are humid, as at night or on damp, overcast days. Submergence by the incoming tide stimulates both crawling and feeding in previously inactive *Littorina littorea* (Newell, 1958a, b) and *L. obtusata* (Williams, 1987). Exposure to air may also stimulate activity but feeding will continue only whilst conditions remain damp (Newell *et al.*, 1971). As the rock dries the animals commonly become inactive and close their opercula. Thus on the upper shore feeding activity is greatly influenced by the state of the tide but on the lower shore conditions permit feeding at all times, even when the tide is out (Newell, 1958a, b; Thamdrup, 1935). The effects of wave action on submerged foraging behaviour is obviously difficult to investigate but it is likely that foraging stops in rough conditions. Adverse conditions (storms, cold temperatures) have been

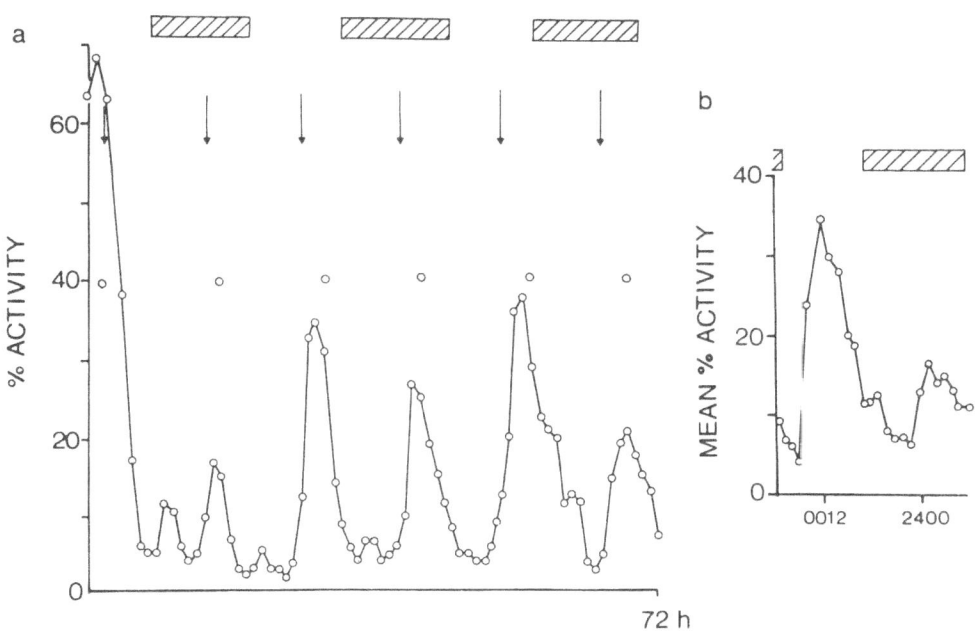

*Fig. 1.* a) The endogenous rhythm of crawling activity shown by 150 freshly collected *Littorina nigrolineata* immersed in seawater under constant illumination. The mean number crawling is expressed as a percentage each hour, and the median point of each active period indicated (o) above the main activity peaks. The times of expected high water are indicated by the arrows and the dark phase of the natural photoperiod shown by the shaded rectangles. b) The mean percentage activity taken over the three days of the experiment shows a strong rhythm of period 12 hours (after Petpiroon & Morgan 1983).

reported to terminate foraging activity in *L. obtusata* when emersed (Bray, 1974; Guiterman, 1970 and pers. obs.).

The different temporal constraints at increasing shore levels can affect feeding rates. Newell *et al.* (1971) studied the rate of radular activity of *L. littorea* feeding on an algal film in the laboratory. They found that specimens from the upper shore had a higher rate of activity than mid shore individuals, which in turn fed more rapidly than those from the lower shore. This phenomenon was confirmed on the shore by Petraitis & Sayigh (1987) but not by Cornelius, 1972. These results may indicate that higher on the shore the snails graze more rapidly to compensate for the shorter immersion and therefore more limited feeding time. The feeding rate of individual snails can be modified by experimentally subjecting them to the tidal regime typical of a different region of the shore (Newell *et al.*, 1971).

## Ecological consequences

*Algal defences versus grazing prowess.*

The benefit gained by an animal from its food depends on its ability to manipulate and ingest the food, and to digest, absorb and assimilate the nutrients. Each of these steps constitutes a potential obstacle to the grazer and provides scope for the development of effective anti-herbivore defences by the alga.

The manner in which the shape and structure of the algal thallus interacts with the body plan and feeding apparatus of the herbivore to influence the efficiency of feeding has been demonstrated by Steneck & Watling (1982) and Watson & Norton (1985b). The importance of a food item which can be manipulated easily and with minimum energy expenditure is illustrated by the preference of *L. littorea* for *Ulva lactuca* over the softer thallus of *Enteromorpha intestinalis*

126

(Watson & Norton, 1985a). While the broad, ulvoid thallus is readily pinned down by the snail's foot to provide a relatively stable food substrate, a single tubular frond of *Enteromorpha intestinalis* is too narrow to carry the entire foot and if inflated with gas has a strong inherent buoyancy which often frustrates efforts to hold it in place.

For some seaweeds structural defences constitute the first and most effective barrier to grazing (Table 6). Grazing of the erect, coralline alga *Corallina officinalis* is precluded by its tough calcium carbonate reinforced thallus. The articulated joints between calcareous segments constitute the only potential weak point. As the radula is wider than the uncalcified joints the alga's defences are impregnable (Watson & Norton, 1985a). Other inedible algae, such as *Mastocarpus stellatus* (Stackh.) Guiry and *Chondrus crispus*. although not calcified are dauntingly tough (Bertness *et al.*, 1983, Watson & Norton, 1985b) and consuming them may cause wear to the snails teeth (Fig. 2).

Subjective assessments of thallus toughness may be misleading. The toughness of algal tissue has been measured in terms of resistance to penetration, scratch resistance, abrasion resistance and even indirectly by examining the wear on the radular teeth (Bertness *et al.*, 1983; Watson & Norton, 1985b). Unfortunately the degree of radula wear does not necessarily correlate with for example scratch resistance of the tissue (Bert-

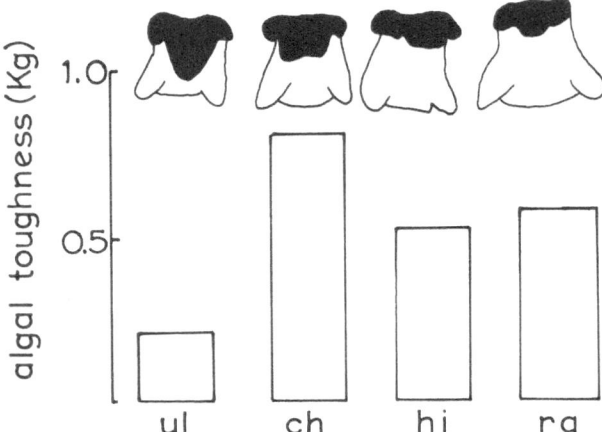

Fig. 2. The degree of wear on rachidian denticles of *Littorina littorea* radula when fed exclusively on different species of algae. ul = *Ulva lactuca*, ch = *Chondrus crispus*, hi = *Hildenbrandia rubra*, ra = *Ralfsia verrucosa*. Toughness is measured as the weight required for a metal spike to score the surface (after Bertness *et al.*, 1983)

ness *et al.*, *loc. cit.*) and toughness rankings based on puncture resistance and abrasion resistance may be markedly different (Table 6). This is because they measure different things. Puncture resistance is dependent on the strength of the outermost cortical layer, whereas resistance to abrasion is related to general cell size in the cortex, which influences the ratio of wall material to cytoplasm (Watson & Norton, 1985b). The measure of toughness that is of most ecological relevance seems to depend on the species of grazer being considered. For *L. littorea*, seaweed edbility is inversely correlated with puncture strength but is not in any way related to abrasion resistance (Watson & Norton, 1985b). Edibility rankings for *L. mariae* on the other hand, are inversely correlated with the abrasion resistance of the thalli but are not influenced by its puncture resistance. This reflects the basic differences described above in the radulae of *L. littorea* and flat periwinkles, *L. obtusata* and *L. mariae*. *L. littorea* can employ its sharp denticular cusps to tear the thallus, a mode of action which relies heavily on the initial penetration of the tissue. *L. obtusata* on the other hand uses the cusps to abrade the thallus. *Fucus spp.* the preferred foods for the latter species, are much less resistant to abrasion than non-

Table 6. Two assessments of the relative 'toughness' of the thalli of a variety of seaweeds. All figures represent a percentage of the value obtained for the toughest species, designated 100%.

| Species | Puncture strength | Abrasion resistance |
|---|---|---|
| *Porphyra umbilicalis* | 13 | 13 |
| *Ulva lactuca* | 25 | 44 |
| *Fucus spiralis* | 30 | 91 |
| *Fucus vesiculosus* | 30 | 100 |
| *Pelvetia canaliculata* | 48 | 18 |
| *Fucus serratus* | 53 | 72 |
| *Mastocarpus stellatus* | 63 | 13 |
| *Ascophyllum nodosum* | 100 | 20 |

preferred species such as *Ascophyllum* and *Ulva lactuca* (Watson & Norton, 1987).

All parts of a seaweed's thallus are not necessarily equally accessible, nor equally palatable to littorinid grazers, herbivores may attack the tougher parts of the thallus only after consuming the more tender portions. D.P. Cheney (unpublished) observed that in early spring *L. littorea* may consume the margins of the frond of *Fucus vesiculosus* L. leaving the midrib intact. Our own observations (Watson & Norton, 1985b) indicate that *L. littorea* grazes the thallus of *P. canaliculata* Dcne & Thur. selectively, preferring the frond tips to the basal tissue. This may relate to variations in the toughness of the thallus, for the tips of plants of *Pelvetia canaliculata* are more easily punctured than is their basal tissue (Watson & Norton, 1985b). Similarly the vegetative and reproductive tissue of the same plant of *Fucus* may be differentially susceptible to grazing (Manley & Norton, unpublished). It is however difficult to test whether swollen mucilaginous reproductive receptacles are more or less tough than the vegetative thallus. The unpublished observations of D.P. Cheney are interesting in this respect. He found that although vegetative plants of *Chondrus crispus* were not consumed by *L. littorea*, reproductive plants were, but only after they had shed their spores. Perhaps the open ostioles allow ingress through the otherwise impenetrable cuticle.

The effects of texture and taste can be differentiated by comparing the rankings of intact plants and extracts of the plants impregnated into agar, starch gel or filter paper (Geiselman, 1980; Geiselman & McConnell, 1981; Manley & Norton, unpublished; Bertness *et al.*, 1983; Watson, 1983; Imrie *et al.*, 1989 and this volume). In spite of the difficulty of quantifying the results obtained by some of these methods (see Ragan & Glombitza, 1986) such treatments clearly rendered some inedible algae attractive (e.g. *Cladophora rupestris* L.). Whereas extracts from *Laurencia pinnatifida* (Huds.) Lamour. *Ascophyllum nodosum*, *Ralfsia verrucosa* (Aresch.) J.Ag. *Hildenbrandia rubra* (Sommerf.) Menegh. and *Chondrus crispus* remained as unpalatable as the

intact plants. Similarly, it seems unlikely that seaweed germlings could possess effective physical defences, yet *L. littorea* finds germlings of *Ascophyllum* as repellent as the adult plants, indicating that chemical factors are important (Table 3). The general correspondence of preferences obtained with intact plants and plant extracts (Imrie *et al.*, 1989), with a few exceptions (see above), suggest that chemical cues are primarily responsible for determining food preferences in *L. littorea*.

It has been claimed that polyphenols produced by *Ascophyllum* may act as grazer repellents (Geiselman & McConnell, 1981). Such compounds are not only produced by *Ascophyllum* but also excreted into the surrounding water, especially during spring and summer (Sieburth & Jensen, 1968; Sieburth & Tootle, 1981), when the water above the algal beds becomes significantly enriched with phenolics. In laboratory experiments *L. littorea* is not only repelled by exudates of *Ascophyllum* (Table 5), but also the presence of *Ascophyllum* extract reduces the apparent palatability of other more favoured algae (Manley & Norton, unpublished). Geiselman & McConnell (1981) found that *Ascophyllum* contained exceptionally high concentrations of polyphenols in comparison with *Fucus vesiculosus*. Our results indicate that the concentration in *Ascophyllum* is comparable with that in several other fucoids (Fig. 3, see also Ragan & Glombitza, 1986), although in general the levels of phenols in fucoids are significantly higher than in other brown algae (E.M. Bell & D.P., Cheney pers. commun.). Nonetheless, phenolic compounds are well-known grazer repellants in higher plants and are also found in a variety of both brown and red algae. They owe their efficacy as repellants to their astringency which is assumed to result from cross-linking between the polyphenols and proteins or glycoproteins in the mouth of the herbivore (Goldstein & Swain, 1965). The cross-linking precipitates the grazer's salivary proteins, or immobilizes enzymes, rendering the plant tissue unpalatable (Bate-Smith, 1973). We and independently E.M. Bell & D.P. Cheney (pers. commun.) both found no correlation between the

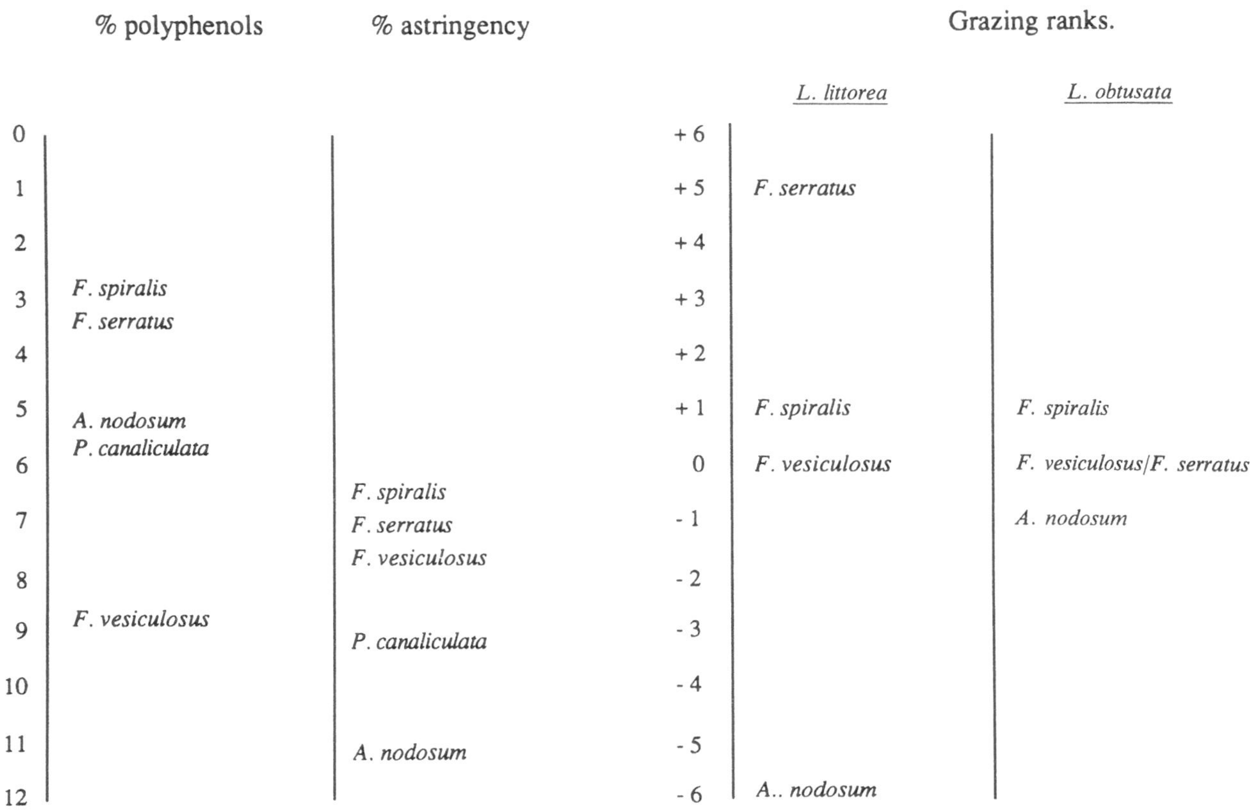

% polyphenols     % astringency           Grazing ranks.

| % polyphenols | % astringency | Grazing ranks | L. littorea | L. obtusata |
|---|---|---|---|---|
| 0 | | + 6 | | |
| 1 | | + 5 | F. serratus | |
| 2 | | + 4 | | |
| 3 F. spiralis / F. serratus | | + 3 | | |
| 4 | | + 2 | | |
| 5 A. nodosum | | + 1 | F. spiralis | F. spiralis |
| 6 P. canaliculata | | 0 | F. vesiculosus | F. vesiculosus/F. serratus |
| 7 | F. spiralis / F. serratus | - 1 | | A. nodosum |
| 8 | F. vesiculosus | - 2 | | |
| 9 F. vesiculosus | P. canaliculata | - 3 | | |
| 10 | | - 4 | | |
| 11 | A. nodosum | - 5 | | |
| 12 | | - 6 | A.. nodosum | |

*Fig. 3.* Polyphenol analyses and grazing preference ranks for vegetative algal tissue. Rankings are based on levels of significance (Wilcoxon signed-rank Test) between tissue consumed in pairwise choice experiments.

total polyphenol content of the tissue as determined by the Folin-Denis method (Tempel, 1973) and astringency as measured by haemanalysis (Bate-Smith 1973; Fig. 3). This is almost certainly because it is the high molecular weight polyphenols that have the greatest protein-binding capacity and therefore the highest astringency and presumably confer the greatest defence against grazers. These findings cast doubt on the value of the widely used Folin-Denis test as an indicator of the concentration in plant tissues of grazer repellent phenols. Our assessments clearly indicate that the tissues of *Ascophyllum* and *Pelvetia canaliculata* are very markedly more astringent than those of other intertidal fucoids (Fig. 3) which correlates well with the apparent repugnance exhibited by *L. littorea* for these plants. However, *Ascophyllum* is more readily consumed by *L. obtusata*. It may be that *L. obtusata*, a herbivore living predominantly amidst the *Ascophyllum* canopy, has gained immunity in this particular 'arms race' overcoming defences effective against the more generalist feeder, *L. littorea*.

Van Alstyne (1988) has recently shown that there are higher phenol levels in *Fucus distichus* (de la Pyl.) Pow. plants grazed by *Littorina sitkana* compared to ungrazed plants. Experimental wounding induced 20% elevated phenol levels. Initially *L. sitkana* were attracted to damaged plants but after two weeks they preferred undamaged controls, in which subsequent grazing damage was 50% more. Elevated phenols occurred in the whole plant not just the injured part. These results help to explain spatial and temporal variation in phenol levels. Higher levels in summer may correlate with greater grazing activity and initial grazing damage may stimulate defences against further grazing damage (see Fig. 4).

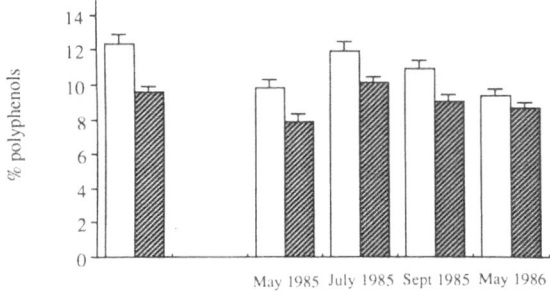

a) Natural grazing damage b) Experimental wounding

*Fig. 4.* a) Mean phenolic concentrations ($\pm$ S.E.) of naturally grazed and ungrazed *Fucus distichus*. Grazing damage was inflicted by *Littorina sitkana*. b) Mean phenolic concentrations ($\pm$ S.E.) of damaged and uninjured *Fucus distichus* when wounded by clipping (Clear blocks represent damaged plants – whether natural or clipped; hatched blocks represent undamaged – control plants) (after Van Alstyne, 1988).

Plantlets of fucoids are particularly vulnerable to grazing by *L. littorea* (Lubchenco, 1983). This may be because they are too small to have effective physical defences, but also because the astringency of faster growing plants (such as juveniles) is lower than in slower growing adults (E.M. Bell & D.P. Cheney, pers. commun.). Similarly, in fucoids the vunerable tips and receptacles appear not to be preferentially enriched with polyphenols when compared with other parts of the thallus and may even have lower astringency (E.M. Bell & D.P. Cheney, pers. commun.).

### Effects on individuals and populations of algae

Invertebrate grazing can influence the fitness and longevity of the algal victim quite profoundly. The nature of the herbivore/plant relationship depends on the form of damage sustained by the alga (Lubchenco & Gaines, 1981). Grazing may be lethal or even beneficial (e.g. the removal of epiphytes from the thallus of fucoids by *L. mariae*). More often it is sustainable, causing no long term damage to algal populations.

Some of the most palatable seaweeds are opportunists that counteract vulnerability with very rapid growth rates and immense reproduc-

tive potential. The sporadic, patchy distribution of *Enteromorpha* and *Ulva* may render them too unpredictable to sustain large populations of specialist grazers and the plants occur mainly as escapes from epilithic grazing resulting from refuges presented by irregularities in the rock surface, and by snails missing plants that may have been encountered had the animals moved randomly (Petraitis, 1983). *Littorina littorea* however, is an opportunist and in general makes good use of the periodic availability of such a valuable food source. Although it will locally deplete patches, sufficient escapes occur to maintain the population on any particular shore.

At the other end of the scale, slower growing but perennial seaweeds such as *Fucus* and *Ascophyllum*, occupy the same site for long periods and must therefore, develop effective physical or chemical defences. Thus a compromise exists between growth rate and palatability each of which offers an alternative escape from grazing. Faster growing ephemeral algae tend to fall prey to generalist herbivores such as *L. littorea*, whereas the better defended plants will attract more specialist grazers, such as *L. obtusata*.

It has been suggested that the seasonal phenology of some seaweeds may be governed not only by seasonal fluctuations in conditions such as temperature or photoperiod, but also by changes in the grazing activity of *L. littorea* (Lubchenco & Cubit, 1980). In New England the erect plants of *Petalonia fascia* (O. F. Mull) O. Kuntz and *Scytosiphon lomentaria* (Lyngb.) Link are prevalent in autumn and winter, whereas their *Ralfsia* alternative phase dominates in summer. As the erect plants are preferred foods for *L. littorea* whereas the *Ralfsia* crusts are not (Lubchenco, 1978), the snails may indeed influence the abundance of either phase. However, D.C. Cheney (pers. commun.) found that adding snails to pools made no difference to the timing of the decline of *Scytosiphon*. Similarly, the seasonality of *Ralfsia verrucosa* Aresch. is not controlled by grazing snails although they have some influence on its success (Bertness *et al.*, 1983).

The fate of a plant population may be governed both by the environmental factors influencing the rate of grazing by the snails and the rate of growth and recovery by the plants. For example, at low salinities in an estuarine environment, the foraging activity of *L. littorea* is greatly reduced. In culture, lawns of germlings of the euryhaline alga *Enteromorpha intestinalis* have been subjected to grazing by *L. littorea* at a range of salinities. The snails demolished the algal lawns at all salinities down to 17%, (Norton, 1986). Below this salinity, the impact of the grazers was reduced. In nature, wherever the salinity falls below 12% on a regular basis littorinids are replaced as the dominant herbivores by amphipods. Although the grazing activity of the latter is not affected to the same extent by reduced salinity, the amphipods do less damage to the plants. It is possible that some algae thrive in estuaries because of the reduced grazing pressure there. In the Firth of Clyde the outermost limit of *Fucus ceranoides* L. coincides with the innermost limit of the littorinid grazers (Norton, 1986). Transplantation and feeding experiments suggest that this alga may be particulary susceptible to littorinid grazing on the open shore (Manley & Norton, unpublished).

*Lacuna vincta* (Montagu) sometimes aggregates at very high densities for short periods during which it can inflict enormous damage on its food plants. Concentrations of 277 snails per plant of *Laminaria saccharina* (L.) Lamour. (Fralick *et al.*, 1974) and 205 snails on each square metre of the laminae of *Saccorhiza polyschides* (Lightf.) Batt., (Norton 1971) were sufficient to completely demolish the plants. Similarly a population explosion averaging 328 snails $m^{-2}$ and reaching a maximum of 1570 $m^{-2}$ removed 79% of the net production of *Fucus edentatus* De la Pyl. in a site in eastern Canada (Thomas & Page, 1983).

*Effects on algal communities*

Work in various geographical regions has indicated that on rocky shores the removal of littorinid snails results in an increase in the abundance of algae and changes in community composition.

This has been demonstrated for *Lacuna vincta* (Thomas & Page, 1983), *Littorina keenae* (Castenholz, 1961; Foster, 1964); *L. littorea* (eg. Menge, 1975; Lubchenco, 1978, Lubchenco & Menge, 1978; Lein, 1980, 1984), *L. sitkana* (Behrens, 1976) and *L. scutulata* (Behrens, 1974; Nicotri, 1977). *Littorina neglecta* removed from the leaves of *Zostera* has a similar effect (Robertson & Mann, 1982). The ecological effects of snails inhabiting unstable substrata have been less well studied although it is claimed that by grazing on sand binding plants of *Spartina* and by 'bulldozing' sediment, *L. littorea* can convert sandy habitats to rocky ones (Bertness, 1984). In mangroves *Bembicium auratum* reduces standing crop of the microalgal ilm (Branch & Branch, 1981a).

Littorinid snails are so characteristic of the upper shore that they have been used as one of the organisms to define the limits of the supralittoral fringe world wide (Stephenson & Stephenson, 1949; Lewis, 1964; as the littoral fringe). In temperate regions littorinids are often the most abundant grazers in this zone. Species such as *Littorina saxatilis* and *Melarhaphe neritoides* are particularly abundant on broken shores or where there is an abundance of crevices and pits that offer some protection from desiccation and wave action (e.g. Raffaelli & Hughes, 1978). In the vicinity of such refuges littorinids can exert considerable influence on the macroalgae present (Hawkins & Hartnoll, 1983; Fig. *16*, see also Branch & Branch, 1981b) and would be expected to exert a considerable effect on the microfloral component of the community. Filling in refuge holes with cement resulted in a more uniform cover of algae (Hawkins, unpublished). Snails that inhabit shore levels too high to support algal growth graze predominantly on lichens, e.g. *Melarhaphe neritoides* (Colman, 1940) and *Nodilittorina antipodum* ( = *N. unifasciata*, Branch & Branch, 1981a) – see also Fletcher (1980).

The supply of algal food may vary with seasonal changes in growing conditions, but it is also claimed that winter increases in the abundance of diatoms and of some seaweeds result from the decrease in the activities of littorinids (Castenholz, 1961; Menge, 1975). Although it has not been established that individual snails consume less in winter than in summer, on some shores increased wave action may wash away some littorinids and coupled with lower temperatures restrict the activities of others (Lewis, 1964; Behrens, 1974; V. Chow, unpublished cited in Cubit, 1984). Further controlled experimental work needs to be done on the relative roles of seasonally varying physical factors acting directly both on the growth of algae (particularly micro-algae), and on the grazing activity of littorinids. On balance the evidence supports the hypothesis that the direct effects of physical factors are the most important structuring agency in the littoral fringe. Competition during the seasonal successions of species, and localized grazing effects by littorinids although influencing the temporal and spatial patterns within the littoral fringe community do not normally determine its overall ceiling of productivity nor structure.

Several species of littorinids are found in the eulittoral zone (sensu Lewis, 1964). We best understand their role in algal communities on the intensively studied shores of New England (Menge, 1975; Menge, 1976; Lubchenco & Menge, 1978; Menge & Lubchenco, 1981). According to Lubchenco and Menge fucoid algae are prevented from establishing themselves in the eulittoral of exposed shores by competition from mussels, which dominate there in the absence of effective predation by *Nucella*. *L. littorea* is virtually absent which accounts for the blooms of ephemeral algae at such sites. Epiphytic ephemeral algae, which proliferate on fucoids in exposed conditions due to absence of epiphytic grazers, such as *L. obtusata*, may also increase the rate of dislodgement of any fucoids which do manage to grow (Menge, 1975).

At moderately exposed sites, *L. littorea* are able to prevent colonization of fucoids by grazing their propagules and germlings from the open rock (see also Keser & Larson, 1984), but this only occurs between late spring and early autumn when *L. littorea* are active. Crevices and barnacles enhance the chances of early survival of fucoids partially because the feeding rate of *L. littorea* is lower over barnacles than over smooth rock (Petraitis & Sayigh, 1987). *L. littorea* is also thought to accelerate succession by selectively removing *Enteromorpha* and other ephemeral algae that can delay the dominance of *Fucus* (Lubchenco & Menge, 1978).

Moderately sheltered areas are covered by canopies of macroalgae in the midshore with turfs of *Chondrus crispus* downshore where predation by *Nucella* is thought to prevent mussels out-competing perennial algae. On these shores the grazers, principally *L. littorea*, do not control recruitment of perennial algae. Algal recruitment under the canopy is limited, even in grazer exclusion cages, presumably because of the shading effects of the dense canopy (Schonbeck & Norton, 1980).

Lubchenco (1982) also showed that grazing by *L. littorea* excluded fucoids from tide pools. In an earlier study (Lubchenco, 1978), algal species diversity on emergent rock was shown to be inversely related to the grazing pressure of *L. littorea*. In pools, however, algal diversity was greatest at intermediate levels of grazing because low grazing pressure allowed dominance by *Enteromorpha*, a preferred food, and high grazing pressure led to dominance by a turf of *Chondrus crispus* a slow growing but grazer-resistant species proliferating vegetatively.

Lubchenco & Menge (1978) suggested that similar processes to those above were responsible for structuring the community in the low intertidal: mussels are competitively superior in the absence of predators on exposed shores; in moderate shelter mussel predation by *Nucella* and *Asterias* allows *Chondrus* to predominate. This dominance is reinforced by *Littorina* grazing which removes competing ephemerals, and influences the competitive interactions between *Fucus distichus* and *Chondrus* (Lubchenco, 1980; Cheney, 1982). In addition, Bertness *et al.* (1983) have shown that *L. littorea* can affect the cover of encrusting algae such as *Hildenbrandia rubra* and *Ralfsia verrucosa*.

Petraitis (1987) has recently shown that in very sheltered bays the above generalizations do not hold. On such shores, mussels and barnacles are the most common organisms in the rocky lower intertidal zone, despite abundant predators. Perennial algae are rare and *L. littorea* are extremely abundant. *Littorina* grazing was shown to prevent establishment of *Fucus* and stands of *Fucus* established in grazer exclusions were demolished by *L. littorea* within a few months when cages were removed. *Littorina* grazing also reduced barnacle settlement and hence indirectly enhanced mussel recruitment (Petraitis *loc. cit.*). Thus *L. littorea* was particularly important in this community.

In Europe there has been much less work on *L. littorea* although Lein (1980, 1984) working on the inner reaches of Norwegian fjords has shown that *L. littorea* can have an important effect on *Enteromorpha* and on *Fucus* germlings. Clokie & Boney (1980) were able to control *Enteromorpha* growth on a shore in the Firth of Clyde by importing *L. littorea*. In many respects these results echo those of Petraitis (1987) on very sheltered shores in New England.

On moderately exposed barnacle dominated shores of the Isle of Man grazing by the tiny *L. neglecta* has been implicated in the differences in algal colonization of *Patella* exclusion areas. On rock from which both barnacles and *L. neglecta* have been removed *Fucus* colonizes slowly after a succession of other algae, but over barnacles *Fucus* appears directly; perhaps because *L. neglecta* selectively removes the ephemeral algae (Hawkins, 1981). Recent work has shown that green algae and *Fucus* will grow on barnacle covered rock maintained without limpets and *L. neglecta* on a freshly primed seawater circulation bench, whilst *Fucus* predominates on rocks without limpets but with *L. neglecta* left in place (R.G. Hartnoll, pers. commun.). This needs confirmation in the field but again suggests that selective feeding by *L. neglecta* can influence the course of succession when the main grazer, *Patella*, is removed.

Grazers such as *L. mariae* and *L. obtusata* that live amongst the fucoid canopy rather than on the rock surface are unlikely to exert much direct influence on the shore community as a whole although they may affect the population dynamics

of the canopy plants, and influence the structure of the epiphytic subcomponent of the community (Williams, this volume). Potentially damaging population explosions like those recorded of *Lacuna vincta* (Fralick *et al.*, 1974) seem to occur very rarely in these species. Perhaps as flat periwinkles live and breed on their host plant, local density dependent feedback and predation may control population size. The tolerance to algal defences of *L. obtusata*, their homing behaviour (Williams, 1987) and their removal of potentially deleterious epiphytes hints at a co-evolutionary association which benefits both partners.

The influence of littorinids on the community structure of epilithic microalgal film is regrettably, but not surprisingly, less well explored than their effect on macroalgae. On the upper shore a density of 440 *L. keenae* per square metre can reduce the biomass of microalgae (measured as chlorophyll *a* content) by about 10% per month (Foster, 1964), and about 300 snails m$^{-2}$ can keep the rock surface clear of diatoms (Fig. 5), although a higher density may be required to clear an established stand of diatoms (Castenholz, 1961). *Littorina littorea* also reduces both the diversity and biomass of microorganisms on glass slides (measured both as dry matter dm$^{-2}$ and organic C dm$^{-2}$) as compared to ungrazed controls (Hunter & Russell-Hunter, 1983). The reduction

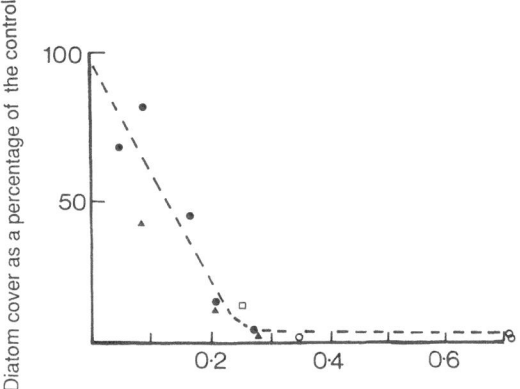

*Fig. 5.* The relationship between the biomass of *Littorina keenae* and diatom cover during recolonization of sterilized pools (after Castenholz, 1961). The different symbols represent different experiments.

in biomass increased with snail density over the range 13–504 snails m$^{-2}$. However, the nutritional quality of the microflora improved, for the carbon per unit dry mass increased with snail grazing, and the C:N ratio on grazed surfaces decreased over the range of snail densities tested (Hunter & Russell-Hunter *loc. sit.*). An assessment of the productivity of the microflora based on intake by snails indicated that moderate levels of grazing increased microalgal productivity by 88% over the controls (Hunter & Russell-Hunter *op. cit.*). The same authors conclude that the impact of snail grazing on the microflora would depend on the prior composition of the community. Grazing would reduce diversity providing the community was not dominated by one or very few species. If it were, and the dominants could escape grazing, then increased grazing would have little effect on the already limited microalgal diversity. If however the dominants could not escape grazing, then their removal by low to moderate levels of grazing would result in the release of competitively inferior species and therefore to increased algal diversity.

Several species of *Littorina* can consume both microalgae and macroalgae and there have been few studies of the conditions under which they eat one or the other, or which they prefer. However, from an analysis of the gut contents of snails living on the shore at Roscoff in north-west France, Sacchi *et al.* (1981) concluded that at least for *L. littorea*, *L. nigrolineata* and *L. saxatilis* seaweeds are eaten preferentially and that the microphagous habit may prevail only in habitats poorly colonized by macroalgae.

Many of the above studies demonstrate the importance of *Littorina* grazing in structuring shore communities in some regions. In particular, on many shores on the north eastern coast of the United States *L. littorea* is by far the most important grazer and is a major ecological influence in the intertidal zone. By contrast, in Europe *L. littorea* is rarely considered to have a significant influence on shore vegetation and attention has focused on limpets rather than littorinids. Field experiments and the consequences of oil spills have revealed the importance of *Patella* in prevent-

ing the growth of *Fucus* on moderately to very exposed shores (reviewed by Hawkins & Hartnoll, 1983). On the shore of New England *Patella* are absent and *Acmaea* seem to have little impact on the algae.

Undoubtedly where *L. littorea* is sufficiently abundant it may be a significant grazer, for each snail can ingest 1.27–3.5% of its dry body weight daily (Grahame, 1973). Although the density of *L. littorea* varies considerably from shore to shore (Fig. 6), perhaps due to localized differences in recruitment (e.g. Hawkins & Hiscock, 1983), it seems to be much more abundant on the shores of the north eastern United States than on European coasts. In Britain, for example, although in damp gullies the snails occasionally reach very high densities, elsewhere, even on favourable shores, they very rarely exceed 200 snails m$^{-2}$. Usually their density is very much lower than this, and on many rocky shores they are hard to find (e.g. Watson & Norton, 1985a, and Fig. 6).

In contrast, on the other side of the Atlantic typical densities of *L. littorea* for open rock are 126–315 snails m$^{-2}$ (Vineyard Sound, Massachucetts, Hunter & Russell-Hunter, 1983) >200 m$^{-2}$ (Pemaquid Point & Bennet Neck, Maine, Keser & Larson, 1984), 400 m$^{-2}$, (Long Island, New York, Petraitis, 1983) and in tide pools 100–1000 m$^{-2}$ is typical (Nahant, Massachusetts, D.P. Cheney, pers. comm.). Such high abundances are particularly interesting in view of the relatively recent introduction of *L. littorea* to North America (Carlton, 1982), and demonstrations of its deleterious effects on native snails (Brenchley, 1982, Behrens & Mansour, 1987).

The need for comparative work on both sides of the Atlantic using similar methods to test similar hypotheses is obvious. Moreover, the shores on either side of the North Atlantic with their similar fauna and flora are ideal for work of this nature. Some embayed shores in western Scotland and Norway are perhaps more similar to those in New England, but the contrasts elsewhere are striking. New England provides localities lacking *Patella* but with *L. littorea*, whose grazing is strongly seasonal, Iceland lacks both

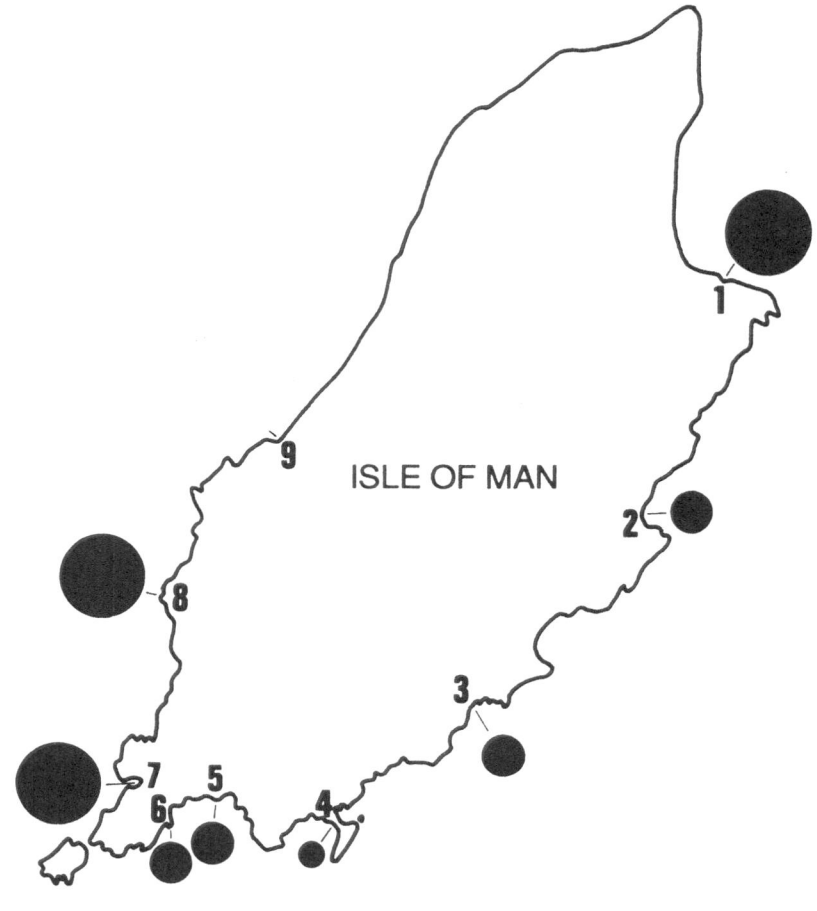

Key

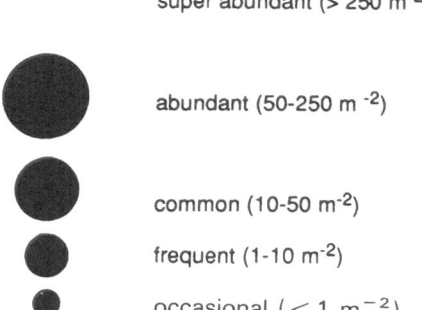

super abundant (> 250 m⁻²)

abundant (50-250 m⁻²)

common (10-50 m⁻²)

frequent (1-10 m⁻²)

occasional (< 1 m⁻²)

*Fig. 6.* Abundance of *Littorina littorea* on favourable boulder or broken rock shores of similar moderate exposure around the Isle of Man (July, 1988). Abundance was assessed by counts in >11 random 0.5 × 0.5 m. quadrats and converted to the abundance scale shown. The sample sites were: 1. Port-e-Vullen; 2. Laxey; 3. Port Soderick; 4. Derbyhaven; 5. Kentraugh; 6. Port St. Mary; 7. Port Erin swimming pool; 8. Niarbyl; 9. White Strand.

*Patella* and *L. littorea* whereas Europe offers shores with an abundance of various species of *Patella* and a variety of littorinids and trochids, and with mild conditions in winter that allow grazing all year round.

## Conclusions

1. The versatility of the littorinid feeding apparatus and feeding behaviour enables them to exploit a variety of micro and macroalgal food sources in a variety of habitats.
2. Littorinids select from this broad menu primarily in response to chemical cues.
3. The attractiveness of algae varies with both their food and habitat value. Preferences are different in epilithic and epiphytic grazers.
4. In response to grazing, algae have evolved structural and chemical defences and life history traits to avoid intense grazing pressure. Polyphenol compounds have been shown to be particularly important.
5. Littorinids can be important organisms structuring shore communities. Many of the important processes, such as escapes from grazing and removal of competitively superior species accelerating succession are a direct consequence of their feeding behaviour.
6. Much current knowledge relates to very few species of *Littorina* inhabiting northern temperate seas, and concentrates largely on their relationships with macroalgae. There is a need to study the feeding and diet of a much wider variety of species from different geographical regions, and to examine in more detail the microalgae on which many of them feed.
7. The contrasting grazer regimes on shores in different geographical regions, especially the North Atlantic, present an ideal natural experiment by which the ecological role of littorinids can be assessed. We hope that this review will help to stimulate such comparative studies.

## Acknowledgement

The review benefitted greatly from the helpful comments of Dr D.P. Cheney to whom we are greatly indebted.

## References

Ankel, W. E., 1937. Wie frisst Littorina? 1. Radula. Beweg. Fressp. Senckenberg. 19: 317–333.
Bate-Smith, E. C., 1973. Haemanalysis of tannins: The concept of relative astringency. Phytochemistry 12: 907–912.
Bebout, B., 1986. The role of marine fungi in the nutrition of the saltmarsh periwinkle, Littorina. Ph.D. thesis Univ. N. Carolina Chapel Hill.
Behrens, S. Y., 1974. Ecological interactions of three Littorina (Gastropoda, Prosobranchia) along the West Coast of North America. Ph.D. thesis, University of Oregon.
Behrens, S. Y., 1976. Range extension in Littorina sitkana Phillippi, 1845, and range contraction in Littorina planaxis Phillippi, 1847. Veliger 19 368.
Behrens, S. Y. & R. A. Mansour, 1987. Growth inhibition of native Littorina saxatilis (Olivi) by introduced L. littorea (L.). J. exp. mar. Biol. Ecol. 105: 187–196.
Bertness, M. D., 1984. Habitat and community modification by an introduced herbivorous snail. Ecology 65: 370–381.
Bertness, M. D., P. O. Yund & A. F. Brown, 1983. Snail grazing and the abundance of algal crusts on a sheltered New England rocky beach. J. exp. mar. Biol. Ecol. 71: 147–164.
Branch, G. M., 1979. Food as a limiting resource for intertidal herbivores. S. African J. Sci. 75: 562.
Branch, G. M., 1984. Competition between marine organisms: Ecological and evolutionary implications. Oceanogr. Mar. Biol. Ann. Rev. 22: 429–593.
Branch, G. M. & M. L. Branch, 1981a. Competition in Bembicium auratum (Gastropoda) and its effect on microalgal standing stock in mangrove muds. Oecologia (Berl.) 46: 106–114.
Branch, G. M. & M. L. Branch, 1981b. Experimental analysis of intraspecific competition in an intertidal gastropod, Littorina unifasciata. Aust. J. mar. Freshwat. Res. 32: 573–589.
Bray, C. J., 1974. A study of the mobility of L. obtusata. MSc thesis in Ecology. University of Wales.
Brenchley, G. A., 1982. Predation on encapsulated larvae by adults: effects of introduced species on the gastropod Ilyanassa obsoleta. Mar. Ecol. Progr. Ser. 9: 235–262.
Brenchley, G. A., 1987. Herbivory in juvenile Ilyanassa obsoleta Neogastropoda. Veliger 30: 167–172.
Calow, P., 1974. Some observations on locomotory strategies and their metabolic effects in two species of freshwater gastropods, Ancylus fluviatilis Mull and Planorbis contortus. Linn. Oecologia 16: 149–161.

136

Carefoot, T. H., 1980. Studies on the nutrition and feeding preferences of *Aplysia*: development of an artificial diet. J. exp. mar. Biol. Ecol. 42: 241–252.

Carlton, J. T., 1982. The historical biogeography of *Littorina littorea* on the Atlantic coast of North America and implications for the structure of New England intertidal communities. Malacol. Rev. 15: 146.

Castenholz, R. W., 1961. The effect of grazing on marine littoral diatom populations. Ecology 42: 783–794.

Chapman, M. G., 1986. Assessment of some controls in experimental transplants of intertidal gastropods. J. exp. mar. Biol. Ecol. 103: 181–201.

Cheney, D. P., 1982. The role of *Littorina littorea* grazing on recruitment and distribution of algae in the lower intertidal zone. Malacol. Rev. 15: 147.

Clokie, J. J. P. & A. D. Boney, 1980. The assessment of changes in intertidal ecosystems following major reclamation work: framework for interpretation of algal-dominated biota and the use and misuse of data. In J. H. Price, D. E. G. Irvine & W. F. Farnham (eds), The Shore Environment Vol. 2, 609–675.

Colman, J. C., 1940. On the faunas inhabiting intertidal seaweed. J. mar. biol. Ass. U.K. 24: 129–184.

Cornelius, P. F. S., 1972. Thermal acclimation of some intertidal invertebrates. J. exp. mar. Biol. Ecol. 9: 43–53.

Croll, R. P., 1983. Gastropod chemoreception. Biol. Rev. 58: 293–319.

Cubit, J. D., 1984. Herbivory and the seasonal abundance of algae on a high intertidal rock shore. Ecology 63: 1905–1917.

Dahl, A. L., 1964. Macroscopic algal foods of *Littorina planaxis* Philippi and *Littorina scutulata* Gould. Veliger 7: 139–143.

Dinter, I. & P. J. Manos, 1972. Evidence of a pheromone in the marine periwinkle *Littorina littorea*. Veliger 15: 45–47.

Fletcher, A., 1980. Marine and maritime lichens of rocky shores: their ecology, physiology and biological interactions in J. P. Price, D. E. G. Irvine & W. F. Farnham (eds), The Shore Environment Vol. 2: 789–842.

Foster, M. S., 1964. Microscopic algal food of *Littorina planaxis* Phillipi and *Littorina scutulata* Gould. Veliger 7: 149–152.

Fralick, R. A., K. W. Turgeon & A. C. Mathieson, 1974. Destruction of kelp populations by *Lacuna vincata* (Montagu). Nautilus 88: 112–114.

Fretter, V. & A. Graham, 1962. British Prosobranch Molluscs. Ray Soc. London. 755 pp.

Frid, C. L. J. & R. James, 1988. Interactions between two species of saltmarsh gastropod, *Hydrobia ulvae* and *Littorina littorea*. Mar. Ecol. Prog. Ser. 43: 173–179.

Frings, H. & C. Frings, 1965. Chemosensory bases of food-finding and feeding in *Aplysia juliana* (Mollusca, opisthobranchia). Biol. Bull. 128: 211–217.

Garrity, S. D., 1984. Some adaptations of gastropods to physical stress on a tropical rocky shore. Ecology 65: 559–574.

Geiselman, J. A., 1980. Ecology of chemical defenses of algae against the herbivorous snail, *Littorina littorea* in the New England rocky intertidal community. Ph.D. thesis Woods Hole Oceanographic Institution, Massachusetts Institute of Technology: 209 pp.

Geiselman, J. A. & O. J. McConnell, 1981. Polyphenols in brown algae *Fucus vesiculosus* and *Ascophyllum nodosum*: chemical defenses against the marine herbivorous snail, *Littorina littorea*. J. chem. Ecol. 7: 1115–1133.

Gendron, R. P., 1977. Habitat selection and migratory behaviour of the intertidal gastropod, *Littorina littorea* (L.). J. anim. Ecol. 46: 79–92.

Goldstein, J. L. & T. Swain, 1965. The inhibition of enzymes by tannins. Phytochemistry 4: 185–192.

Grahame, J., 1973. Assimilation efficiency of *Littorina littorea* (L.) (Gastropoda Prosobranchiata). J. anim. Ecol. 42: 383–389.

Guiterman, J. D., 1970. The population biology of *Littorina obtusata* (L.) (Gastropoda Prosobranchiata). Ph.D. thesis. University of Wales.

Haseman, J. D., 1911. The rhythmical movements of *Littorina littorea* synchronous with ocean tides. Biol. Bull. 21: 113–121.

Hawkins, S. J., 1981. The influence of season and barnacles on the algal colonization of *Patella vulgata* exclusion areas. J. mar. biol. Ass. U.K. 61: 1–15.

Hawkins, S. J. & R. G. Hartnoll, 1983. Grazing of intertidal algae by marine invertebrates. Oceanogr. mar. biol. Ann. Rev. 21: 195–282.

Hawkins, S. J. & K. Hiscock, 1983. Some anomalies on Lundy in the distribution of common eulittoral prosobranchs with planktonic larvae. J. moll Stud. 49: 86–88.

Hawkins, S. J., D. C. Watson, A. S. Hill, S. Hutchinson, S. Harding, M. A. Kyriakides & T. A. Norton, 1989. A comparison of feeding mechanisms in microphagous herbivorous gastropods in relation to resource partitioning. J. moll. Stud. 55: 151–165.

Hunter, R. D. & W. D. Russell-Hunter, 1983. Bioenergetic and community changes in intertidal aufwuchs grazed by *Littorina littorea* Ecology 64: 761–769.

Hylleberg, J. & J. T. Christensen, 1978. Factors affecting the intra-specific competition and size distribution of the periwinkle *Littorina littorea* (L.). Natura jutl. 20: 193–202.

Imrie, D. W., S. J. Hawkins & C. R. McCrohan, 1989. The olfactory–gustatory basis of food preference in the herbivorous prosobranch, *Littorina littorea* L.. J. moll. Stud. 55: 217–225.

Keser, M. & B. R. Larson, 1984. Colonization and growth dynamics of three species of *Fucus*. Mar. Ecol. Progr. Ser. 15: 125–134.

Kohlmeyer, J. & B. Bebout, 1986. On the occurrence of marine fungi in the diet of *Littorina angulifera* and observations on the behaviour of the periwinkle. P.S.Z.N.I. mar. Ecol. 7: 333–343.

Land, M. F., 1968. Functional aspects of the optical and retinal organisation of the mollusc eye. Symposia of the Zoological Society, Lond. 23: 75–96.

Lein, T. E., 1980. The effects of *Littorina littorea* (Gastropoda) grazing on littoral green algae in the inner Oslofjord, Norway. Sarsia 65: 87–92.

Lein, T. E., 1984. A method for the experimental exclusion of *littorina littorea* L. (Gastropoda) and the establishment of fucoid germlings in the field. Sarsia 69: 83–86.

Lewis, J. R., 1964. The Ecology of Rocky Shores. English Universities Press, London: 323 pp.

Little, C., 1989. Factors governing patterns of foraging activity in littoral marine herbivorous gastropods. J. moll. Stud. 55: 273–284.

Little, C., G. A. Williams, D. Morritt, J. A. Perrins & P. Sterling, 1988. Foraging behaviour of *Patella vulgata* L. in an Irish sea-lough. J. exp. mar. Biol. Ecol. 120: 1–21.

Lubchenco, J., 1978. Plant species diversity in a marine intertidal community: Importance of herbivore food preference and algal competitive abilities. Am. Nat. 112: 23–39.

Lubchenco, J., 1980. Algal zonation in a New England rocky intertidal community – an experimental analysis. Ecology 61: 333–344.

Lubchenco, J., 1982. Effects of grazers and algal competitors on fucoid colonization in tide pools. J. Phycol. 18: 544–550.

Lubchenco, J., 1983. *Littorina* and *Fucus*: Effects of herbivores, substratum heterogeneity and plant escapes during succession. Ecology 64: 1116–1123.

Lubchenco, J. L. & J. D. Cubit, 1980. Heteromorphic life histories of certain marine algae as adaptations to variations in herbivory. Ecology 61: 676–687.

Lubchenco, J. L. & S. D. Gaines, 1981. A unified approach to marine plant-herbivore interactions. I. Populations and communities. Annu. Rev. Ecol. Syst. 12: 405–437.

Lubchenco, J. & B. A. Menge, 1978. Community development and persistence in a low rocky intertidal zone. Ecol. Monogr. 48: 67–94.

Luckens, P. A., 1974. Removal of intertidal algae by herbivores in experimental frames and on shores near Auckland. N.Z. J. mar. Freshwat. Res. 8: 637–654.

Mclean, R. F., 1967. Measurements of beachrock erosion by some tropical marine gastropods. Bull. mar. Sci. 17: 551–561.

McQuaid, C. D., 1981. The establishment and maintenance of vertical size gradients in populations of *Littorina africana knysnaensis* (Phillippi) on an exposed rocky shore. J. exp. mar. Biol. Ecol. 54: 77–89.

Menge, J. L., 1975. Effect of herbivores on the community structure on the New England rocky intertidal region: distribution, abundance and diversity of algae. Ph.D. Thesis, Harvard University. 165 pp.

Menge, B. A., 1976. Organization of the New England rocky intertidal community: Role of predation, competition and environmental heterogeneity. Ecol. Monogr. 49: 355–369.

Menge, B. & J. Lubchenco, 1981. Community organization in temperate and tropical rocky intertidal habitats: prey refuges in relation to consumer pressure gradients. Ecol. Monogr. 51: 429–450.

Naylor, E., 1985. Tidal rhythmic behaviour of marine animals. In M. S. Laverack (ed.), Physiological Adaptations of Marine Animals. Symposia of the Society for Experimental Biology, Symposium XXXIX. The Company of Biologists Ltd.: 63–93.

Newell, G. E., 1958a. The behaviour of *Littorina littorea* (L.) under natural conditions and its relation to position on the shore. J. mar. biol. Ass. U.K. 37: 229–239.

Newell, G. E., 1958b. An experimental analysis of the behaviour of *Littorina littorea* (L.) under natural conditions and in the laboratory. J. mar. biol. Ass. U.K. 37: 241–266.

Newell, G. E., 1965. The eye of *Littorina littorea*. Proc. zool. Soc. Lond. 144: 75–86.

Newell, R. C., 1970. Biology of Intertidal Animals. Lagos Press. London, 555 pp.

Newell, R. C., V. I. Pye & M. Ahsanullah, 1971. Factors affecting the feeding rate of the winkle *Littorina littorea*. Mar. Biol. 9: 138–144.

Nicotri, R. E., 1977. Grazing effects of four marine intertidal herbivores on the microflora. Ecology 58: 1020–1032.

North, W. J., 1954. Size distribution, erosive activities and gross metabolic efficiency of the marine intertidal snails *Littorina planaxis* and *L. scutulata*. Biol. Bull. 106: 185–197.

Norton, T. A., 1971. An ecological study of the fauna inhabiting the sublittoral marine algae *Saccorhiza polyschides* (Lightf.) Batt. Hydrobiologia 37: 215–231.

Norton, T. A., 1986. The ecology of macroalgae in the Firth of Clyde. In J. A. Allen, P. R. O. Barnett, J. M. Boyd, R. C. Kirkwood & J. C. Smyth (eds.), The Environment of the Estuary and Firth of Clyde. Proc. Roy. Soc. Edinb. 90B: 225–269.

Odum, E. P. & A. E. Smalley, 1959. Comparison of population energy flow of a herbivorous and deposit-feeding invertebrate in a salt marsh ecosystem. Proc. natn. Acad. Sci. U.S.A. 45: 617–622.

Petpiroon, S. & E. Morgan, 1983. Observations on the tidal activity rhythm of the periwinkle *Littorina nigrolineata* (Gray). Mar. Behav. & Physiol. 9: 171–192.

Petraitis, P. S., 1982. Occurrence of random and directional movements in the periwinkle *Littorina littorea* (L.). J. exp. mar. Biol. Ecol. 59: 207–217.

Petraitis, P. S., 1983. Grazing patterns of the periwinkle and their effect on sessile intertidal organisms. Ecology 64: 522–533.

Petraitis, P. S., 1987. Factors affecting rocky intertidal shores of New England: Herbivory and predation in sheltered bays. J. exp. mar. Biol. Ecol. 109: 117–136.

Petraitis, P. S. & L. Sayigh, 1987. *In situ* measurement of radula movements of three species of *Littorina* (Gastropoda; Littorinidae). Veliger 30.

Raffaelli, D. G., 1985. Functional feeding groups of some intertidal molluscs defined by gut contents analysis. J. moll. Stud. 51: 233–239.

Raffaelli, D. G. & R. N. Hughes, 1978. The effect of crevice size and availability on populations of *Littorina rudis* and *Littorina neritoides*. J. anim. Ecol. 47: 71–83.

Ragan, M. A. & K. W. Glombitza, 1986. Phlorotannins, brown alga polyphenols. Progress in Phycol. Res. 4: 129–241.

Reimchem, T. E., 1974. Studies on the biology and colour

138

polymorphism of two sibling species of marine gastropod (*Littorina*). Ph.D. thesis, University of Liverpool. 389 pp.

Robertson, A. I. & K. H. Mann, 1982. Population dynamics and life history adaptations of *Littorina neglecta* Bean in an eelgrass meadow (*Zostera marina* L.) in Nova Scotia. J. exp. mar. Biol. Ecol. 63: 151–171.

Sacchi, C. F., A. O. Ambrogi & D. Voltolini, 1981. Recherches sur le spectre trophique compare de *Littorina saxatilis* (Olivi) et de *L. nigrolineata* (Gray) (Gastropoda, Prosobranchia) sur le greve de Roscoff. Cah. Biol. mar. 22: 83–88.

Santelices, B. & R. Ugarte, 1987. Algal life-history strategies and resistance to digestion. Mar. Ecol. Prog. Ser. 35: 267–275.

Sieburth, J. McN. & A. Jensen, 1968. Studies on algal substances in the sea. I. Gelbstoff (humic material) in terrestrial and marine waters. J. exp. mar. biol. Ecol. 2: 174–189.

Sieburth, J. McN. & J. L. Tootle, 1981. Seasonality of microbial fouling on *Ascophyllum nodosum* (L.) Le Jol., *Fucus vesiculosus* L., *Polysiphonia lanosa* (L.) Tandy and *Chondrus crispus* Stackh. J. Phycol. 17: 57–64.

Steneck, R. S. & L. Watling, 1982. Feeding capabilities and limitations of herbivorous molluscs: a functional group approach. Mar. Biol. 68: 299–319.

Stephenson, T. A. & A. Stephenson, 1949. The universal features of zonation between tidemarks on rocky coasts. J. Ecol. 37: 289–305.

Schonbeck, M. W. & T. A. Norton, 1980. Factors controlling the lower limits of fucoid algae on the shore. J. exp. mar. Biol. Ecol. 43: 131–150.

Tempel, A. S., 1973. Tannin measuring techniques: A review. J. chem. Ecol. 8: 1289–1298.

Thamdrup, H. M., 1935. Beitrage zur Okologie der Wattenfauna. Meddr. Kommn. Havunders, Serie. Fiskeri 10: 1–125.

Thomas, J. D., 1986. The chemical ecology of *Biomphalaria glabrata* (Say): Sugars as attractants and arrestants. Comp. Biochem. Physiol. 83: 457–460.

Thomas, J. D., J. Osfosu-Barko & R. L. Patience, 1983. Behavioural responses to carboxylic and amino acids by *Biomphalaria glabrata* (Say), the snail host of *Schistosoma mansoni* (Sambon), and other freshwater molluscs. Comp. Biochem. Physiol. 75: 57–76.

Thomas, J. D., P. R. Sterry, H. Jones, M. Gubala & B. M. Grealy, 1986. The chemical ecology of *Biomphalaria glabrata* (Say): Sugars as phagostimulants. Comp. Biochem. Physiol. 83a: 461–475.

Thomas, M. L. H. & F. H. Page, 1983. Grazing by the gastropod *Lacuna vincta* in the lower intertidal area of Musquash Head, New Brunswick, Canada. J. mar. biol. Ass. U.K. 63: 725–736.

Townsend, C. R. & R. N. Hughes, 1981. Maximising net energy returns from foraging. In C. R. Townsend & P. Calow (eds), Physiological Ecology: an Evolutionary Approach to Resource Use. Blackwell Scientific Publs. Oxford. 86–108.

Uhazy, L. S., R. D. Tanaka & A. J. MacInnis, 1978. *Schistosoma mansoni*: identification of chemicals that attract or trap its snail vector *Biomphalaria glabrata*. Science 201: 924–926.

Underwood, A. J., 1979. The ecology of intertidal gastropods. Adv. Mar. Biol., 16: 111–210.

Underwood, A. J. & K. E. McFadyen, 1981. Ecology of the intertidal snail *Littorina acutispira* Smith. J. exp. mar. biol. Ecol. 66: 169–197.

Underwood, A. J. & M. G. Chapman, 1985. Multifactorial analyses of directions of movement of animals. J. exp. mar. Biol. Ecol. 91: 17–43.

Van Alstyne, K. L., 1988. Herbivore grazing increases polyphenolic defenses in the intertidal brown alga *Fucus distichus*. Ecology. 69: 655–663.

Van Dongen, A., 1956. The preference of *Littorina obtusata* for Fucaceae. Archs. néerl. zool. 11: 373–386.

Warren, J. H., 1985. Climbing as an avoidance behaviour in the salt marsh periwinkle *Littorina irrorata* (Say). J. exp. mar. Biol. Ecol. 89: 11–28.

Watson, D. C., 1983. Seaweed palatability and selective grazing by littoral gastropods. Ph.D. Thesis, University of Glasgow. 187 pp.

Watson, D. C. & T. A. Norton, 1985a. Dietary preferences of the common periwinkle *Littorina littorea*. J. exp. mar. Biol. Ecol. 88: 193–211.

Watson, D. C. & T. A. Norton, 1985b. The physical characteristics of seaweed thalli as deterrents to littorine grazers. Bot. mar. 28: 383–387.

Watson, D. C. & T. A. Norton, 1987. The habitat and feeding preferences of *Littorina obtusata* (L.) and *Littorina mariae* Sacchi et Rastelli. J. exp. mar. Biol. Ecol. 112: 61–72.

Williams, G. A., 1987. Niche partitioning in *Littorina obtusata* and *Littorina mariae*. Ph.D. thesis, University of Bristol.

Wright, J. R. & R. G. Hartnoll, 1981. An energy budget for a population of the limpet *Patella vulgata*. J. mar. biol. Ass. U.K. 61: 627–646.

Wright, J. R., 1977. The construction of energy budgets for three intertidal gastropods, *Patella vulgata*, *Littorina littoralis* and *Nucella lapillus*. Ph.D. thesis, University of Liverpool.

Hydrobiologia **193**: 139–146, 1990.
K. Johannesson, D. G. Raffaelli and C. J. Hannaford Ellis (eds), Progress in Littorinid and Muricid Biology.
© 1990 Kluwer Academic Publishers.

# *Littorina mariae* – a factor structuring low shore communities?

Gray A. Williams
*Department of Marine Biology, University of Liverpool, Port Erin, Isle of Man, UK (present address:
Department of Botany, University of Hong Kong, Pokfulam Road, Hong Kong)*

*Key words:* Fucus serratus, herbivory, epiphytes, community structure

## Abstract

The role of littorinids in structuring communities is discussed. On hard substrates preferential grazing of the dominant algae often enables competitively inferior species to utilize the rock substrate. Investigations of these systems has led to the proposal of a number of factors that should be analyzed in order to assess the effect of herbivores on community structure. Some of these factors have been investigated for *Littorina mariae* Sacchi & Rastelli. *L. mariae* is a micro-epiphytic grazer browsing the surface covering of epiphytes off the alga *Fucus serratus* (L.). The relationship between the host alga, *F. serratus*, and *L. mariae* is far more intricate than that between winkles and hard substrates as the alga itself is a dynamic resource. *L. mariae* is spatially and temporally linked to *F. serratus*. *L. mariae* is found almost exclusively on the alga and is positively attracted by extracts of the alga. The life history of the winkle is closely synchronized with that of the alga (the winter decrease of *L. mariae* populations is associated with the seasonal die-back of the host alga). On analysis of the factors considered important to assess herbivory, and examples of the effects of other epiphytic herbivores on algal success, it is suggested that *L. mariae* could potentially play an important role in structuring the community of it's host fucoid and that this may influence larger scale community structure. Experimental manipulations are proposed to evaluate these hypotheses.

## Introduction

Current research on herbivory has attempted to integrate botanical and zoological schools of thought and to move away from previous views that algae were simply fodder for herbivores and herbivores a nuisance to the algae (Lubchenco & Gaines, 1981; Hawkins & Hartnoll, 1983). It is only recently that herbivory has been viewed in the context of community structure and that biological interactions have been considered (Vadas, 1985). Similar trends are now being applied to littorinid biology. The relationships between littorinids and algae are now being assessed at the community level and the importance of littorinid grazing in structuring communities realized.

Recent research has shown most winkles to be generalist feeders but to have strict preferences for certain algae when given the choice (e.g. *L. littorea* (L.) – Watson & Norton, 1985; *L. obtusata* and *L. mariae* – Watson & Norton, 1987). The importance of grazing by these species in structuring communities has not been studied in detail on

British shores, although a role for certain species has been suggested, for example *Littorina neglecta* Bean might clear barnacle cases of microalgae (Hawkins & Hartnoll, 1983). In contrast the role of littorinids in structuring communities has been thoroughly investigated in New England, USA.

In a detailed investigation of winkle feeding preferences and their effect on algal communities, grazing by *L. littorea* has been shown to influence community structure on hard substrates (Menge, 1975; Lubchenco, 1978, 1980). Preferential grazing by the winkle can remove the competitively dominant alga and release space for colonization by competitively inferior algal species. Here community structure is clearly affected by the feeding preferences of the winkle and the competitive hierarchy of the algal assemblage. The relationship between the winkle and algae species is still that of consumer and resource, but the effects are more subtle than simple fodder usage. In Lubchenco's work it was the inert rock surface, which is a finite resource, that is freed for colonization. Many smaller grazers are epiphytic and the substrate which they utilise is a dynamic and renewable resource – the algal tissue. Thus algal substrates represent differing constraints on the role of the herbivore in structuring the community.

It is probable that the juveniles of most winkles are important micrograzers. Many are mobile members of the epiphyton, that is they live most of their life on an algal substrate, a dynamic and renewable resource, and they graze on the host algae or on the other members of the epiphyton. The epiphyton can be an important food resource (Menge, 1975; Medlin, 1980; Cattaneo, 1983; D'Antonio, 1985) and their energetic value to the consumer may be higher than macroalgal tissue (Zimmerman *et al.*, 1979). The effects of grazers in this system are likely to be more complex than the relationship between a herbivore and food resource since, as an epiphyte, the herbivore relies on the algae for food (macroalgal or microalgal material) and substrate (Nicotri, 1980).

Sessile epiphytes themselves have a negative effect on many host plant species, by affecting light penetration (Wing & Clendenning, 1971) and photosynthesis (Oswald *et al.*, 1984). Also the loading effect of epiphytes is sufficient to weigh the host down in the water column and increase detachment rates (Menge, 1975), the risk of axis breakage (D'Antonio, 1985) and predation by fish (Dixon *et al.*, 1981). Coralline algae have been shown to die when covered with epiphyte films (Steneck, 1982) and epiphytes decreased the growth rate and reproductive success of foliose red algae (D'Antonio, 1985).

Plants have many ways of overcoming the negative effects of epiphytic loading. Many species have an unstable surface, thereby decreasing the attachment potential for the epiphyte. Epidermis shedding has been recorded in *Ascophyllum nodosum* (L.) Le Jol. (Filion-Myklebust & Norton, 1981) and crustose coralline algae (Johnson & Mann, 1986); *Enteromorpha* has a number of thickened outer layer of cells which are actually worn away taking epiphytes with them; and *Halidrys siliquosa* (L.) Lyngb. casts off the outer layer of cell walls (the meristoderm) to decrease the epiphyte load (Moss, 1982). Other algae exude antifoulant chemicals which discourage epiphyte settlement, e.g. *Himanthalia elongata* (L.) S.F. Gray. (Kitching, 1987); *Laminaria digitata* (Huds.) (Al-Ogilvy & Knight Jones, 1977); for a detailed review see McNeill Sieburth (1968).

Many algae may rely upon herbivores to remove fouling organisms from their frond surface. Crustose corallines can die through fouling by epiphytes in the absence of limpet grazing (Steneck, 1982), and grazing by amphipods and littorinids (*Littorina scutulata* Gould) has been shown to control epiphyte growth and thereby increase plant fitness (Brawley & Adey, 1981 a & b, D'Antonio, 1985). The interactions between the host plant and herbivore are therefore intricate, and the life histories of many species of epiphyte grazers have been shown to be closely linked to epiphyte abundance and the life history of the algal host (Cattaneo, 1983; Hicks, 1979; Trotter & Webster, 1984).

On European shores two species of flat winkle, *Littorina obtusata* and *L. mariae*, spend their entire life histories as epiphytes on macroalgae. *L. obtusata* lives almost exclusively on *Ascophyllum nodosum* on sheltered shores (Williams, 1987 and unpubl. data). On British shores *Ascophyllum* has a low epiphyte load (Round, 1984) which is maintained by the alga frequently shedding its epidermis. *L. obtusata* uses the alga itself as a food source, excavating the fucoid thallus (Williams, 1987; Watson & Norton, 1987); and also as a substrate for egg deposition (Goodwin, 1978). The relationship between *L. obtusata* and *Ascophyllum* thus appears to be one of consumer and resource. On American shores *Ascophyllum* has been recorded as having a large covering of epiphytes in the summer months (Menge, 1975) and *L. obtusata* is recorded as grazing micro-epiphytically. It is probable that juvenile (but not adult) *L. obtusata* also use microalgae on the algal surface as a food source in Britain. The relationship between *L. mariae* and *F. serratus* is different to that between *L. obtusata* and *Ascophyllum*. *L. mariae* grazes epiphytically off *F. serratus* fronds which support a rich community of epiphytes. The aim of this paper is to review the relationship between *L. mariae* and its host alga *F. serratus* with particular reference to the winkles effect on epiphyte populations and consequent host plant fitness. Data have been collected from previously published work (where acknowledged) and also from unpublished work conducted at Sawdern Point in West Wales (Ordnance Survey Grid reference SM 888032) and St. Michaels Island on the Isle of Man (O.S. Grid reference SC 296675).

## The alga – epiphyte – herbivore community

### The host alga

The life history of *F. serratus* on the Isle of Man is well documented (Knight & Parke, 1950). Plants usually live for 3 + years. Growth is seasonal reaching a maximum in the summer. Reproduction takes place usually in the second year of growth and gametes are released in autumn/winter. All the fronds that bore reproductive tissue defoliate; the tissue dying back to the internode. Subsequent growth is from the remaining vegetative apices. As growth is from apical meristems there is a strict age gradient along the plant representing old tissue at the base to young developing tissue at the apex. The surface area available for epiphytes therefore varies both seasonally, due to storm damage and defoliation, and within individual plants, as a reflection of past reproduction and secondary thickening of stipe material.

### The epiphytes

*Fucus serratus* is host to many animal and plant species. Hagerman (1966) recorded 164 species of epiphytes on *F. serratus* plants from Sweden and Boaden *et al.* (1975) recorded 79 taxa of animals on *F. serratus* in Strangford Lough. The majority of epiphytes are sessile filter-feeders – bryozoans, serpulids, hydroids and tunicates – and there are numerous examples of space limiting their distribution and abundance on the fronds of *F. serratus* (Stebbing, 1973; Boaden *et al.*, 1976; O'Connor *et al.*, 1979, 1980). The distribution patterns of the various species are dictated primarily by physical conditions (which show differences both between sites and within the alga itself) and by differential settlement and growth along the frond of the plant. As the resource is renewable through growth there is selection to partition the available space by settling on the meristematic areas of the plant (Ryland & Stebbing, 1971). *F. serratus* is also host to a rich assemblage of microalgae (D.M. Paterson & G.A. Williams unpubl. data) which form the primary food source, together with the settling stages of other species, for epiphytic grazers. The component species of this community vary in space and time according to environmental and seasonal effects and community structure is dictated by factors other than those controlling rocky substrates (Seed & O'Connor, 1981).

### The herbivore

In order to understand the role of *L. mariae* in structuring this community the factors proposed by Menge (1975) to estimate the effects of herbivores on plants in the community should be determined. These are:

(1) The feeding mode of the herbivore
(2) Size of the herbivore relative to the plant/food
(3) The mobility of the herbivore
(4) The selectivity of the herbivore
(5) The diet of the herbivore (specialist or generalist)
(6) How the herbivore's food choice affects algal competitive hierarchies
(7) Is the herbivore limited by food or predation?
(8) Relative distributions of the plant and herbivore
(9) Relative length of plant/herbivore life cycles.
For the purpose of ascertaining the role of *L. mariae* these factors can be broadly divided into three groups; Food and Selection (incorporating factors 1,2,4,5); Spatial Distribution (within algal stands-factors 2,3,8 and within the plant-factors 2,3,4,6); and the Temporal Distribution of the winkle and host species (factors 7,8,9).

*Food and Selection*

Recent work has shown *L. mariae* to be principally a micro-epiphytic grazer browsing the surface covering of microalgae, detritus and other organic material off the fronds of *F. serratus* (Watson, 1983; Williams, 1987). Faecal analysis has shown the diet to be composed primarily of diatoms and detritus (Williams, unpubl. data) and it is assumed that there is little selection at the micro-epiphyte level and that *L. mariae* ingests whatever is on the surface of the algae. The diet may involve some macroalgal material but it is not an important constituent of the diet, although attractiveness trials have shown *F. serratus* to be actively preferred to other macroalgae by *L mariae* (Watson & Norton, 1987).

*Spatial distribution patterns- within algal stands*

*L. mariae* reaches an adult size of approximately 9–12 mm on sheltered shores and as such it

spends its entire life history in the canopy of *F. serratus*. The zonation patterns of the two species overlap on shores of varying exposures (Williams, 1987; Watson & Norton, 1987). All the stages of the winkles life cycle are spent on *F. serratus*: from egg mass deposition, hatching and growth to sexual maturity (Williams, 1987 and unpubl. data). Movement of the species in the *F. serratus* canopy is random in direction and limited in distance, in the order of 30 cm/day (Williams, 1987 and unpubl. data), which will often be within the frond structure of a single plant. Movement between plants will be dependent on the density of the algal stand and has been shown to be preferentially directed towards *F. serratus* plants as opposed to other fucoid species (Watson, 1983).

*Spatial distribution patterns- within plant variation*

The surface area of the plant, and therefore the space available for colonization and grazing, decreases from the apex of the plant to the holdfast as the plant develops and grows (Fig. 1). The basal parts of the plant loose their foliage and thicken to form the characteristic stipe, whilst the distal parts continue growing from their apical meristems into frond and reproductive tissue. The

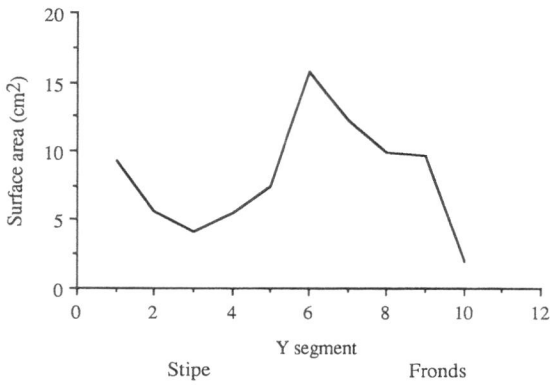

Fig. 1. Variation in surface area (cm$^2$) of *Fucus serratus* fronds progressing from the holdfast to the apex of the plant. (Data from 11 plants collected from Sawdern Point, West Wales. The longest frond of the plant was divided into 'Y' segments (after Boaden *et al.*, 1975), the surface area of each segment traced and the area measured using a digitizer.)

distribution and abundance of sessile and mobile epiphytes reflect this gradient by showing zonation patterns along the plant. On sheltered shores, where *L. mariae* is not an important constituent of the community (e.g. Strangford Lough in Northern Ireland, Boaden *et al.*, 1975), the zonation pattern is dictated by competitive interactions and differential settlement and growth patterns of the bryozoan guild on *F. serratus* (O'Connor *et al.*, 1979, 1980, Boaden *et al.*, 1976); analogous to algal interactions on hard substrates. However, on shores where *L. mariae* is present epiphyte loading is restricted towards the basal part of the algae (Fig. 2), and this is negatively associated with winkle, and other mobile gastropod distribution which is predominantly on the plant fronds (Fig. 3). This may represent a spatial partitioning of the plant enforced on the bryozoan by the nonselective grazing of the herbivores affecting bryozoan settlement. It may also represent preferential settlement by the bryozoan on the stipe of the plant at more exposed sites; due to the risk of damage or breakage of the fronds by wave action. This is not thought to be the case in the present situation as the shore at St. Michaels Island is relatively sheltered.

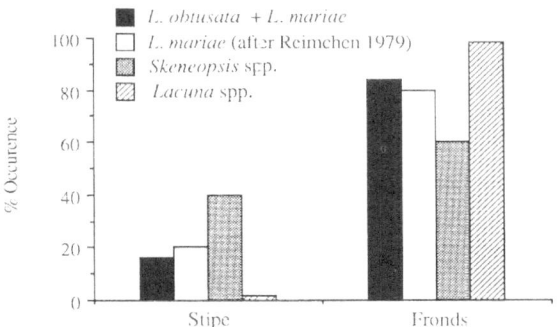

Fig. 3. Distribution of gastropods along the length of *F. serratus* fronds. Data from five plants collected from St. Michaels Island (clear bars after Reimchen, 1979).

*Temporal distribution patterns*

At Sawdern in West Wales the population dynamics of *L. mariae* coincide with annual fluctuations of *F. serratus* cover (Figs. 4 and 5) (Williams, 1987 and unpubl. data). In early spring apical growth of *F. serratus* starts as the egg masses of *L. mariae* begin to hatch. The proportion of adult winkles remaining from the previous year decreases in the population as the numbers of juveniles increase. Both winkles and algae grow through the summer months but at the onset of winter there is a die back in *F. serratus* due to defoliation and storm damage. The winkle population at this time is beginning to be dominated by adults (Fig. 5) which breed and lay egg masses on the plant. This seasonal pattern

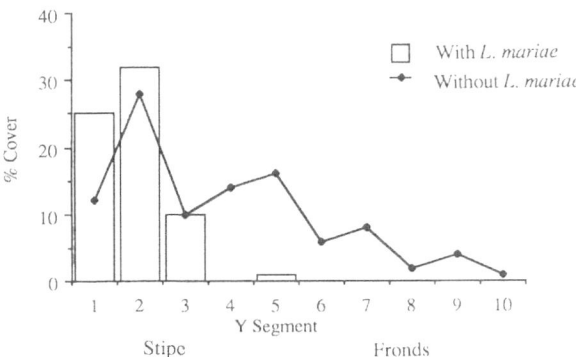

Fig. 2. Zonation of the dominant sessile epiphyte *Flustrellidra hispida* (Fabricius) along the length of *F. serratus* fronds. The bars (with *L. mariae*) represent the zonation in a community where *L. mariae* is abundant (data from 16 plants collected from St. Michaels Island) and the line (without *L. mariae*) represents a community where *L. mariae* is sparse (data from Strangford Lough; after Boaden *et al.*, 1975).

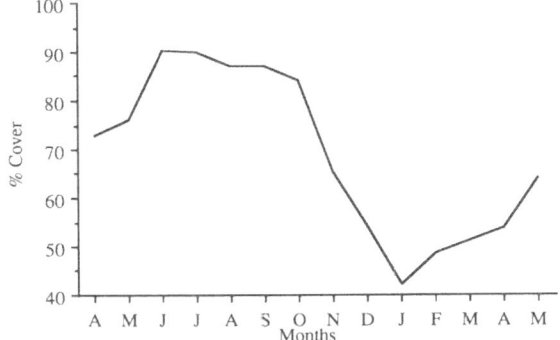

Fig. 4. Annual variation in cover of *F. serratus* at Sawdern Point. Data from monthly samples of ten fixed 0.25 m² quadrats at Sawdern Point (Williams, 1987).

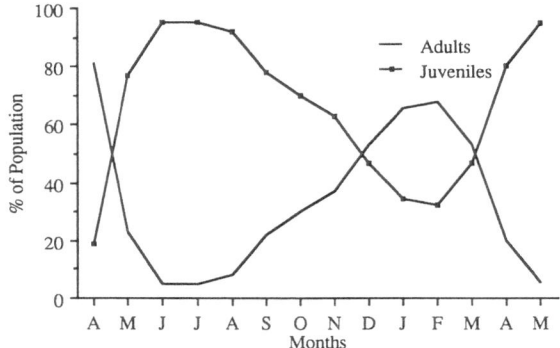

*Fig. 5.* Annual variation in the proportion of adults and juvenile *L. mariae* in the population at Sawdern Point. Data as in Fig. 4.

may reflect the availability of microalgae for *L. mariae*; Cattaneo (1983) has shown a significant association between micrograzer life histories and epiphyte availability. The importance of the reduction in frond tissue (and hence surface area on which to graze) may account for the winter losses of *L. mariae*, although it is likely that predation is also an important factor (Williams, 1987). Close synchronization between epiphytic gastropods and their host algae have also been recorded for *Lacuna pallidula* (da Costa) on *F. serratus* (Smith, 1973). This is probably an important factor in assessing the relationship between epiphytic grazers and algal hosts.

## Discussion

Many of the factors important in assessing the impact of a herbivore on community structure (see above list) seem to be fairly well understood for *L. mariae*. Those factors that are poorly understood require further research effort. Of particular interest is whether the winkle is limited by food or predation. It is possible that the temporal variability of the *F. serratus* fronds limits *L. mariae*; or that microalgal biomasses decrease in the winter. Reimchen (1974) and Williams (1987) have suggested that predation plays an important role in the population dynamics of *L. mariae*. Reimchen (1979) proposed that visual predation by blennies accounted for colour poly-

morphism in *L. mariae*, and that selection varied along the plant as the stipe and frond area represented different cryptic backgrounds. These backgrounds were utilized by different winkle colour morphs. Reimchen's work was conducted on relatively unfouled *F. serratus* and the implications may vary between different sites. These differences in the algal morphology are irrevocably linked as the difference between stipe and frond also reflects a variation in patch quality (the stipe having a low surface area and so being a poor area to graze).

The close relationship between the biologies of *L. mariae* and *F. serratus* may imply a mutual selection and evolutionary development. This is in contrast to the view of an evolutionary 'arms race' between some herbivores (such as *L. obtusata*) and their food plants (*Ascophyllum*) where the plants appear to produce secondary chemical defenses to deter grazing and the herbivores adapt to render these defenses ineffectual. In the case of *L. mariae* and *F. serratus* the winkle gains the benefit of a substance off which to graze and live out its life history. But does the plant gain by the interaction?

Epiphyte grazing has been demonstrated to be beneficial to plants in freshwater systems where there is a high epiphyte loading detrimental to the host plants success (Brönmark, 1989) and several examples in the marine environment have been given above. Furthermore grazing can structure the microfloral assemblage found on the host plant (Nicotri, 1980; Brönmark, 1989). An increase in plant success, due to the removal of epiphytes, could result in the maintenance of dense stands of algae (as a result of reduced losses due to epiphyte loading and increased plant reproductive success), this will have consequent large scale effects on community structure with dense stands of algae occupying most of the free space available, dependant upon physical constraints. Dense stands of algae with a wide spectrum of ages are thought to inhibit the foraging efficiency of predators (Brawley & Adey, 1981 b) and this in turn could limit predation on grazer populations, creating a positive feedback.

Does *L. mariae* in fact structure low shore

communities? Evidence from factors which have been shown to be important in hard substrate communities (Menge, 1975; Lubchenco, 1978) add weight to this idea. The feeding methods of the winkle; its distribution relative to the alga stands and individual variation; and the temporal cycles of the winkle and alga all point towards a close relationship between the winkle and plant. It is possible therefore that *L. mariae* could influence community structure on a large scale. The main questions to be answered concern the importance of predation in the interaction and whether any beneficial effects of the winkle grazing off the host algae can be demonstrated. Both these questions can only be answered by experimental field and laboratory experiments. The role of *L. mariae* may be seen by investigating the settlement and growth of epiphytes both in the presence and absence of grazing pressure from *L. mariae* and monitoring the subsequent growth and success of *F. serratus*. The effect of predators, especially shore crabs, can be investigated using caging and tethering techniques, allowing the indirect influence of predators on plant fitness to be assessed.

## Acknowledgements

This work was funded by the Natural Environment Research Council. I am grateful to C. Little, D. Raffaelli & K. Johannesson for critically reading the manuscript and improving the text. Thanks also to the many people that helped with the fieldwork.

## References

Al-Ogily, S. M. & E. W. Knight-Jones, 1977. Antifouling role of antibiotics by marine algae and bryozoans. Nature 265: 728–729.

Boaden, P. J. S., R. J. O'Connor & R. Seed, 1975. The composition and zonation of *F. serratus* community in Strangford Lough, Co. Down. J. exp. mar. Biol. Ecol. 17: 111–136.

Boaden, P. J. S., R. J. O'Connor & R. Seed, 1976. The fauna of a *Fucus serratus* L. community: ecological isolation in sponges and tunicates. J. exp. mar. Biol. Ecol. 21: 249–267.

Brawley, S. H. & W. H. Adey, 1981a. The effect of micro-grazers on algal community structure in a coral reef microcosom. Mar. Biol. 61: 167–177.

Brawley, S. H. & W. H. Adey, 1981b. Micrograzers may affect macroalgal density. Nature 292: 177.

Brönmark, C., 1989. Interactions between epiphytes, macro-phytes and freshwater snails: review and prospects for future research. J. moll. Stud. 55: 299–311.

Cattaneo, A., 1983. Grazing on epiphytes. Limnol. Oceanogr. 28: 124–132.

D'Antonio, C., 1985. Epiphytes on the rocky intertidal red alga *Rhodomela latrix* (Turner) C. Agardh: negative effects on the host and food for herbivores. J. exp. mar. Biol. Ecol. 86: 197–218.

Dixon, J., S. C. Schroeter & J. Kastendiek, 1981. Effects of the encrusting bryozoan, *Membranipora membranacea*, on the loss of blades and fronds by the giant kelp, *Macrocystis pyrifera* (Laminariales). J. Phycol. 17: 341–345.

Filion-Myklebust, C. & T. A. Norton, 1981. Epidermis shedding in the brown seaweed *Ascophyllum nodosum* (L.) Le Jolis and its ecological significance. Mar. Biol. Lett. 2: 42–51.

Goodwin, B. J., 1978. The growth and breeding cycle of *Littorina obtusata* (Gastropoda: Prosobranchiata) from Cardigan Bay. J. moll. Stud. 44: 231–242.

Hagerman, L., 1966. The macro- and microfauna associated with *Fucus serratus* L., with some ecological remarks. Ophelia 3: 1–43.

Hawkins, S. J. & R. G. Hartnoll, 1983. Grazing of intertidal algae by marine invertebrates. Oceanogr. mar. Biol. A. Rev. 21: 195–282.

Hicks, G. R. F., 1979. Pattern and strategy in the reproductive cycles of benthic harpacticoid copepods. In E. Naylor & R. G. Hartnoll (eds), Cyclic Phenomena in Marine Plants and Animals. 13th European Marine Biology Symposium: 139–147.

Johnson, C. R. & K. H. Mann, 1986. The crustose coralline alga *Phymatolithon* Foslie inhibits the overgrowth of seaweeds without relying on herbivores. J. exp. mar. Biol. Ecol. 96: 127–146.

Kitching, J. A., 1987. The flora and fauna associated with *Himanthalia elongata* (L.) S. F. Gray in relation to water current and wave action in the Lough Hyne Marine Nature Reserve. Estuar. coast. shelf Sc. 25: 663–676.

Knight, M. & M. Parke, 1950. A biological study of *Fucus vesiculosus* (L.) and *F. serratus* (L.). J. mar. biol. Ass. UK 29: 439–514.

Lubchenco, J., 1978. Plant species diversity in a marine intertidal community: importance of herbivore food preferences and algal competitive abilities. Am. Nat. 112: 23–39.

Lubchenco, J., 1980. Algal zonation in the New England rocky intertidal community: An experimental analysis. Ecology 61: 333–344.

Lubchenco, J. & S. D. Gaines, 1981. A unified approach to marine plant-herbivore interactions. I. Populations and communities. Annu. Rev. Ecol. Syst. 12: 405–437.

146

McNeill Sieburth, J., 1968. The influences of algal antibiosis on the ecology of marine micro-organisms. In M. R. Droop & E. J. Ferguson Woods (eds), Advances in Microbiology of the Sea. 1. Academic Press, New York. 63–94.

Medlin, L. K., 1980. Effects of grazers on epiphytic diatom communities. 6th Diatom Symposium: 339–412.

Menge, J. L., 1975. Effects of herbivores on community structure of the New England rocky intertidal region: Distribution, abundance and diversity of algae. Ph.D. thesis, Harvard University.

Moss, B., 1982. The control of epiphytes by *Halidrys siliquosa* (L.) Lyngb (Phaeophyta, Cystoseiraceae). Phycologia 21: 185–191.

Nicotri, M. E., 1980. Factors involved in herbivore food preference. J. exp. mar. Biol. Ecol. 42: 13–26.

O'Connor, R. J., R. Seed & P. J. S. Boaden, 1979. Effects of environment and plant characteristics on the distribution of bryozoa in a *F. serratus* community. J. exp. mar. Biol. Ecol. 38: 151–178.

O'Connor, R. J., R. Seed & P. J. S. Boaden, 1980. Resource space partitioning by bryozoa of a *F. serratus* L. community. J. exp. mar. Biol. Ecol. 45: 117–137.

Oswald, R. C., N. Telford, R. Seed & C. M. Happey-Wood, 1984. The effects of encrusting bryozoans on the photosynthetic activity of *Fucus serratus* L.. Estuar. coast. shelf Sc. 19: 697–702.

Reimchen, T. E., 1974. Studies on the biology and colour polymorphism of two sibling species of marine gastropod (*Littorina*). Ph.D. thesis, University of Liverpool.

Reimchen, T. E., 1979. Substratum heterogeneity, crypsis, and colour polymorphism in an intertidal snail (*Littorina*). Can. J. Zool. 57: 1070–1085.

Round, F. E., 1984. The Ecology of Algae, Cambridge University Press.

Ryland, J. S. & A. Stebbing, 1971. Settlement and orientated growth in epiphytic and epizoic bryozoans. In D. J. Crisp (ed.), 4th European Marine Biology Symposium: 105–123.

Seed, R. & R. J. O'Connor, 1981. Community organization in marine algal epifaunas. Annu. Rev. Ecol. Syst. 12: 49–74.

Smith, D. A., 1973. The population biology of *Lacuna pallidula* (da Costa) and *Lacuna vincta* (Montagu) in North East England. J. mar. biol. Ass. UK 53: 493–520.

Stebbing, A. R. D., 1973. Competition for space between the epiphytes of *F. serratus* (L.). J. mar. biol. Ass. UK 53: 247–261.

Steneck, R. S., 1982. A limpet-coralline alga association: adaptations and defences between a selective herbivore and its prey. Ecology 63: 507–522.

Trotter, D. B. & J. M. Webster, 1984. Feeding preferences and seasonality of free-living marine nematodes inhabiting the kelp *Macrocystis integriflora*. Mar. Ecol. Prog. Ser. 14: 151–157.

Vadas, R. L., 1985. Herbivory. In M. M. Littler & D. S. Littler (eds), Handbook of Phycological methods. Ecological field methods: Macroalgae. Cambridge University Press: 531–572.

Watson, D. C., 1983. Seaweed palatability and selective grazing by littoral gastropods. Ph.D. thesis, Glasgow University.

Watson, D. C. & T. A. Norton, 1985. Dietary preferences of the common periwinkle, *Littorina littorea* (L.). J. exp. mar. Biol. Ecol. 88: 193–211.

Watson, D. C. & T. A. Norton, 1987. The habitat and the feeding preferences of *Littorina obtusata* and *L. mariae* Sacchi et Rastelli. J. exp. mar. Biol. Ecol. 112: 61–72.

Williams, G. A., 1987. Niche partitioning in *Littorina obtusata* and *L. mariae*. Ph.D. Thesis, Bristol University.

Wing, B. L. & K. A. Clendenning, 1971. Kelp surfaces and associated invertebrates. Beih. Nova Hedwigia. 32: 19–41.

Zimmerman, R., R. Gibson & J. Harrington, 1979. Herbivory and detritivory among gammeridean amphipods from a Florida seagrass community. Mar. Biol. 54: 41–47.

*Hydrobiologia* **193**: 147–154, 1990.
*K. Johannesson, D. G. Raffaelli and C. J. Hannaford Ellis (eds), Progress in Littorinid and Muricid Biology.*
© 1990 *Kluwer Academic Publishers.*

# Field observations on the feeding habits of *Littorina scutulata* Gould and *L. sitkana* Philippi (Gastropoda, Prosobranchia) of southern Vancouver Island (British Columbia, Canada)

Domenico Voltolina[1] & Cesare F. Sacchi[2]
[1] *Centro de Investigación Científica y de Educación Superior de Ensenada. Grupo de Acuicultura. Apdo. Postal 2732. Ensenada, Baja California. México*; [2] *Universitá di Pavia, Dip. di Genetica, Sez. Ecologia, Palazzo Botta, Pavia, Italy (address for reprint requests)*

*Key words:* periwinkles, food selectivity, vertical microdistribution

## Abstract

The microdistribution of *Littorina scutulata* and *L. sitkana* on the rocky shores of Vancouver Island can be related to their feeding characteristics. Their gut contents, studied with light and fluorescence microscopy, showed that both species are opportunistic herbivores. However, *L. scutulata* can feed on the sparse epi- and endolithic microflora of the supralittoral zone, while *L. sitkana* is confined to special microenvironments, with abundant epilithic growth.

## Introduction

Among the many faunistic similarities between the North Atlantic and the North Pacific intertidal areas, the case of periwinkles is particularly noteworthy. The species of the group *Littorina scutulata* Gould have a position which is in many ways analogous to that of *Melaraphe neritoides* (L). Such similarities are even greater between the oviparous species of the *Littorina saxatilis* (Olivi) complex and their Pacific counterpart, *Littorina sitkana* Philippi (Sacchi & Voltolina, 1987).

We have previously examined the feeding habits of the *Littorina saxatilis* complex in Venice (Sacchi *et al.*, 1977a) and in Roscoff (Sacchi *et al.*, 1977b, 1981). Here we extend our observations to *L. sitkana* and to *L. scutulata* sensu Murray (1979). *L. plena* (Gould), the other species of the

*scutulata* complex, was rare or completely absent from our samples, and its feeding habits were not investigated.

The general aims of this study as well as details of the sampling stations are given in Sacchi & Voltolina (1987). In the present paper we aim to show by an examination of gut contents and radular structure that, like many other Littorinids (e.g. Sacchi *et al.*, 1977a and b, 1981; Jensen, 1981), both species are in nature opportunistic omnivores. Therefore, studies on their food preferences (e.g. Van Dongen, 1956; Watson & Norton, 1987) are not likely to explain their microdistributions, which we relate instead to their different radular structures and their consequent abilities to forage on the food sources available, as shown for other herbivorous molluscs by Steneck & Watling (1982) and partly by Watson & Norton (1987).

148

## Materials and methods

A total of 55 samples were obtained from April to September 1985 from all the stations indicated in Fig. 1, with the exception of station 7. Additional observations were also carried out in some of these stations in 1987.

The habitats we sampled represent several different types of hard substrate: Rocky shores with bare rock surfaces of the supralittoral, both smooth (1, 10, 14b and 15) and with deep crevices (4, 11, 14 and 17). Intertidal rocks, bare (16, 19, 21 and 29) or with macroalgal growth (1 and 2). Tidal pools of the infralittoral, with macroalgae (5 and 22) or bare (2, 5, 17 and 20), and a spray pool

in the high supralittoral (3). Pebbly beaches were only sampled in the intertidal area, since no specimens were found in the supralittoral. Bare pebbles were found at stations 8, 9, 12 and 18, while at stations 8 to 15, 25 and 26 the pebbles were accompanied by an abundant macroalgal growth. Finally, snails were collected at station 27 on semi-submerged woodwork in a fishing harbour. When the same station is given for more than one habitat, this may refer either to samples taken in different seasons, or at different locations within the site.

At each station, twenty to fifty organisms were collected at random and immediately preserved in a small amount of formalin. At the end of sam-

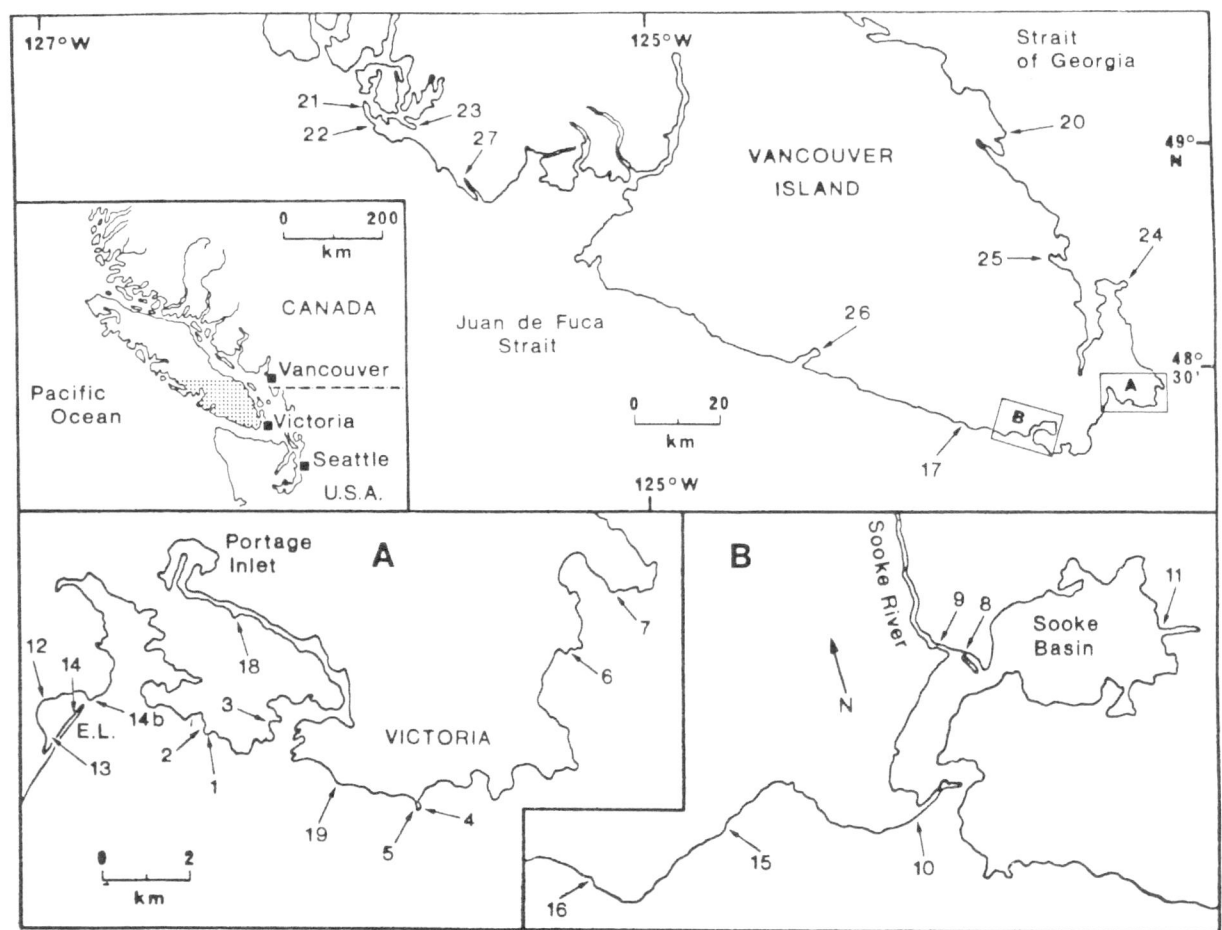

*Fig. 1.* Location of sampling stations. Maps A and B are drawn to the same scale. E.L.: Esquimalt Lagoon. *L. sitkana* was absent from stations 1, 3 and 11. (From: Sacchi & Voltolina, 1987).

pling the formalin was diluted to approximately 10% with filtered seawater, and a sample of the substrate, including surrounding macrophytes, was added to the container.

In the laboratory, all *Littorina* from each sample were scraped with a short brush to collect epiphytic growth, since on several occasions specimens of either species were observed actively feeding on other shells, rinsed and put in a new container with clean 10% formalin. Small pebbles and rock fragments underwent the same treatment and were then discarded. Scrapings and rinses were returned to the original container. Macrophyte fragments were picked from the container and their cell type and structure, as well as amount and type of epiphytic organisms were observed under the light microscope.

The original container was then shaken a few times and a few subsamples (from 5 to 10% of the original) studied using ordinary light and occasionally fluorescence microscopy. The latter technique was mainly used for samples with high detrital content, since it allowed the recognition of the live fraction of the plant material present. It was found especially useful when examining radulas or digestive tract contents, as the presence of recently detached or ingested fresh plant material was evident from its intense red self-fluorescence. This allowed us to determine whether the snails had been feeding on plant litter, almost invariably abundant in our samples, or on living micro- and macroalgae.

Finally, two random subsamples of *Littorina*, representing at least 10% and up to 100% of each species were taken from each sample. Their radulas were extracted and studied immediately, and their gut contents were blended in a dimpled slide with the aid of few drops of distilled water and a fine needle. These were examined using the techniques mentioned above. The amount of material present on radulas and digestive tracts, as well as the identity of any material recognized were noted and compared to those found in the surrounding environment.

The frequency notations used in the tables (a: absent rr: 1 to 4; r: 5 to 10; +: 10 to 20; + +: 20 to 50; + + +: > 50) refer to one hour of continuous observations on subsamples varying in number and in dilution. Consequently, they cannot be taken as indicative of absolute concentrations. In the case of macrophytes they are only used as a subjective evaluation of the amounts seen.

## Results

In the intertidal areas where both species were present, no differences were noted between the

*Table 1.* Results of floristic and gut content observations at station 19 (Intertidal, bare rock with fissures). Dates of collection: 1: April 6; 2: May 18; 3: June 9, 1985. p: planktonic.

| | | Substrate | L. scutulata | L. sitkana |
|---|---|---|---|---|
| Cyan. | *Oscillatoria* sp. | + [1,2,3] | + [1,2]; + + [3] | + + [1,2,3] |
| Chlor. | Germlings – plantulae | + + [1]; r [3] | + + [1]; + [3] | + + [1]; r [3] |
| Bac. | *Amphora coffeaeformis* | rr [2]; a [3] | r [2]; rr [3] | a [2,3] |
| | *Cocconeis scutellum* | r [1]; rr [2]; + [3] | + [1]; a [2]; + + [3] | + [1,2]; + + [3] |
| | *Gyrosigma* sp. | r [1]; a [2,3] | + [1]; r [2]; a [3] | r [1]; a [2,3] |
| | *Licmophora* sp. | a [1]; r [2,3] | + [1,2]; r [3] | r [1,2,3] |
| | *Melosira moniliformis* | r [2,3]; + [3] | r [1]; + [2]; + + [3] | a [1]; + [2]; + + [3] |
| | *Navicula* sp. | + [1]; rr [2,3] | + [1]; r [2,3] | + [1,2]; r [3] |
| | *Synedra* sp. | a [1]; a+; + + + [3] | rr [1]; + + [2]; + + + [3] | a [1]; + + [2]; + + + [3] |
| | *Thalassiosira* sp. (p) | rr [1,2]; + [3] | r [1,2]; + + [3] | a [1]; r [2]; + [3] |
| Others | Pedunculate ciliates | r [1,2,3] | r [1,2]; a [3] | rr [1,2]; r [3] |
| | Organic debris | + + [1,2]; + + + [3] | + + + [1,2,3] | + + + [1,2,3] |
| | rock particles | | + + [1,2,3] | r to + [1,2,3] |

species, in the type of food they ingested (see e.g. Tables 1 and 2). As a general rule, plant litter and microalgae seemed to form a large fraction of the diet of both species. Several other organisms were also recorded in guts. Pedunculate and free-living protozoa were fairly common, and copepods, nematodes and annelids were also occasionally noticed in the digestive tracts of both species. During the spring diatom bloom, and in the areas characterized by summer red tides, large quantities of pelagic diatoms and dinoflagellates washed ashore to form a thick coating along the tideline.

*Table 2*. Results of floristic and gut content observations at station 12 (intertidal, muddy sand and pebbles). Dates of collection: 1: April 6 (phytoplankton bloom); 2: May 18; 3: June 9; 4: July 11 (red tide). p: planktonic; n.f.: non fluorescent – dead.

| | | Substrate | *L. scutulata* | *L. sitkana* |
|---|---|---|---|---|
| **Macro** | | | | |
| Chlor. | *Enteromorpha intestinalis* | + + +[1 to 4] | + + +[1 to 4] | + + +[1 to 4] |
| | *Enteromorpha* sp. | + + +[1 to 4] | + + +[1 to 4] | + + +[1 to 4] |
| Anth. | *Zostera marina* (n.f.) | + + +[1 to 4] | + + +[1 to 4] | + + +[1 to 4] |
| **Micro** | | | | |
| Cyan. | Chroococcales | + + +[4] | + + +[4] | + + +[4] |
| | *Lyngbya* sp. | + + +[3,4] | + + +[3,4] | + + +[3,4] |
| | *Nostoc* sp. | + +[2] | + + +[2] | + + +[2] |
| | *Oscillatoria* sp. | r[1]; + + +[2,3,4] | +[1]; + + +[2,3,4] | +[1]; + + +[2,3,4] |
| Chlor. | Germlings – plantulae | + + +[1,4]; + +[2,3] | + + +[1 to 4] | + + +[1 to 4] |
| Bac. | *Achnanthes brevipes* | + +[1,2,3]; + + +[4] | + + +[1 to 4] | + + +[1 to 4] |
| | *Achnanthes* sp. | + +[1 to 4] | + + +[1 to 4] | + + +[1 to 4] |
| | *Amphora coffeaeformis* | +[1,2]; + + +[3,4] | +[1,2]; + + +[3,4] | + +[1,2]; + + +[3,4] |
| | *Amphora* sp. | + +[1,2,3]; +[4] | + + +[1,2]; + +[3,4] | +[1]; + + +[2,3]; +[4] |
| | *Biddulphia* sp. | r[1,2]; +[3,4] | +[1,2]; + +[3,4] | +[1]; + +[2,3,4] |
| | *Chaetoceros affine* (p, n.f.) | + +[1,2] | + + +[1,2] | + + +[1,2] |
| | *Chaetoceros sociale* (p, n.f.) | + + +[1] | + + +[1] | + + +[1] |
| | *Gyrosigma* sp. | +[3] | t[3] | +[3] |
| | *Melosira moniliformis* | + + +[1 to 4] | + + +[1 to 4] | + + +[1 to 4] |
| | *Melosira nummuloides* | + + +[1,2,3]; + +[2] | + + +[1 to 4] | + + +[1 to 4] |
| | *Navicula grevilleana* | + + +[1,2,4]; + +[3] | + + +[1 to 4] | + + +[1 to 4] |
| | *Navicula* sp. | + + +[1]; + +[2,3,4] | + + +[1 to 4] | + + +[1 to 4] |
| | *Navicula* sp. | + +[1,2,4]; +[3] | + + +[1,2,4]; + +[3] | + + +[1 to 4] |
| | *Navicula* sp. | +[1,2] | +[1,2] | + +[1,2] |
| | *Navicula* sp. | +[1 to 4] | + +[1 to 4] | +[1,3] + +[2,4] |
| | *Navicula* sp. | r[1,3] | +[1,3] | +[1,3] |
| | *Nitzschia closterium* (p) | + +[2] | + + +[2] | + + +[2] |
| | *Nitzschia* sp. | +[1,3]; r[4]; a[2] | +[1,3]; r[2,4] | +[1,3]; rr[2,4] |
| | *Skeletonema costatum* (p, n.f.) | + + +[1] | + + +[1] | + + +[1] |
| | *Synedra affinis* | + + +[3,4] | + + +[3,4] | + + +[3,4] |
| | *Synedra* sp. | + +[1 to 4] | + + +[1 to 4] | + + +[1 to 4] |
| Pyrr. | *Prorocentrum minimum* (p, n.f.) | + + +[4] | + + +[4] | + + +[4] |
| Others | Ciliates | +[1,3,4]; a[2] | + +[1]; r[2,3]; a[4] | +[1,2]; r[3,4] |
| | Nematodes | rr[1]; r[2,3]; + +[4] | a to rr[1 to 4] | rr to r[1 to 4] |
| | Polychaetes | a to rr[1 to 4] | a to r[1 to 4] | rr to rr[1 to 4] |
| | Copepods | +[1,2]; rr[3,4] | +[1]; r[2,3,4] | + +[1]; +[2]; r[3,4] |
| | Zostera fibers | + + +[1 to 4] | + + +[1 to 4] | + + +[1 to 4] |
| | Litter, organic debris | + + +[1 to 4] | + + +[1 to 4] | + + +[1 to 4] |

This coating, along with the accompanying bacteria and silt material, constituted the bulk of the diet of the periwinkles present in the area.

Where macroalgae were present, most of the specimens collected from their surfaces showed evidence of having fed not only on the epiphytic community, but also on the macroalga itself. None of the algae seemed immune from the abrasive action of *Littorina* radulas, since fragments of Chlorophyta (*Enteromorpha*, *Ulva*), Phaeophyta (*Fucus*, *Pelvetiopsis*, *Nereocystis*), even of crustose or coralline Rhodophyta (*Lithothamnion*, *Corallina*) were frequent in the digestive tracts and on the radula surfaces of both species. However, neither *L. scutulata* collected on *Phyllospadix* sp. (Station 22, June 1985; Station 17, May 1987)

nor *L. sitkana* kept in aquaria with freshly collected specimens of *Phyllospadix* sp. and *Zostera marina* L., showed evidence of having fed on these seagrasses. The distinctive fibres of Zosteraceae were only found in the gut contents of animals collected in areas rich in plant debris, such as along the shores of Esquimalt lagoon (Stations 12, 13 and 14) where the decomposing blades of *Zostera marina* L. constitute a large fraction of the plant litter (Table 2).

The rocks of the supralittoral zone were mostly populated by *L. scutulata*, especially if the surface was smooth. The gut contents mirrored the taxonomic composition of the scarce microflora (mostly Cyanophyta and lichens), and the abundance of rock particles is testimory to the abra-

*Table 3.* Results of floristics and gut content observations on samples collected from smooth, bare rock of the supralittoral, at stations 1, 3, 14, 14b and 15. S: sunny; SH: shaded; D: dry; W: wet.

| Stn. | Date | Condition | Microflora | Substrate | L. scutulata | L. sitkana |
|------|------|-----------|------------|-----------|--------------|------------|
| 1 | April 6, 85 | D, SH | Chroococcales | + + + | + + | − |
| | | | Lichens | + | + | − |
| | | | *Navicula* sp. (n.f.) | r | + | − |
| | | | *Cocconeis* sp. (n.f.) | r | + | − |
| | | | rock particles | | + + | − |
| 1 | May 18, 85 | D, S | Chroococcales | + | a | rr |
| | | | Hormogonales | a | a | rr |
| | | | *Achnanthes* sp. | r | rr | a |
| | | | *Cocconeis* sp. (n.f.) | rr | rr | a |
| | | | *Melosira* sp. (n.f.) | a | rr | a |
| | | | *Navicula* sp. (n.f.) | a | a | rr |
| | | | rock particles | | + | rr |
| 1 | June 9, 85 | D, S | Chroococcales | + | a | − |
| | | | Hormogonales | r | rr | − |
| | | | *Cocconeis* sp. (n.f.) | rr | rr | − |
| | | | *Navicula* sp. (n.f.) | rr | a | − |
| | | | rock particles | | + | − |
| 1 | June 30, 85 | D, S | Chroococcales | + + | a | − |
| | | | Lichens | + | rr | − |
| | | | rock particles | | r | − |
| 1 | Sept. 2, 85 | D, S | Chroococcales | + + | a | − |
| | | | Lichens | + | a | − |
| | | | rock particles | | rr | − |
| 1 | Sept. 2, 85 | W | Chroococcales | + + | + + + | − |
| | | | Lichens | + | + + | − |
| | | | rock particles | | + + | − |

*Table 3.* (continued)

| Stn. | Date | Condition | Microflora | Substrate | *L. scutulata* | *L. sitkana* |
|------|------|-----------|------------|-----------|----------------|--------------|
| 10 | July 7, 85 | D, SH | Chroococcales | + + | + + | + |
| | | | Hormogonales | r | + | r |
| | | | *Synedra* sp. (n.f.) | r | a | a |
| | | | rock particles | | + | a |
| 14 | June 4, 85 | D, SH | Chroococcales | + + + | + + + | – |
| | | | Hormogonales | + + | + + + | – |
| | | | Lichens | + | + + | – |
| | | | Diatom frustules | + + + | + + + | – |
| | | | rock particles | | + + + | – |
| 14b | July 12, 85 | D, SH | Chroococcales | + + + | + + + | + |
| | | | Hormogonales | + + | + + + | r |
| | | | Diatom frustules | rr | + | a |
| | | | rock particles | | + + + | rr |
| 15 | July 14, 85 | D, SD, S | Lichens | r | r | – |
| | | | Diatom frustules | a | rr | – |
| | | | rock particles | | rr | – |
| 15 | July 14, 85 | D, SH | Chroococcales | + + | + + + | + |
| | | | Lichens | r | + | r |
| | | | Diatom frustules | + | + | r |
| | | | rock particles | | + + | rr |
| 15 | Aug. 30, 85 | W | Chroococcales | + + | + + + | + |
| | | | Lichens | + + | + | rr |
| | | | rock particles | | + + | rr |

sive power of the radular surface of this species (Table 3). In this high shore environment day-feeding activity seemed limited. The gut contents of specimens collected on dry, sunny surfaces were scarce, their dull orange fluorescence showing that they had been ingested many hours before. Occasional red fluorescence was however noted, especially if the specimens had been collected on the shaded side of a rock, indicating more recent ingestion of food.

Early morning sampling of snails, especially on dewy surfaces, yielded abundant, intensely red fluorescent gut contents, as did the guts of snails sampled during or shortly after rain, irrespective of the time of day (Table 3).

Only occasionally was *L. sitkana* found in this environment. Their gut contents, even in specimens collected on wet surfaces, were sparse and

rock particles were practically absent (Table 3).

The situation was quite different on rocks provided with deep crevices or fissures. While *L. scutulata* was equally well represented, on the dry, sunny areas and in the crevices, *L. sitkana* seemed to congregate in the deeper more humid microhabitats, where it was often observed actively feeding. This was clearly seen at Station 17 (July 1985 and May 1987) and in areas close to Stations 1 and 4 (May 1987, Table 4). In these, we collected several specimens of *L. sitkana* from deep fissures where the abundant moisture maintained a rich growth of a filamentous, *Spirogyra* – like green alga, which constituted all of the abundant, intensely red fluorescent gut contents examined.

In addition to the lack of living Zosteraceae, the only other case in which the gut contents of

*Table 4.* Results of floristic and gut content observations on samples collected from fissures in rocks of the supralittoral, at or near stations, 1, 4 and 17.

| Stn. | Date | Description and microflora | Substrate | *L. scutulata* | *L. sitkana* |
|---|---|---|---|---|---|
| 1 | May 30, 1987 | Fissure with rainwater | | | |
| | | Chroococcales | + + + | + + + | + |
| | | Hormogonales | + + + | + + + | + + + |
| | | *Spirogyra* (?) | + + + | + + + | + + + |
| | | rock particles | | r | a |
| 4 | May 30, 1987 | Fissure with rainwater (approx. salinity 14‰) | | | |
| | | Chroococcales | + + | + + + | + + |
| | | Hormogonales | + + + | + + + | + + + |
| | | *Spirogyra* (?) | + + + | + + + | + + + |
| | | *Melosira nummuloides* | + + | + + + | + + + |
| | | *Synedra* sp. | + | + + | + + |
| | | *Cocconeis* sp. | + | + | + |
| | | rock particles | | a | a |
| 17 | July 14, 1985 | Fissure with rainwater | | | |
| | | Chroococcales | + + + | + + + | + + + |
| | | *Spirogyra* (?) | + + + | + + + | + + + |
| | | rock particles | | r | a |
| 17 | May 25, 1987 | Fissure with rainwater | | | |
| | | Chroococcales | + | r | + |
| | | Hormogonales | + + + | + + + | + + + |
| | | *Spirogyra* (?) | + + + | + + + | + + + |
| | | rock particles | | a | a |

periwinkles were found to differ greatly in composition from the food available in the surrounding environment was the sample collected in a spray pool, high above the intertidal line at Station 3 (June 1985). The pool contained a small *Fucus* sp. in an advanced state of decomposition, on which all of the *L. scutulata* were present. The rocky bottom was covered with a rich growth of naviculoid diatoms, but none of these was found in the gut contents which were predominantly non-fluorescent and undifferentiated, with occasional cells of *Synedra* sp., (the only epiphytic diatoms recorded on *Fucus*).

## Discussion

Our results indicate that the two species have similar feeding habits in the intertidal zone, where

their distributions frequently overlap. Previous authors have focussed their attention on the living components of the diet of *L. scutulata*, and concluded that this species feeds on the natural microalgal communities (e.g. Castenholz, 1961; Foster, 1964), but that it is also able to profit from any macroalgal material available (Dahl, 1964). Our data demonstrate that this species seems to be more of an opportunistic omnivore, which can feed equally well on micro- or macroalgae, and even on an exclusively litter-based diet, as Jensen (1981) has shown for juvenile *L. scutulata*.

The efficiency of the 'moderate' radula of *L. scutulata*, described by Rosewater (1979) as 'a successful food gathering organ... in the spray zone of rocks' is clearly apparent in our results. (See also Sacchi & Voltolina, 1987). Our data also agree with Castenholz (1961) regarding the importance of blue-green algae (and lichens) as a source of food for *L. scutulata* in the supralittoral fringe. In addition, our studies show that *L. scutulata* does not depend exclusively on periodic or occasional events, such as dew or rain, for its ability to forage on bare rocks, but that it is able to take advantage of the limited food resources of that type of environment even on dry days, at least when shielded from the direct action of the sun.

The feeding habits of *L. sitkana*, as far as we are aware, have never been studied in detail. The only noteworthy reference we could find is in Rosewater (1979) who mentioned the 'rhomboidal' radula of *L. sitkana* as a useful tool in feeding on macroalgae. In our studies *L. sitkana* could be regarded as an opportunistic omnivore, in the same way as *L. scutulata*, feeding on any food source which is available. It is, however, in the supralittoral zone that differences between the two species become evident (Sacchi & Voltolina, 1987). The occasional specimens of *L. sitkana* found on bare rocks are probably strays, which would starve given their inability to utilize epi- or endolithic microalgae (as shown by the absence of rock particles in their gut contents), unless they can find a more suitable microenvironment, such as a damp, deep crevice with abundant microalgal growth. To this factor we would attribute the absence of *L. sitkana* from this zone, since an equal or even superior resistance to desiccation would seem of little use, unless accompanied by a suitable food gathering ability.

As far as we know, fluorescence microscopy has not been used previously for studies of this kind. It would seem to have great potential, especially when used in combination with light microscopy, given the ease with which the observer can decide on the relative importance of different food sources.

## Acknowledgements

This research was partially supported by funds from the Italian National Research Council (C.N.R.), through a grant to Prof. C.F. Sacchi. The field work was carried out while one of us (D.V.) was the recipient of an N.S.E.R.C. Visiting Fellowship at Royal Roads Military College (Victoria, B.C., Canada), supported by funds from the Department of National Defense (D.N.D. – Canada).

## References

Castenholz, R. W., 1961. The effect of grazing on littoral diatom populations. Ecology 42: 783–794.

Dahl, A. L., 1964. Macroscopic algal foods of *Littorina planaxis* Philippi and *Littorina scutulata* Gould (Gastropoda: Prosobranchiata). Veliger 7: 139–143.

Foster, M. S., 1964. Microscopic algal food of *Littorina planaxis* Philippi and *Littorina scutulata* Gould (Gastropoda: Prosobranchiata). Veliger 7: 149–152.

Jensen, J. T., 1981. Distribution, activity and food habits of juvenile *Tegula funebralis* and *Littorina scutulata* (Gastropoda: Prosobranchia) as they relate to resource partitioning. Veliger 23: 333–338.

Murray, T., 1979. Evidence for an additional *Littorina* species and a summary of the reproductive biology of *Littorina* from California. Veliger 21: 469–474.

Rosewater, K., 1979. A close look at *Littorina* radulae. Bull. Malac. Union, Inc. 1979: 5–8.

Sacchi, C. F., A. R. Torelli & D. Voltolina, 1977a. L'alimentation de *Littorina saxatilis* (Olivi) dans la lagune de Venise. Malacologia 16: 241–242.

Sacchi, C. F., P. Testard & D. Voltolina, 1977b. Recherches sur le spectre trophique de *Littorina saxatilis* (Olivi) et *L. nigrolineata* (Gray) sur la gréve de Roscoff. Cah. Biol. mar. 18: 499–505.

Sacchi, C. F., A. Occhipinti Ambrogi & D. Voltolina, 1981. Recherches sur le spectre trophique comparé de *Littorina saxatilis* (Olivi) et de *L. nigrolineata* (Gray) (Gastropoda, Prosobranchia) sur la gréve de Roscoff. Cas de populations vivant au milieu d'algues macroscopiques. Cah. Biol. mar. 22: 83–88.

Sacchi, C. F. & D. Voltolina, 1987. Recherches sur l'ecologie comparée des Littorines (Gastropoda, Prosobranchia) dans l'Ile de Vancouver. Atti Soc. Ital. Sci. nat. 128: 209–234.

Steneck, R. S. & L. Watling, 1982. Feeding capabilities and limitations of herbivorous molluscs: a functional group approach. Mar. Biol. 68: 299–319.

Van Dongen, A., 1956. The preference of *Littorina obtusata* for Fucaceae. Arch. Néerl. Zool. 11: 373–386.

Watson, D. C. & T. A. Norton, 1987. The habitat and feeding preferences of *Littorina obtusata* (L.) and *L. mariae* Sacchi et Rastelli. J. exp. mar. Biol. Ecol. 112: 61–72.

*Hydrobiologia* **193**: 155–182, 1990.
*K. Johannesson, D. G. Raffaelli and C. J. Hannaford Ellis (eds), Progress in Littorinid and Muricid Biology.*
© 1990 *Kluwer Academic Publishers.*

# Effect of crab effluent and scent of damaged conspecifics on feeding, growth, and shell morphology of the Atlantic dogwhelk *Nucella lapillus* (L.)

A. Richard Palmer
*Department of Zoology, University of Alberta, Edmonton, Alberta T6G 2E9 and Bamfield Marine Station, Bamfield, British Columbia VOR 1BO, Canada*

*Key words:* gastropod, *Cancer*, phenotypic variation, plasticity, laboratory experiment, allometry, alarm response, norm of reaction, adaptation

## Abstract

Juvenile *Nucella lapillus* of two different shell phenotypes, exposed shore and protected shore, were maintained in running seawater under each of three experimental conditions for 94 d a) laboratory control, b) exposed to the effluent of crabs (*Cancer pagurus*) fed frozen fish ('fish-crab'), and c) exposed to the effluent of crabs fed live conspecific snails ('snail-crab'). Rates of barnacle consumption and rates of body weight change varied significantly between phenotypes and among experimental conditions. Individuals from the protected-shore consumed consistently fewer barnacles and grew consistently less than those from the exposed shore. Body weight increases in the fish-crab treatments were from 25 to 50% less than those in the controls and body weights in the snail-crab treatment either did not change or actually decreased. The perceived risk of predation thus appears to have a dramatic effect on the rates of feeding and growth of *N. lapillus*.

At the end of the experiment, size-adjusted final shell weights for both phenotypes were consistently higher than controls (no crab) in both the fish-crab and snail-crab treatments. In addition, apertural tooth height, thickness of the lip, and retractability (i.e. the extent to which a snail could withdraw into its shell), with few exceptions all varied in an adaptive manner in response to the various risk treatments. Similar changes in the shell form of starved snails exposed to the same stimuli suggest very strongly that the morphological responses of both phenotypes were not just due to differences in rates of growth. These differences, at least in part, represented a direct cueing of the shell form of *Nucella lapillus* to differences in the perceived risk of predation. Somewhat surprisingly, the extent of phenotypic plasticity appeared to differ between the populations examined. Both field and laboratory evidence suggest that the exposed-shore population was much more labile morphologically than the protected-shore population.

In many instances, particularly among starved snails, the development of antipredatory shell traits was greater in the fish-crab treatment than in the snail-crab treatment. Because the scent of crabs was present in both treatments, these results suggest a) that, at the frequency/concentration used in the experiments, the scent of damaged conspecifics may have been a supernormal stimulus and b) that the morphological response in these treatments might have been greater if the stimulus had been provided at a lower level.

156

## Introduction

The shells of dogwhelks (Thaidinae) can vary enormously among populations within a single species. This variability is a conspicuous feature of species from rocky shores of the northeastern Pacific (*Nucella canaliculata*, *N. emarginata*, *N. lamellosa* and *N. lima*; Kincaid, 1957, 1964; Kitching, 1976; Spight, 1973), Australia (*Dicathais aegrota*; Phillips *et al.*, 1973), New Zealand (*Lepsiella albomarginata*, *L. scobina*; Kitching & Lockwood, 1974), the North Atlantic (*Nucella lapillus*; Colton, 1916, 1922; Crothers, 1985) and South Africa (*Nucella dubia*; Kilburn & Rippey, 1982) and includes variability in shell thickness, shape and sculpture. In *Nucella lapillus*, this variation has persisted at least since the Late Pliocene (Cambridge & Kitching, 1982; Moore, 1985).

Much of the intraspecific variation in the shells of dogwhelks appears to be adaptive. Thicker shells, or those with smaller apertures, are less vulnerable to predation by shell-breaking crabs (Hughes & Elner, 1979; Kitching *et al.*, 1966; Kitching & Lockwood, 1974; Palmer, 1985a). Thinner shells, on the other hand, are less expensive to produce energetically and less likely to limit the maximum rate of growth (Palmer, 1981). Larger apertures are associated with proportionally larger feet which reduce the probability of dislodgement by breaking waves (Etter, 1988; Kitching *et al.*, 1966). Compared to smooth shells, spiral sculpture increases the force required to crush entire shells and may reduce the vulnerability of snails to attack by shell-crushing fish (Palmer, unpublished observations).

Although some of this morphological variation has a genetic basis (Largen, 1971; Palmer, 1985b), a sizeable fraction may also be ecophenotypic (Etter, 1988; Palmer, 1985b; Spight, 1973). *Nucella lamellosa* of the northeastern Pacific exhibit a rather striking range of shell forms in response to environmental cues. In this species, the scents of crabs and of damaged conspecifics both induce the development of larger apertural teeth (Appleton & Palmer, 1988) and heavier shells (Palmer, unpublished observations) com-

pared to controls. Rates of feeding and growth also declined substantially with increasing apparent risk (Appleton & Palmer, 1988). To determine the generality of these responses in dogwhelks I initiated a similar experiment examining the effect of these environmental stimuli on the feeding, growth and shell morphology of the North Atlantic dogwhelk, *Nucella lapillus* (L.).

## Materials and methods

### Collection and measurement

Dogwhelks were collected from two sites on the shores of Anglesey, North Wales, UK. Because of different wave-exposure regimes these two sites harbored different shell phenotypes. One site was an exposed headland at the southwestern edge of Trearddur Bay facing directly west into the Irish Sea ('exposed'; 53° 16′ 00″ N, 4° 37′ 10″ W, Ordnance Survey grid reference SH250779) and the other a boulder and cobble beach towards the north end of the Menai Straits, Trwyn Y Penrhyn ('protected'; 53° 17′ 45″ N, 4° 03′ 10″, O.S. grid reference SH631798).

Small preliminary samples were collected initially on July 10, 1986 to determine the relationship between shell length and wet body weight for each site. Based on this relationship, large numbers of dogwhelks of approximately the same wet body weight were collected from each site on July 20 and taken to the University College of North Wales Marine Science Laboratory at Menai Bridge where they were held immersed in running seawater. Snails were identified individually by writing a number on their shell with a fine-tipped permanent marker and covering it with a clear, cyanoacrylate glue to prevent abrasion.

Prior to the experiment, shells were measured for total length, aperture length and width, body whorl diameter, and thickness of the apertural lip (Fig. 1) to the nearest 0.05 mm using Vernier calipers. Lip thickness was measured either between apertural teeth, if present, or at the location on the lip where the teeth would have devel-

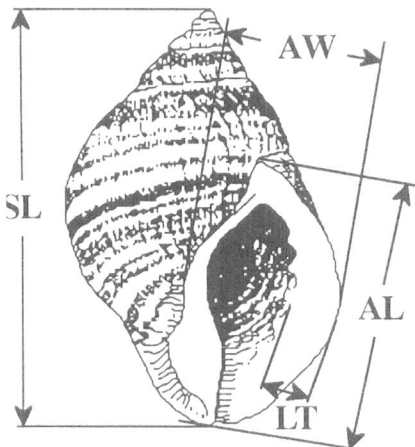

*Fig. 1*. Dimensions measured on the shells of *Nucella lapillus* (mm). SL – shell length, AL – aperture length, AW – aperture width, LT – thickness of the apertural lip. Diameter of the body whorl was measured by aligning the axis of coiling with the jaws of Vernier calipers, placing the aperture of the shell flush against the lower jaw and bringing the upper jaw in contact with the dorsal-most surface of the body whorl.

oped. At the end of the experiment these traits were remeasured and apertural tooth height was also measured. Apertural tooth height was measured as the difference between two measurements of lip thickness at the middle of the apertural lip, one from the tip of an apertural tooth to the outside of the lip, and perpendicular to it, and the other from the adjacent inter-tooth space to the same point on the outside of the lip. Because the snails were small and actively growing, no apertural teeth were present on any of them at the beginning of the experiment.

Initial shell weights and wet body weights were estimated following the procedure of Palmer (1982). Snails were first weighed while suspended in seawater (immersed weight) to estimate shell dry weight using a previously determined regression of destructively sampled shell dry weight (Y, mg) upon immersed weight (X, mg) for each shell phenotype: Exposed – $Y = 1.5707 \pm 0.0048X - 6.75$ ($r^2 = 0.9998$, $N = 28$), Protected – $Y = 1.6036 \pm 0.0021X - 4.68$ ($r^2 = 0.9999$, $N = 29$). Snails were then weighed in air (whole weight) after gently pressing out as much of the extra-visceral water as possible and allowing the shells

to dry. Subtracting estimated shell dry weight from whole weight yielded a non-destructive estimate of wet body weight. Test correlations between estimated wet body weight and destructively sampled wet body weight were high for both phenotypes (Exposed–$r^2 = 0.997$, $N = 28$; Protected–$r^2 = 0.990$, $N = 29$). Immersed weight was measured twice, 24 h apart, for each snail prior to initiation of the experiment.

The unoccupied volume of a shell was measured to compare the body size of a snail to the habitable volume of its shell and hence provide a measure of retractability. Unoccupied volume was measured by placing a live snail aperture up on a small supporting ring on the tray of the balance. The shell was orientated carefully so that the plane of the aperture was as close to horizontal as possible. After orientating the shell, the balance was tared and distilled water was introduced into the aperture with a Pasteur pipette until the water was flush with the lip and the columella. The weight of water added was then recorded as volume of shell unoccupied (1 g = 1 ml). Prior to this procedure, the snail had been pressed gently back into the shell with absorbent tissue, to remove as much of the extra-visceral water as possible, and the shell was allowed to dry. Repeat estimates of unoccupied volume on the same individual, accomplished by refilling the aperture a second time while still on the balance, varied by less than 5% [mean difference = $2.1 \pm 1.38\%$ (mean $\pm$ SD); $N = 21$].

Barnacles (*Semibalanus balanoides*), collected on small stones from the protected site, were provided as food for dogwhelks. Barnacle size was measured to the nearest mm as opercular diameter between the inside margins of the rostral and carinal plates. Crabs (*Cancer pagurus*) were collected by divers from the Menai Straits and their size was measured as maximum carapace width.

*Experimental design*

The basic unit of the experiment consisted of a 20-liter plastic aquarium provided with a continuous supply of running seawater at ambient

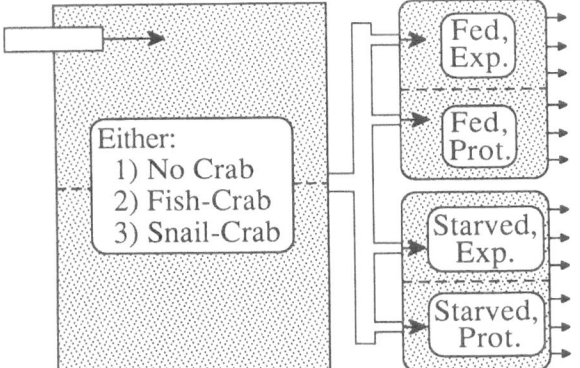

*Fig. 2.* Patterns of water circulation through experimental cages. Exp. – compartment containing snails from the exposed-shore population, Prot. – compartment containing snails from the protected-shore population. Dashed lines indicate permeable partitions between adjacent compartments.

temperature (Fig. 2). To this aquarium was added either a) nothing ('no crab' = laboratory control) or b) two crabs, one male and one female, which were fed frozen fish ( = 'fish-crab') or c) two crabs, one male and one female, fed live, intact *N. lapillus* (15–20 mm shell length) from the exposed site ( = 'snail-crab'). Each of these three risk treatments was replicated twice. The four aquaria containing crabs were subdivided with rigid, but perforated partitions to keep the two crabs separate and to allow seawater to circulate

between compartments. They were also kept covered with black plastic to minimize disturbance to the crabs. The *C. pagurus* were size-matched as much as possible among replicates (Table 1), however, some crabs moulted or died over the course of the experiment. These were replaced within 24 h.

From each experimental aquarium, seawater flowed by gravity into two cages made from plastic freezer containers ($20 \times 16 \times 7$ cm) from which the center of the lids had been removed and replaced with 7 mm plastic mesh. Each cage was further subdivided into two compartments by a plastic mesh partition (Fig. 2). The cages were tilted slightly so that seawater flowed in one side and out the other. Each compartment contained 10 *N. lapillus* from one of the source populations, and the adjoining compartment contained 10 snails from the other. Both compartments of a given cage were either loaded with stones covered with barnacles ('fed') or with bare stones ('starved'). In this manner, snails in the 'starved' treatments were unable to sense food available in the 'fed' treatments. Before placing them in the cage, the stones covered with barnacles were inspected carefully and dead barnacles (i.e. those missing opercular plates) were removed. Upon subsequent examination, all dead barnacles were then assumed to have been eaten.

*Table 1.* Carapace widths and final wet weights of the crabs (*Cancer pagurus*) used over the course of the experiments (July 24–Oct. 26, 1986). Crabs that escaped, moulted or died were replaced as noted. Dashes indicate crabs that survived for the entire experiment. Total snails eaten refers to the total number of *N. lapillus* eaten by both crabs in a particular replicate. Repl. – replicate, m – male, f – female.

| Treatment | Repl. | Initial | | | Replacement | | | Final wet weight (g) | Total snails eaten |
|---|---|---|---|---|---|---|---|---|---|
| | | Carapace width (mm) | Sex | Dates (mo/d) | Carapace width (mm) | Sex | Dates (mo/d) | | |
| Fish-crab | 1 | 107 | m | 7/24–9/16 | 104 | m | 9/16–10/26 | 167.3 | |
| | | 79 | f | 7/24–9/3 | 100 | f | 9/3–10/26 | 172.7 | |
| | 2 | 101 | m | 7/24–8/26 | 125 | m | 8/26–10/26 | 285.5 | |
| | | 110 | f | 7/24–10/26 | – | – | – | 215.0 | |
| Snail-crab | 1 | 76 | m | 7/24–10/26 | – | – | – | 72.3 | 544 |
| | | 84 | f | 7/24–10/13 | 111 | f | 10/13–10/26 | 154.9 | |
| | 2 | 82 | f | 7/24–8/16 | 105 | f | 8/16–10/26 | 171.0 | 703 |
| | | 98 | m | 7/24–10/26 | – | – | – | 153.7 | |

## Execution of the experiment

From the time of collection until initiation of the experiment (4 d), snails were held continuously immersed in cages without food in fresh running seawater which, at least after entering the laboratory seawater system, had no prior contact with crabs. On July 24, following tagging and measurement, all snails were loaded into their respective cages and then the cages were connected to aquaria containing the experimental stimuli.

Snails from the starved and fed treatments were monitored differently. To determine the effect of risk treatments on the short-term rates of shell deposition of starved snails, immersed weight was measured 24 h after initiation of the experiment, at 48 h intervals for the next six days, at 72 h intervals for the subsequent six days, and with declining frequency for the remainder of the experiment. These weighings were conducted as quickly as possible to minimize disturbance. Snails of both phenotypes were removed from an individual cage, held immersed in seawater in a plastic container, weighed and then returned to running seawater in their respective cage within 15 min. At the end of the experiment (October 26), snails were removed from their cages and measured for shell length, lip thickness, apertural tooth height, immersed weight and whole weight.

Snails in the fed treatment were not monitored as frequently to avoid disturbing them. At 10 to 30 d intervals they were measured for shell length and immersed weight. Whole weight was not measured because it disrupted activity of the snails for the subsequent 24–48 h. Barnacles were replaced with fresh ones every 20–30 d. Both eaten and uneaten barnacles removed from the cages were counted and the opercular diameters of eaten barnacles measured.

Flow rates through the aquaria were measured every two to four days and adjusted if necessary to a rate close to 1.7 liters min$^{-1}$ (0.85 l m$^{-1}$ cage$^{-1}$). In addition, the aquaria were inspected daily to insure that the flow of seawater had not been interrupted. Water temperature was also measured daily and ranged from 16 °C at the beginning to 13 °C at the end

of the experiment. At least once a day, the number of snails eaten by crabs in the snail-crab treatments were recorded, and replacement snails were added to bring the total up to five per crab. Over the 94 day duration of the experiment, more than 250 snails were eaten per crab (Table 1). Crabs in the fish-crab treatment were fed roughly 2–4 g of frozen fish (haddock or cod) every third or fourth day, and any uneaten fish was removed at the end of the day.

## Statistical analyses

Statistical analyses were conducted using the microcomputer statistical package Statview 512 + ™ (Abacas Concepts, Berkeley, CA). Because of the design of the experiment, F-values from analysis of variance (ANOVA) were computed according to the procedure suggested by Hartley (1962) and Sokal & Rohlf (1981; p. 395–396) as follows. Sums of squares (SS) were computed via a three-way fully factorial ANOVA (A = risk treatment × B = source population × C = replicate). Because replicates were nested within the main effects, and because the question to be answered was whether the variation among main effects exceeded that between replicate cages rather than that among snails within cages, the mean squares (MS) for main effects (A or B) and their interaction (AB) were tested over the MS for replicates. The appropriate MS for replicates was computed by summing the SS for three terms: a) the dummy main effect 'replicate' (C), b) the two two-way interactions which included this dummy effect (AC and BC), and c) the three-way interaction term (ABC). This sum was then divided by the sum of the degrees of freedom for these terms. This MS for replicates was then tested over the error MS.

Because rates of growth varied among risk treatments, so did the final sizes of snails. Hence to compare morphological traits among groups of different average size the effect of size had to be scaled out. Analysis of covariance (ANCOVA) could not be conducted with confidence on these data for two reasons. First, because the final size

160

ranges of some experimental groups did not overlap, adjusted means would have to have been extrapolated outside the range for which I had data. Second, because the size range within experimental groups was not very large slopes within groups could not be determined with much confidence. To circumvent these difficulties, reference samples of 100 snails each were collected near the end of the experiment (October 16, 17). Both phenotypes were collected from the same sites as those that had yielded the experimental snails. These samples included roughly equal numbers of all sizes of snails from 10 mm shell length up through fully mature adults and were used to define the size-dependence of the traits of interest for each population.

To compare traits of experimental snails at the end of the experiment, the size of the trait in question for a given snail was transformed to that of a standard-sized snail as follows:

$$\log V_i = (\log O_i - \log E_i) + \log V_s$$

where $V_i$ = value of a trait for snail $i$ scaled to that of a standard-sized snail, $O_i$ = original observed value of the trait for snail $i$, $E_i$ = the expected value of that trait for a snail of the same size as $i$ determined from the regression obtained from the appropriate reference sample (see Table 2), and $V_s$ = the average expected value of that trait for the standard-sized snail, also determined from the reference sample regression. For example, the observed final shell weight of 1580 mg for a snail of the protected phenotype of 270 mg wet body weight was transformed to that for a snail of 350 mg wet body weight as follows: $O_i$ = log(1580), $E_i$ = 0.921 (log(270)) + 1.013 (from regression 6b, Table 2), and $V_s$ = 0.921 (log(350)) + 1.013. Hence $V_i$ = 2007 mg. This transformation assumes that both the variance and the effects of the risk treatments were proportional to size. Although means and standard errors of these size-scaled values were graphed in untransformed units, statistical analyses [i.e. $t$-tests comparing adjusted means ($V_i \pm$ SE) vs. expected means ($V_s \pm$ SE), and the relevant $P$ values of Figs. 12 and 13] were conducted on the log-transformed variates.

## Results

### Morphological variation within and between populations

Field-collected snails of both phenotypes differed substantially in most traits examined. For shells of a given length, the aperture was significantly longer and significantly wider for the exposed-shore phenotype (Figs. 3a, b; Regressions 1a, b, and 2a, b in Table 2). Mature snails of both phenotypes had equally thick lips (Fig. 3c) but because of their lower spires, lip thickness at a given length was greater for the exposed-shore

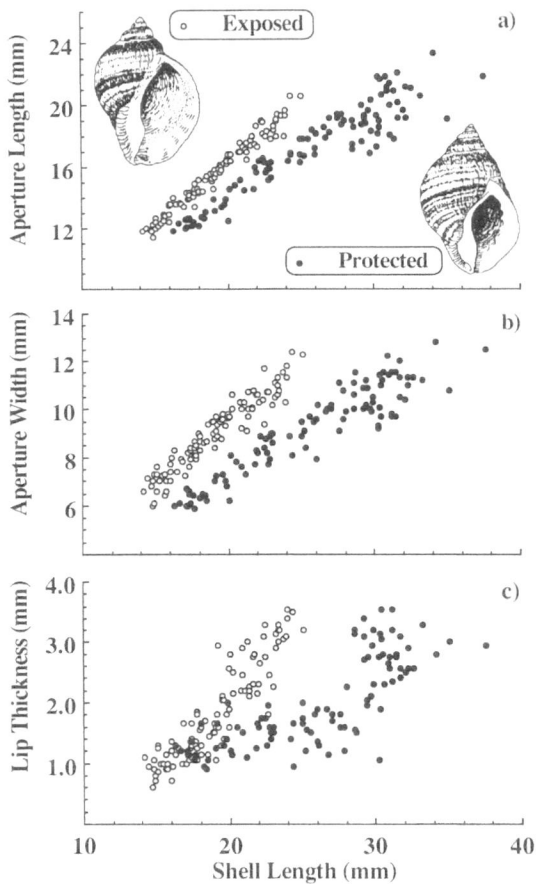

Fig. 3. Variation in aperture length, aperture width and lip thickness as a function of shell length for two populations of Nucella lapillus from shores of different wave exposure. See Table 2 for regression equations (of log-transformed values) and statistics.

*Table 2.* Regression equations (both variables log-transformed; slope and intercept $\pm$ SE) describing morphological differences between the exposed- and protected-shore populations of *Nucella lapillus* used in the experiments. See Figs. 3–5 for scatterplots of these data[§].

| X | Y | Source | Regression | $r^2$ | Comparison of slopes | | | |
| | | | | | Allometry | | a) vs b) or c) vs d) | |
| | | | | | $T_s$ | $P$ | $T_s$ | $P$ |
|---|---|---|---|---|---|---|---|---|
| 1) Shell length | Aperture length | a) ExpII | $Y = 0.991(\pm 0.017)X - 0.078(\pm 0.001)$ | 0.972 | 0.53 | 0.60 | 6.58 | <0.001 |
| | | b) ProtII | $Y = 0.792(\pm 0.025)X + 0.120(\pm 0.002)$ | 0.910 | 8.32 | <0.001 | | |
| 2) Shell length | Aperture width | a) ExpII | $Y = 1.070(\pm 0.035)X - 0.416(\pm 0.002)$ | 0.905 | 2.00 | 0.048 | 3.14 | 0.002 |
| | | b) ProtII | $Y = 0.921(\pm 0.032)X - 0.337(\pm 0.003)$ | 0.895 | 2.47 | 0.015 | | |
| 3) Shell length | Body whorl diameter | a) ExpII | $Y = 1.014(\pm 0.022)X - 0.309(\pm 0.001)$ | 0.957 | 0.64 | 0.53 | 0.03 | 0.97 |
| | | b) ProtII | $Y = 1.013(\pm 0.020)X - 0.345(\pm 0.002)$ | 0.962 | 0.65 | 0.52 | | |
| 4) Shell length | Lip thickness | a) ExpII | $Y = 2.677(\pm 0.126)X - 3.191(\pm 0.008)$ | 0.821 | 13.31 | <0.001 | 7.94 | <0.001 |
| | | b) ProtII* | $Y = 1.343(\pm 0.111)X - 1.619(\pm 0.010)$ | 0.601 | 3.09 | 0.003 | | |
| | | b') ProtII** | $Y = 0.635(\pm 0.144)X - 0.695(\pm 0.010)$ | 0.254 | 2.53 | 0.014 | | |
| 5) Shell length | Body wet wt. | a) ExpII | $Y = 2.948(\pm 0.075)X - 1.227(\pm 0.005)$ | 0.941 | 0.69 | 0.49 | 5.38 | <0.001 |
| | | b) ProtII | $Y = 3.468(\pm 0.061)X - 2.291(\pm 0.006)$ | 0.971 | 7.67 | <0.001 | | |
| | | c) ExpI | $Y = 2.891(\pm 0.174)X - 1.153(\pm 0.010)$ | 0.911 | 0.63 | 0.53 | 0.09 | 0.93 |
| | | d) ProtI | $Y = 2.911(\pm 0.144)X - 1.587(\pm 0.008)$ | 0.938 | 0.62 | 0.54 | | |
| 6) Body wet wt. | Shell wt. | a) ExpII | $Y = 1.275(\pm 0.042)X - 0.216(\pm 0.008)$ | 0.903 | 6.54 | <0.001 | 7.61 | <0.001 |
| | | b) ProtII | $Y = 0.921(\pm 0.020)X - 1.013(\pm 0.007)$ | 0.955 | 3.95 | <0.001 | | |
| | | c) ExpI | $Y = 1.134(\pm 0.083)X - 0.002(\pm 0.015)$ | 0.879 | 1.61 | 0.118 | 1.39 | 0.17 |
| | | d) ProtI | $Y = 0.979(\pm 0.075)X - 0.899(\pm 0.013)$ | 0.864 | 0.28 | 0.78 | | |
| 7) Shell length | Shell wt. | a) ExpII | $Y = 3.974(\pm 0.092)X - 2.056(\pm 0.006)$ | 0.950 | 10.58 | <0.001 | 6.34 | <0.001 |
| | | b) ProtII | $Y = 3.245(\pm 0.069)X - 1.169(\pm 0.006)$ | 0.958 | 3.55 | <0.001 | | |
| | | c) ExpI | $Y = 3.512(\pm 0.175)X - 1.588(\pm 0.011)$ | 0.939 | 2.92 | 0.007 | 1.93 | 0.059 |
| | | d) ProtI | $Y = 3.050(\pm 0.163)X - 0.916(\pm 0.010)$ | 0.928 | 0.31 | 0.76 | | |
| 8) Body wet wt. | Unocc. volume | a) ExpII[†] | $Y = 0.710(\pm 0.040)X - 2.427(\pm 0.008)$ | 0.769 | 7.25 | <0.001 | 0.75 | 0.45 |
| | | b) ProtII | $Y = 0.745(\pm 0.026)X - 2.317(\pm 0.009)$ | 0.892 | 9.77 | <0.001 | | |
| 9) Shell length | Unocc. volume | a) ExpII[†] | $Y = 2.251(\pm 0.100)X - 3.499(\pm 0.007)$ | 0.839 | 7.49 | <0.001 | 4.21 | <0.001 |
| | | b) ProtII | $Y = 2.643(\pm 0.086)X - 4.108(\pm 0.008)$ | 0.907 | 4.15 | <0.001 | | |

[§] Linear dimensions are in mm, weights in mg and volume in ml. All regression equations are for log-transformed values, even though some scatterplots are on linear axes. The SEs tabulated for intercepts actually correspond to the SE of the expected Y at the average X for the sample. $N = 100$ for both populations except for ExpI and ProtI where $N = 28$. $T_s$ – either the value from a T-test for allometry, computed as the difference between observed slopes and those expected theoretically for isometry (1.0 or 3.0 depending on dimensionality), or the value from a T-test comparing the slopes of the two populations sampled [a) vs b) or c) vs. d)], $r^2$ – coefficient of determination, $P$ = exact probability, Source - source population, Exp – exposed-shore phenotype, Prot – protected-shore phenotype. ExpI and Prot I were collected July 10, 1986, ExpII and ProtII were collected from the field near the end of the experiment (Oct. 16/17; see methods).

* Relationship not linear even on log-log plot (see Fig. 3c).

** Regression for linear region of scatter: snails $\leq$ 28 mm shell length ($N = 59$); used to estimate adjusted lip thickness in Fig. 13.

[†] One outlier removed prior to computing regression (see Fig. 5).

snails. In addition to aperture size, the body weight of the animal occupying a shell of a given length was significantly greater for the exposed-shore phenotype (Fig. 4a, Regressions 5a–d). Related to this latter difference, the weight of shell for a given weight of animal was substantially higher for the protected-shore phenotype (Fig. 4b, Regressions 6a–d). Ironically, because of the dif-

ference in shape, the shell weights of the two phenotypes overlapped broadly for a given shell length (Fig. 4c, Regressions 7a–d). The two phenotypes did, however, differ in another interesting way: the volume of unoccupied shell, a measure of retractability, was proportionally larger for the protected- than the exposed-shore phenotype (Fig. 5; Regressions 8a, b). This assumes that the visceral mass extends equally far up the apex of both phenotypes, an assumption which seems justified based upon an examination of animals removed from fractured shells. Curiously, the diameter of the body whorl did not differ between populations for shells of the same length (Regressions 3a, b).

Several traits exhibited significant allometry within populations, and the coefficients of allometry themselves often differed between populations. Aperture length, for example, became proportionally smaller with increasing shell length for the protected- but not the exposed-shore population (Regressions 1a, b, Table 2). Aperture width, on the other hand, became proportionally larger with increasing length in the exposed population but proportionally smaller in the protected one (Regressions 2a, b). In the protected but not the exposed population, body weight increased disproportionally with increasing shell length (Regressions 5a, b), whereas shell weight increased disproportionally with length in both populations (Regressions 7a, b). In contrast, shell

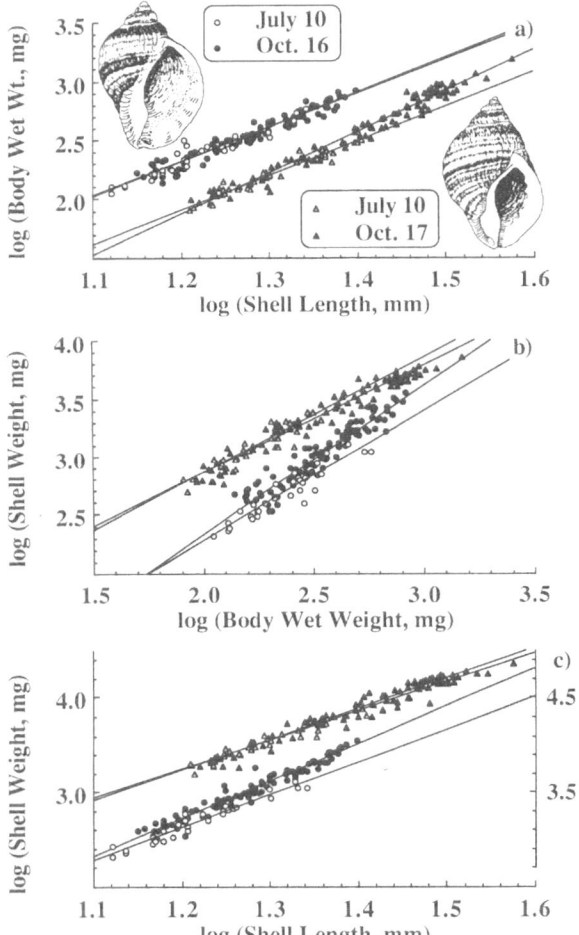

*Fig. 4.* Relationships between log transformed shell length, wet body weight and shell weight for two populations of *Nucella lapillus* from shores of different wave exposure. The samples were collected on two different dates in 1986. See Table 2 for regression equations and statistics. Triangles – protected-shore population, circles – exposed-shore population. Note that points for the protected-shore population have been shifted up 0.5 log units in Fig. 4c to avoid overplotting those of the exposed-shore population and should thus be compared against the axis to the right of figure.

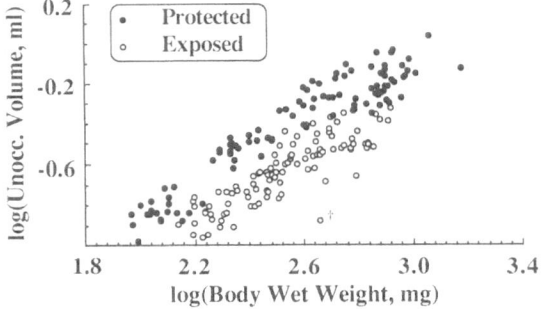

*Fig. 5.* The relation between unoccupied volume of the shell and wet body weight for two populations of *Nucella lapillus* from shores of different wave exposure. See Table 2 for regression equations and statistics. † – this point was eliminated as an outlier prior to computing Regression 8a (Table 2).

weight increased disproportionally with respect to body weight only for the exposed population, while it decreased with respect to body weight in the protected population (Regressions 6a, b). These last two patterns were qualitatively the same for the smaller preliminary samples collected from the same sites three months earlier (Regressions 6c, d and 7c, d), but were less dramatic probably because of smaller sample sizes and narrower size ranges. Finally, the volume of unoccupied shell decreased in proportion to body weight in both populations as size increased (Regressions 8a, b, 9a, b).

The thickness of the apertural lip exhibited the most complicated variation with size (Fig. 3c). In both populations, lip thickness increased disproportionally with increasing shell length (Regressions 4a, b, Table 2), but even log-transformed values varied nonlinearly (most notable for protected-shore snails; data not shown). For both populations, as snails approached maturity, lip thickness increased at a much more rapid rate (Fig. 3c).

*Temporal change within natural populations*

Rather unexpectedly, the field collections of exposed-shore *N. lapillus* on two different dates (July 10 and Oct. 16) differed in relative shell weight. Snails collected in October had significantly heavier shells than those collected in July for a given wet body weight (Fig. 4b) and also for a given shell length (Fig. 4c). Body weight for a given shell length, however, did not differ between dates (Fig. 4a) hence the variable that changed was shell weight (compare Regressions 5a vs. 5c, 6a vs. 6c and 7a vs. 7c, Table 2).

Of interest, no significant differences were observed for these traits between dates in snails from the protected shore (Figs. 4a–c; compare Regressions 5b vs. 5d, 6b vs. 6d and 7b vs. 7d).

*Variation among initial groups*

Because of the many morphological differences outlined above, some attribute had to be used by which animals of comparable size from the two populations studied could be selected for the experimental groups. I chose wet body weight for this purpose because it seemed the least biased measure of overall animal size (see size scaling in discussion for an expanded consideration of this problem). Snails from both populations were sorted by shell length and then individuals were chosen to be close to 195 mg wet weight based on Regressions 5a, b (Table 2). No significant differences in wet body weight were present among risk treatments or between source populations at the beginning of the experiment for either starved or fed snails (Tables 3, 4). Unfortunately, by chance, a significant difference did materialize among replicates in the starved treatments ($P = 0.02$), but this did not occur in the fed treatments. Among the remaining traits (shell length, shell weight and lip thickness) no significant differences existed among risk treatments or replicates for either fed or starved snails, even though, as expected, differences between source populations were highly significant (Tables 3, 4).

*Short-term rates of shell deposition in starved snails*

In the 24 h immediately prior to the experiment, the daily rate of shell deposition ranged from 5 to 8 mg snail$^{-1}$ d$^{-1}$ among groups in the starved series (Fig. 6). Although significant differences existed between the two source populations ($P = 0.016$), none appeared among risk treatments ($P = 0.16$, Table 5). In the first 24 h of the experiment, however, some interesting differences arose among risk treatments (Fig. 6). Compared to baseline values, the daily rate of deposition declined dramatically in the snail-crab treatments: $-26$ and $-46\%$ for the exposed- and protected-shore phenotypes respectively. This decline was less dramatic in the fish-crab treatments ($-13$ and $-27\%$ respectively). In the no-crab treatments the rate of shell deposition either continued to increase ($+19\%$) or declined slightly ($-14\%$) for the exposed- and protected-shore phenotypes respectively. These differences among risk treatments were highly significant ($P = 0.008$, Table 5).

*Table 3.* Initial and final values of four traits of *Nucella lapillus* held under various experimental conditions in the laboratory for 94 d. SE – standard error.

| Source population | Treatment | Replicate | N | Shell length (mm) Initial Mean | SE | Final Mean | SE | Body wet wt. (mg) Initial Mean | SE | Final Mean | SE | Shell weight (mg) Initial Mean | SE | Final Mean | SE | Lip thickness (mm) Initial Mean | SE | Final Mean | SE |
|---|---|---|---|---|---|---|---|---|---|---|---|---|---|---|---|---|---|---|---|
| **Starved** | | | | | | | | | | | | | | | | | | | |
| Exposed | No crab | 1 | 9 | 15.6 | 0.20 | 16.1 | 0.16 | 218 | 7.6 | 155 | 6.9 | 388 | 25.1 | 512 | 33.2 | 0.75 | 0.037 | 1.13 | 0.082 |
| | | 2 | 10 | 15.1 | 0.16 | 15.5 | 0.18 | 188 | 9.1 | 132 | 7.2 | 337 | 23.4 | 432 | 31.9 | 0.75 | 0.043 | 0.97 | 0.054 |
| | Fish-crab | 1 | 9 | 15.3 | 0.17 | 15.8 | 0.22 | 202 | 11.2 | 146 | 9.5 | 341 | 18.6 | 475 | 33.9 | 0.71 | 0.020 | 1.17 | 0.073 |
| | | 2 | 10 | 15.0 | 0.17 | 15.3 | 0.22 | 185 | 7.0 | 127 | 5.2 | 296 | 11.2 | 413 | 23.7 | 0.71 | 0.015 | 1.15 | 0.083 |
| | Snail-crab | 1 | 10 | 15.0 | 0.15 | 15.1 | 0.18 | 185 | 7.0 | 127 | 6.1 | 346 | 20.4 | 394 | 26.2 | 0.75 | 0.042 | 0.91 | 0.063 |
| | | 2 | 10 | 15.1 | 0.12 | 15.2 | 0.17 | 195 | 7.8 | 139 | 5.6 | 323 | 10.5 | 393 | 18.1 | 0.71 | 0.022 | 0.89 | 0.040 |
| Protected | No crab | 1 | 10 | 21.3 | 0.20 | 21.1 | 0.20 | 181 | 9.3 | 139 | 6.9 | 1402 | 74.9 | 1434 | 78.6 | 1.42 | 0.062 | 1.54 | 0.071 |
| | | 2 | 8 | 21.3 | 0.25 | 21.2 | 0.23 | 196 | 7.3 | 164 | 9.6 | 1328 | 46.0 | 1359 | 44.5 | 1.31 | 0.054 | 1.34 | 0.069 |
| | Fish-crab | 1 | 10 | 21.6 | 0.26 | 21.4 | 0.25 | 227 | 11.0 | 172 | 8.9 | 1459 | 104.7 | 1517 | 100.2 | 1.32 | 0.101 | 1.52 | 0.092 |
| | | 2 | 8 | 21.2 | 0.35 | 21.1 | 0.35 | 191 | 13.4 | 144 | 10.0 | 1347 | 64.9 | 1392 | 54.5 | 1.42 | 0.096 | 1.56 | 0.091 |
| | Snail-crab | 1 | 10 | 21.5 | 0.30 | 21.4 | 0.28 | 208 | 9.6 | 159 | 7.7 | 1477 | 76.2 | 1516 | 72.7 | 1.45 | 0.055 | 1.52 | 0.061 |
| | | 2 | 9 | 21.2 | 0.22 | 21.0 | 0.23 | 202 | 8.6 | 154 | 10.8 | 1267 | 61.5 | 1316 | 62.0 | 1.29 | 0.073 | 1.46 | 0.073 |
| **Fed** | | | | | | | | | | | | | | | | | | | |
| Exposed | No crab | 1 | 10 | 14.8 | 0.16 | 22.5 | 0.40 | 182 | 7.4 | 621 | 38.0 | 317 | 20.6 | 1648 | 113.8 | 0.65 | 0.028 | 1.80 | 0.153 |
| | | 2 | 10 | 15.0 | 0.19 | 21.6 | 0.56 | 187 | 8.5 | 574 | 37.1 | 318 | 15.4 | 1510 | 115.8 | 0.67 | 0.033 | 1.78 | 0.142 |
| | Fish-crab | 1 | 9 | 14.9 | 0.19 | 22.1 | 0.64 | 182 | 8.7 | 485 | 34.4 | 333 | 20.8 | 1777 | 145.6 | 0.74 | 0.059 | 2.13 | 0.201 |
| | | 2 | 10 | 15.1 | 0.19 | 22.5 | 0.41 | 193 | 9.2 | 509 | 28.7 | 333 | 15.6 | 2047 | 85.5 | 0.72 | 0.023 | 2.66 | 0.091 |
| | Snail-crab | 1 | 10 | 15.3 | 0.20 | 15.7 | 0.17 | 209 | 14.6 | 179 | 10.8 | 372 | 19.0 | 496 | 23.4 | 0.78 | 0.032 | 1.10 | 0.080 |
| | | 2 | 9 | 14.7 | 0.16 | 16.5 | 0.22 | 174 | 5.9 | 204 | 4.5 | 304 | 14.5 | 590 | 39.5 | 0.64 | 0.024 | 1.12 | 0.092 |
| Protected | No crab | 1 | 10 | 21.3 | 0.23 | 23.1 | 0.60 | 202 | 10.3 | 375 | 38.1 | 1467 | 68.6 | 1997 | 140.7 | 1.44 | 0.087 | 1.43 | 0.082 |
| | | 2 | 8 | 21.2 | 0.17 | 23.9 | 0.58 | 187 | 7.3 | 360 | 44.9 | 1360 | 87.9 | 2095 | 109.6 | 1.39 | 0.079 | 1.48 | 0.104 |
| | Fish-crab | 1 | 9 | 21.2 | 0.19 | 22.5 | 0.43 | 196 | 7.5 | 265 | 19.4 | 1402 | 81.4 | 1875 | 134.7 | 1.29 | 0.067 | 1.74 | 0.084 |
| | | 2 | 8 | 20.8 | 0.21 | 23.3 | 0.78 | 190 | 5.5 | 305 | 34.6 | 1271 | 46.9 | 2117 | 195.7 | 1.29 | 0.062 | 1.82 | 0.081 |
| | Snail-crab | 1 | 10 | 21.5 | 0.16 | 21.2 | 0.17 | 206 | 6.3 | 175 | 6.1 | 1457 | 59.1 | 1511 | 53.4 | 1.42 | 0.090 | 1.59 | 0.090 |
| | | 2 | 10 | 21.6 | 0.26 | 21.3 | 0.27 | 213 | 11.9 | 184 | 9.8 | 1589 | 106.4 | 1690 | 114.9 | 1.57 | 0.078 | 1.83 | 0.083 |

*Table 4.* Results from ANOVA on initial traits of *Nucella lapillus* used in the experiments. See Table 3 for trait means. Treatments – No crab, Fish-crab, Snail-crab. Source populations – exposed, protected. df – degrees of freedom, MS – mean squares, P – exact probability. $MS_{replicates}$ was tested over $MS_{error}$, all other MS were tested over $MS_{replicates}$ (see methods for computation of $MS_{replicates}$ and F values).

| Source of variation | df | Shell length | | Body wet wt. | | Shell weight | | Lip thickness | |
|---|---|---|---|---|---|---|---|---|---|
| | | MS | P | MS | P | MS | P | MS | P |
| **Starved snails** | | | | | | | | | |
| Main effects | | | | | | | | | |
|   Treatment | 2 | 0.18 | 0.73 | 292 | 0.88 | 1046 | 0.98 | 0.0018 | 0.95 |
|   Source pop. | 1 | 1071.54 | <0.001 | 858 | 0.55 | $3043 \times 10^4$ | <0.001 | 11.5067 | <0.001 |
| Interaction | 2 | 0.33 | 0.58 | 2668 | 0.36 | 15657 | 0.75 | 0.0049 | 0.88 |
| Replicates | 6 | 0.55 | 0.99 | 2166 | 0.02 | 52667 | 0.09 | 0.0384 | 0.29 |
| Error | 101 | 0.445 | | 797 | | 27900 | | 0.0308 | |
| **Fed snails** | | | | | | | | | |
| Main effects | | | | | | | | | |
|   Treatment | 2 | 0.77 | 0.29 | 1497 | 0.39 | 91077 | 0.19 | 0.0854 | 0.16 |
|   Source pop. | 1 | 1113.24 | <0.001 | 3549 | 0.16 | $3393 \times 10^4$ | <0.001 | 13.8481 | <0.001 |
| Interaction | 2 | 0.61 | 0.36 | 358 | 0.78 | 77410 | 0.23 | 0.1438 | 0.07 |
| Replicates | 6 | 0.50 | 0.24 | 1363 | 0.13 | 40586 | 0.24 | 0.0344 | 0.43 |
| Error | 102 | 0.371 | | 801 | | 30084 | | 0.0344 | |

Except for a brief increase following day three, the rates of shell deposition declined roughly exponentially over the remainder of the experiment (Fig. 6). Following the measurements of day three, bare stones were added to otherwise empty cages to provide a more natural substratum for the starved snails. This appeared to result in an

*Table 5.* Results from ANOVA on rates of shell deposition of starved *Nucella lapillus* held under various conditions in the laboratory (Fig. 6). Baseline – rate of deposition over 24 h immediately prior to the experiment. % Change in rate – percent change in the rate of deposition from baseline to that of the first 24 h of the experiment. Abbreviations and analyses as in Table 4.

| Source of variation | df | Baseline | | % Change in rate | |
|---|---|---|---|---|---|
| | | MS | P | MS | P |
| Main effects | | | | | |
|   Treatment | 2 | 14.113 | 0.16 | 14077 | 0.008 |
|   Source pop. | 1 | 62.085 | 0.016 | 13797 | 0.013 |
| Interaction | 2 | 0.029 | 0.99 | 820 | 0.53 |
| Replicates | 6 | 5.609 | 0.63 | 1148 | 0.65 |
| Error | 101 | 7.679 | | 1649 | |

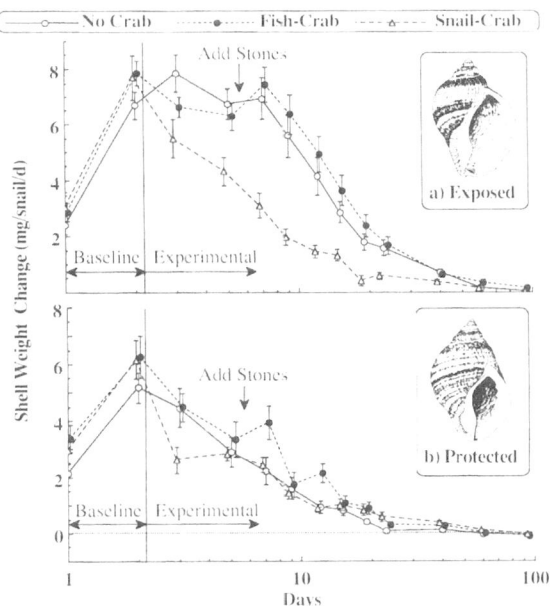

*Fig. 6.* Changes in the rate of shell deposition (mean ± SE) as a function of time for *Nucella lapillus* from shores of different wave exposure. Snails were held without food under three different experimental conditions in the laboratory. Note that time is plotted on a logarithmic scale. Add stones – bare stones added to cages. See Table 1 for sample sizes.

*Table 6.* Results from ANOVA on final values and total change for traits of starved *Nucella lapillus* held under different conditions in the laboratory (Fig. 7). Abbreviations and analyses as in Table 4.

| Source of variation | df | Body weight change | | Shell weight change | | Final tooth ht. | | Lip thickness change | |
|---|---|---|---|---|---|---|---|---|---|
| | | MS | P | MS | P | MS | P | MS | P |
| Main effects | | | | | | | | | |
| Treatment | 2 | 348 | 0.18 | 12785 | 0.016 | 0.04812 | 0.044 | 0.2660 | 0.022 |
| Source pop. | 1 | 4018 | 0.002 | 82530 | <0.001 | 0.03733 | 0.085 | 0.9351 | 0.002 |
| Interaction | 2 | 704 | 0.061 | 11976 | 0.019 | 0.02781 | 0.11 | 0.1295 | 0.086 |
| Replicates | 6 | 153 | 0.32 | 1451 | 0.50 | 0.00868 | 0.062 | 0.0343 | 0.26 |
| Error | 101 | 128 | | 1627 | | 0.00417 | | 0.0261 | |

increase in the rate of shell deposition for both phenotypes in the fish-crab treatments as well as for the exposed phenotype in the no-crab treatment (Fig. 6).

### Differences in final shell morphology among starved snails

Snails of the exposed-shore phenotype lost significantly more body weight over the 94 d of the experiment than those of the protected-shore phenotype (Fig. 7a, Table 6). Risk treatment had no overall effect on body weight loss, although the interaction between source population and risk treatment was nearly significant ($P = 0.061$, Table 6). Shell weight gain, however, not only varied significantly among risk treatment groups ($P = 0.016$) and between source populations ($P < 0.001$), but the interaction between these main effects was also significant ($P = 0.019$; Fig. 7b, Table 6). For both populations, the total

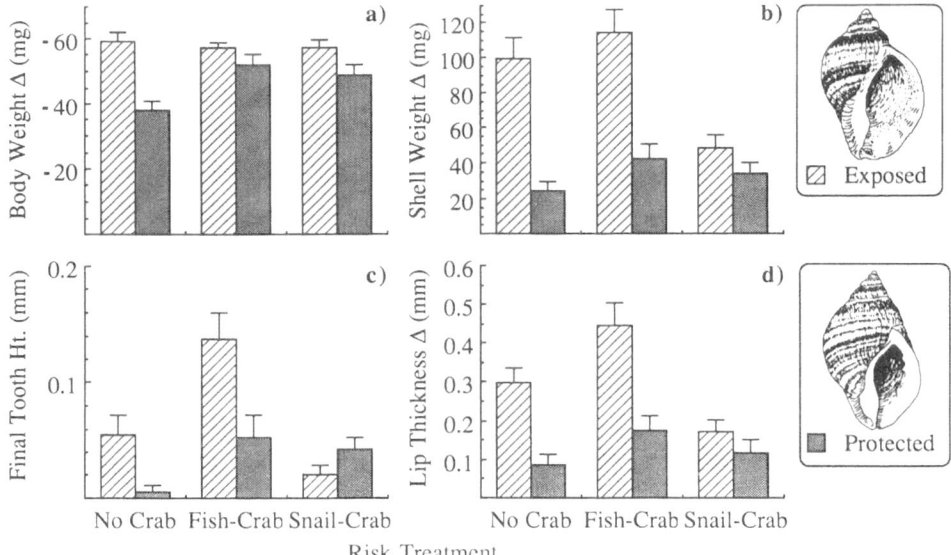

*Fig. 7.* Total change (Δ) in wet body weight, shell weight, apertural tooth height, and lip thickness of *Nucella lapillus* from shores of different wave exposure (mean ± SE). Snails were held without food under three different experimental conditions in the laboratory for 94 d. All snails lacked apertural teeth at the beginning of the experiment, thus final tooth height is synonymous with the change in tooth height. See Table 3 for sample sizes and initial and final values.

shell added was highest in the fish-crab treatments. Exposed-shore snails, however, added consistently more shell than those from the protected shore.

The rank order of final apertural tooth height and change in lip thickness of starved snails paralleled that for total shell weight change among risk treatments (Fig. 7c, d). For both traits, risk treatment had a statistically significant effect, and source population had a significant or nearly significant effect (Table 6). As observed for shell weight gain, the increase in tooth height and lip thickness was the greatest in the fish-crab treatments. Note, however, that the teeth which developed in these immature snails were not very large for either population in any treatment.

## Differences in feeding and growth

Rates of barnacle consumption differed significantly between source populations and also varied substantially among risk treatments (Fig. 8, Table 7). In all risk treatments fewer barnacles were consumed by the protected- than the exposed-shore snails. Among risk treatments, the rate of barnacle consumption was highest in the no-crab treatment: 2.09 and 1.48 barnacles snail$^{-1}$ d$^{-1}$ for the exposed- and protected-shore phenotypes respectively. The feeding rate declined by more than 25% in the fish-crab treatment (to 1.48 and 1.11 barnacles snail$^{-1}$ d$^{-1}$ respectively)

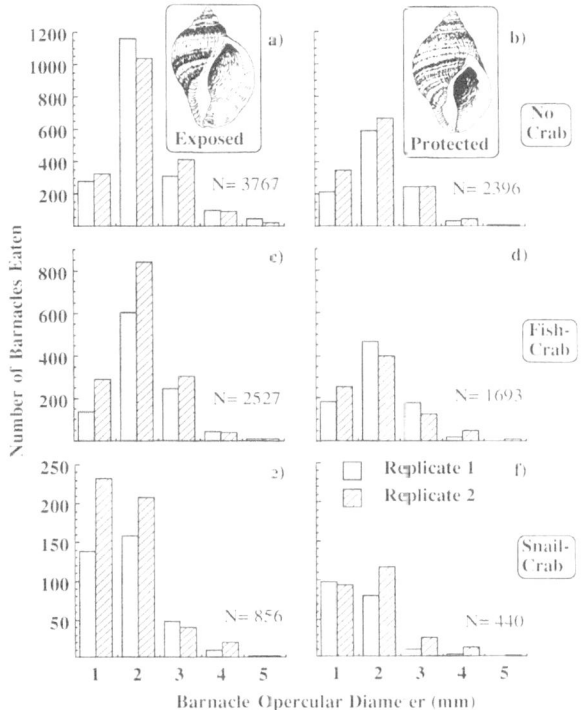

*Fig. 8.* Size-frequency distributions of barnacles eaten by *Nucella lapillus* from shores of different wave exposure. Snails were held under three different experimental conditions in the laboratory for 94 d. See Table 3 for the number of snails in each cage. Note the different scales of the vertical axes among risk treatments. N – total barnacles eaten of all sizes.

and by more than 75% in the snail-crab treatment (to 0.50 and 0.24 barnacles snail$^{-1}$ d$^{-1}$).

Not surprisingly, differences in the rates of

*Table 7.* Results from ANOVA on numbers of barnacles consumed by and % change in wet body weight of fed *Nucella lapillus* held under various conditions in the laboratory (see Figs. 8, 9; see Table 3 for final body weights). Abbreviations and analyses as in Table 4. – not applicable.

| Source of variation | Total # barnacles eaten | | | % Change in body weight | | |
|---|---|---|---|---|---|---|
| | df | MS | P | df | MS | P |
| Main effects | | | | | | |
| Treatment | 2 | 1500530 | <0.001 | 2 | 279702 | <0.001 |
| Source pop. | 1 | 572470 | 0.002 | 1 | 235675 | <0.001 |
| Interaction | 2 | 57297 | 0.16 | 2 | 42118 | 0.001 |
| Replicates | 6 | 22800 | | 6 | 1737 | 0.65 |
| Error | – | – | – | 101 | 2693 | |

168

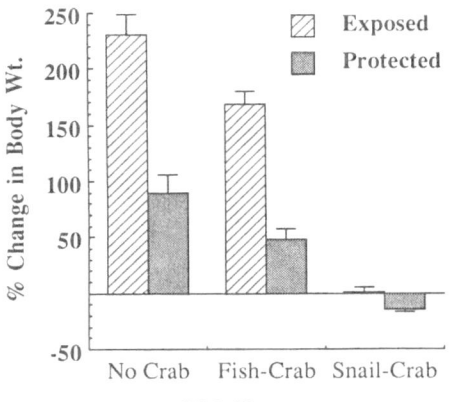

Fig. 9. Percent change in wet body weight of *Nucella lapillus* from shores of different wave exposure (mean ± SE). Snails were held with food (barnacles) under three different experimental conditions in the laboratory for 94 d. See Table 3 for sample sizes and initial and final values.

body growth paralleled those of feeding rate (Fig. 9). For both phenotypes, total body weight change declined in the order: no crab → fish-crab → snail crab ($P < 0.001$, Table 7). Snails from the protected-shore population, however, gained less than half as much in body weight as those from the exposed shore in the no-crab and fish-crab treatments and actually lost weight in the snail-crab treatment. The greater range of rates of body growth among exposed- compared to protected-shore snails was also reflected in a significant interaction between risk treatment and source population ($P < 0.001$, Table 7).

In contrast to the differences in total body weight change, changes in shell length over time were effectively the same for the no-crab and fish-crab treatments for both shell phenotypes (Figs. 10a, c, Table 8). In the snail-crab treatment, however, shell length either increased only slightly for the exposed phenotype or decreased slightly (presumably due to dissolution and abrasion of the apex while handling the shells) for the protected phenotype.

Rates of shell deposition also differed somewhat from rates of body weight gain (Figs. 10b, d, Table 8). Even though they gained less in body weight, snails in the fish-crab treatments added either more shell material than (exposed) or the same amount as (protected) those in the no-crab treatment. Shell weight gains in snails from the snail-crab treatment were slight for both phenotypes.

*Differences in final shell morphology among fed snails*

Final apertural tooth height of fed snails not only varied among risk treatments, but the pattern of variation differed between source populations (Fig. 11). Apertural tooth height was highest in the fish-crab treatment for the exposed phenotype but highest in the snail-crab treatment for the protected phenotype. Because of the magnitude of this interaction ($P = 0.037$, Table 8), neither of the main effects was significant statistically when both source populations were analyzed together (Table 8), even though the effect of risk treatment was highly significant for both source populations

Table 8. Results from ANOVA on final shell length (Fig. 10a, c), final shell weight (Fig. 10b, d) and final apertural tooth height (Fig. 11) of fed *Nucella lapillus* held under various conditions in the laboratory. Abbreviations and analyses as in Table 4.

| Source of variation | df | Final shell length | | Final shell weight | | Final tooth ht. | |
|---|---|---|---|---|---|---|---|
| | | MS | P | MS | P | MS | P |
| Main effects | | | | | | | |
| Treatment | 2 | 206.87 | <0.001 | 8517938 | <0.001 | 0.0312 | 0.13 |
| Source pop. | 1 | 165.52 | <0.001 | 8068268 | <0.001 | 0.0255 | 0.17 |
| Interaction | 2 | 55.49 | 0.002 | 2257363 | 0.005 | 0.0631 | 0.037 |
| Replicates | 6 | 2.41 | 0.31 | 155493 | 0.26 | 0.0104 | 0.13 |
| Error | 101 | 2.01 | | 119502 | | 0.0061 | |

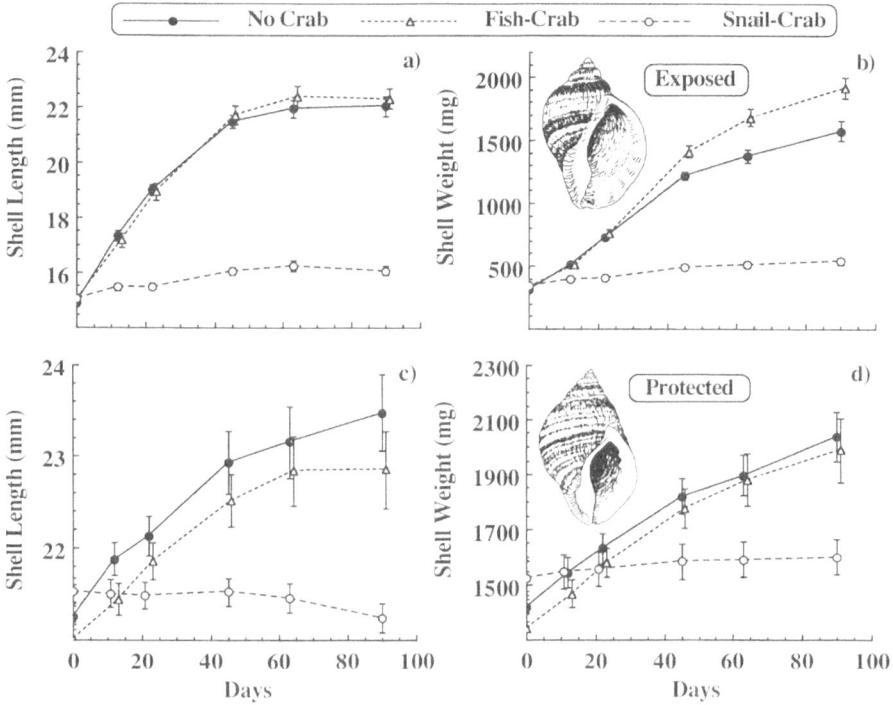

Fig. 10. Changes in shell length and shell weight as a function of time by *Nucella lapillus* from shores of different wave exposure. Snails were held with food (barnacles) under three different experimental conditions in the laboratory for 94 d. See Table 3 for sample sizes and initial and final values. Note that some points have been shifted slightly right or left to avoid overplotting. The actual dates on which the measurements were taken lie underneath the points for the no-crab treatment (solid circles). Each point represents a mean ± SE. Where error bars are not present they are less than the diameter of the symbol.

when each was analyzed separately ($P = 0.003$ and $P < 0.001$ for exposed and protected respectively from 1-way ANOVA).

Both relative shell weight and retractability

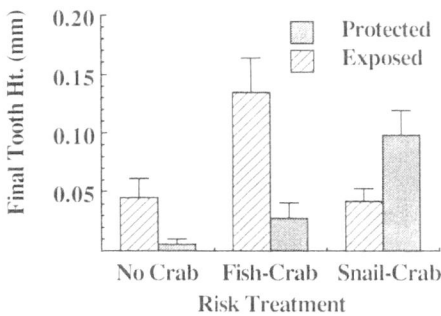

Fig. 11. Final apertural tooth height (mean ± SE) of *Nucella lapillus* from shores of different wave exposure. Snails were held with food (barnacles) under three different experimental conditions in the laboratory for 94 d. See Table 3 for sample sizes.

varied consistently (with one exception) among risk treatments for both shell phenotypes (Fig. 12). Relative shell weight and retractability were lowest in the no-crab, intermediate in the fish-crab and highest in the snail-crab treatments. The only exception to this pattern occurred in exposed-shore snails: retractability in the snail-crab treatment did not differ from controls (no crab).

The precise pattern of variation, and the degree of change relative to references shells collected from the field at the beginning of the experiment, however, depended upon the metric used to standardize 'size'. For shells of a standard length for both phenotypes, shell weight increased significantly (or nearly so) in all groups relative to the reference samples (Fig. 12a). This increase in relative shell weight was highest in the snail-crab treatment, slightly less in the fish-crab treatment and lowest in the no-crab treatment. For snails of

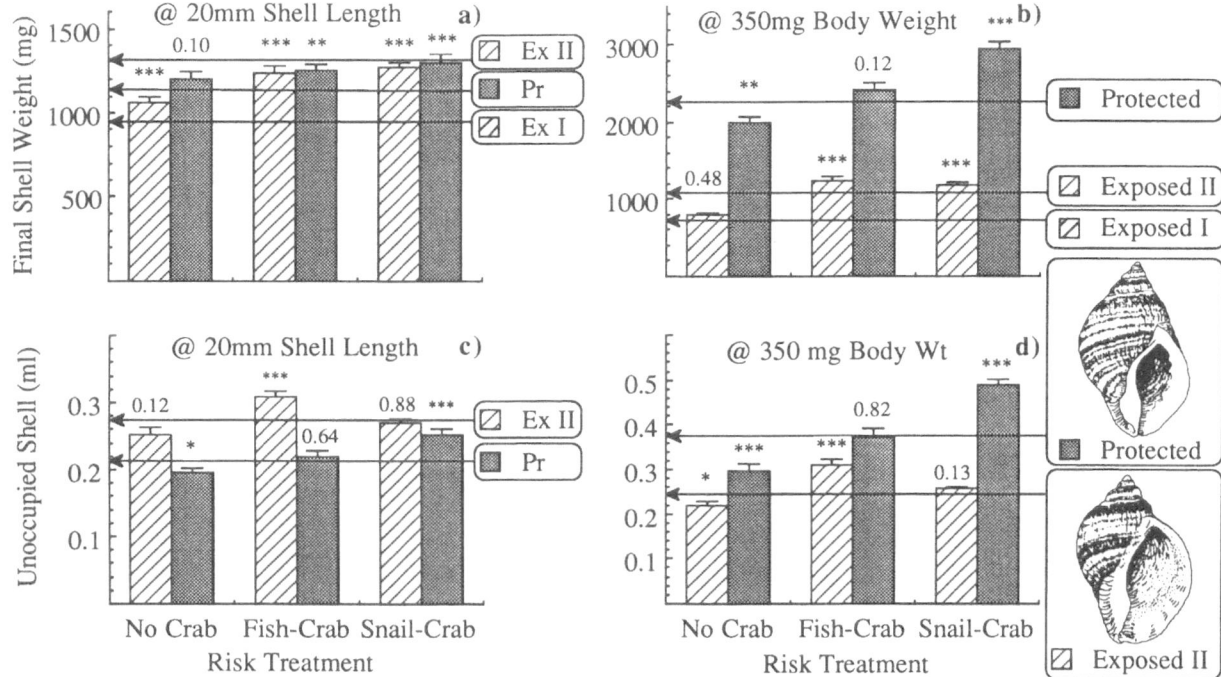

*Fig. 12.* Final shell weights (a,b), and unoccupied volume of shells (c,d) for *Nucella lapillus* from shores of different wave exposure (mean ± SE). Snails were held with food (barnacles) under three different experimental conditions in the laboratory for 94 d. For each trait, the final values have been expressed for a standard-sized snail. The two figures for each trait show the results of using different size metrics (shell length or body weight) to standardize size. See Table 3 for sample sizes, and initial and final values, and see methods for the procedure used to transform these values to those for a standard-sized snail. Arrows indicate the values of these traits for reference shells collected from the field. Exposed I – exposed-shore snails collected in mid July, Exposed II – exposed-shore snails collected from the field in mid October, Protected – protected-shore snails collected in mid October (see methods). Asterisks above bars indicate the significance level of the difference between the experimental group and reference shells (Exposed I, or Protected) from T-tests (* – < 0.05, ** – < 0.01, *** – < 0.001), otherwise exact $P$ values are given. Although means and SE are displayed on a linear scale, P values were computed from log-transformed values.

a standard wet body weight, however, although the rank order of response among risk treatments was the same as that observed for snails of a standard shell length, the departures of relative shell weight from that of the references shells were different (Fig. 12b). Relative shell weight was again significantly higher than that of the reference shells in the snail-crab treatment. In the fish-crab treatment, relative shell weight also increased, but only significantly so for the exposed-shore phenotype. In contrast, relative shell weight in the no-crab treatment either did not change compared to reference shells (exposed-shore phenotype) or actually decreased (protected-shore phenotype).

Because the shell weight of exposed-shore snails collected from the field increased between the time the experiment was started (mid July) and the time it ended (end of October; compare solid vs. open circles Figs. 4b, c), the interpretation of change depended upon which field-collected sample was used as the frame of reference. When compared to shells collected at the end of the experiment (Exposed II), relative shell weight did not change very much in the fish-crab and snail-crab treatments (Figs. 12a, b) whereas it was significantly lower in the no-crab treatment. Hence, although the increase relative to initial shell weight was rather dramatic in the fish-crab and no-crab treatments, this change was within the natural range of temporal variation in the exposed-shore population.

The effect of risk treatment on retractability (unoccupied volume of the shell) was rather insensitive to the method used to standard size (Fig. 12c, d). The main consequence of using different metrics to standardize size was a change in the ranking of the field reference values for the two populations (retractability relative to length was higher for exposed-shore snails whereas retractability relative to body weight was lower). For snails of the exposed-shore phenotype, retractability increased significantly in the fish-crab treatment, remained the same in the snail-crab treatment and decreased slightly in the no-crab treatment. In contrast, retractability of the protected-shore phenotype increased substantially in the snail-crab treatment, whereas it did not change in the fish-crab treatment and actually decreased significantly in the no-crab treatment.

Adjusted final lip thickness also varied rather substantially among groups (Fig. 13). Compared to the no-crab treatment, the lip thickness of both phenotypes increased significantly in both the fish-crab and snail-crab treatments. For the exposed-shore phenotype, this increase was greater in the snail-crab than the first-crab treatment whereas for the protected-shore phenotype the increase was about the same in these two treatments.

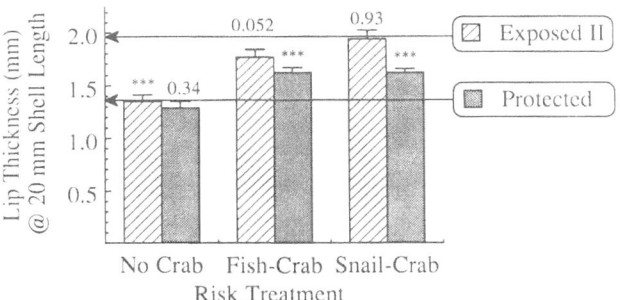

*Fig. 13.* Final thickness (mean ± SE) of the apertural lip of shells of *Nucella lapillus* from shores of different wave exposure. Snails were held with food (barnacles) under three different experimental conditions in the laboratory for 94 d. The final values have been expressed for a standard-sized snail. See Table 3 for sample sizes and initial and final values. See methods for the procedure used to transform these values to those for a standard-sized snail, and Fig. 12 for an explanation of symbols and abbreviations.

### Trajectories of shell form over time

Because shell weights and shell lengths were measured repeatedly, the change in these traits relative to each other could be examined over time. The exposed-shore groups exhibited the most interesting morphological trajectories (Fig. 14a). Individuals in the snail-crab treatment increased only slightly in shell length, but their shell weight relative to length increased steadily over the duration of the experiment and it was consistently higher than at the beginning. Of some interest, among snails in both the fish-crab and no-crab treatments, shell weight relative to length

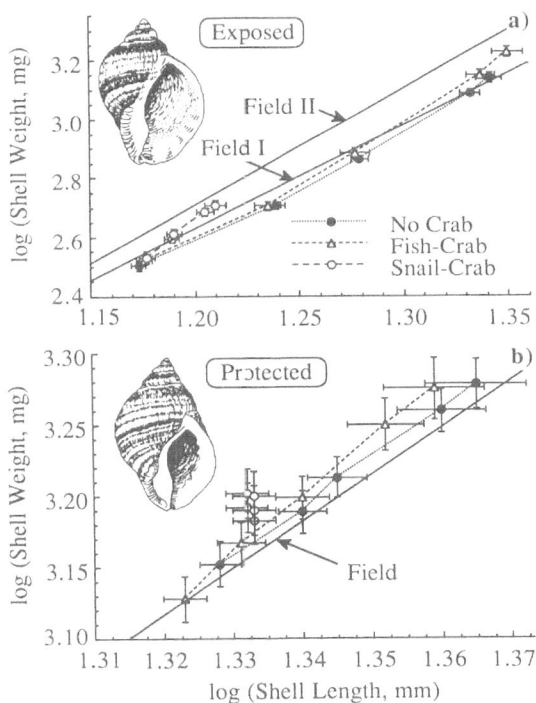

*Fig. 14.* Trajectories of shell weight vs. shell length for *Nucella lapillus* from shores of different wave exposure. Snails were held with food under three different experimental conditions in the laboratory. See Table 3 for sample sizes and initial and final values. Each point corresponds to the average shell length (± SE) and average shell weight (± SE) on a particular date, starting with those at the initiation of the experiment. Solid lines describe the static relationship between these variables for samples of reference shells collected from the field. Field I – snails collected in mid July, Field & Field II – snails collected from the field in mid October.

actually decreased in the early portion of the experiment and then increased towards the end so that the final shell weights relative to shell length were either not significantly different from (no crab) or were significantly higher than (fish-crab) those initially. Curiously, compared to the heavier-shelled snails collected from the field at the end of the experiment (e.g. see Fig. 4c), the shell weights from all risk treatments were lower.

Snails of the protected-shore phenotype grew considerably less than those of the exposed-shore and consequently the within-group variation was large relative to the total change (Fig. 14b). The shell weight of snails from all three risk treatments increased relative to shell length over the duration of the experiment. This increase was greatest in the snail-crab and fish-crab treatments and least in the no-crab treatment. Unlike the exposed-shore population, the shell form of field-collected snails from the protected shore did not change between the beginning and end of the experiment, hence these experimental snails were only compared to a single reference sample.

## Discussion

### Morphological differences between field populations

The observation that the shells of *Nucella lapillus* vary with wave exposure is not new; this phenomenon is widespread and well documented (see Crothers, 1985 for a review). The differences in shell shape and thickness between the two populations reported above are consistent with those described by others (e.g. see Seed, 1978). *N. lapillus* from more wave-exposed shores have wider apertures, lower spires and relatively thinner shells than those from protected shores. This variation appears to be maintained by the opposing selection pressures of wave action on exposed shores and crab predation on protected shores (Kitching *et al.*, 1966).

Three aspects of the shell variation observed in these natural populations seem worthy of note. First, because the wet body weight for a given shell length was much lower for snails from the protected shore (Fig. 4a), shell length will not be a reliable predictor of body size when comparing *N. lapillus* populations of different shell morphology (see size scaling below). Second, the unoccupied volume of the shell was larger by more than 50% for snails of the same body weight from the protected shore. This difference seems likely to be adaptive since animals from the protected shore would thus be able to retract substantially further into their shell than those from the exposed shore. For both populations, however, the capacity to retract into the shell declined allometrically with increasing size (Regressions 8a, b, Table 2). This allometric shift may reflect an ontogenetic increase in the amount of visceral mass relative to foot as animals approach and then reach maturity.

Third, a number of traits exhibited interesting allometric variation within populations. For example, relative aperture width increased with increasing size in the exposed-shore population, and both relative aperture width and length declined with increasing size in the protected-shore population (Regressions 1 and 2, Table 2). Hence, within each population, allometric changes amplified the differences observed between populations which are known to be adaptive (Etter, 1988; Kitching *et al.*, 1966). These data suggest that patterns of allometry themselves may be adaptive, although they do not reveal whether this allometric variation is genetically or environmentally determined. In addition, the significant positive allometry of shell weight relative to length exhibited by both populations could be adaptive, or it could reflect a tradeoff between growth rate and shell thickness. Heavier shells may be more advantageous to mature animals if they are long-lived. On the other hand, because the rate of shell production may limit the maximum rate of body growth (Palmer, 1981), juveniles may have to sacrifice some of the added defense a heavier shell might provide to enable them to grow more rapidly. The initial decline in shell weight relative to length during the period of rapid growth, followed by the increase in shell weight relative to length in the same individuals as growth slowed towards the end of the experiment (exposed-shore phenotype in both the no-crab and fish-crab treatments; compare Figs. 10a and 14a) illustrates nicely the dependence of relative shell weight upon growth rate.

## Short-term rates of shell deposition

The high initial rates at which starved, intermediate-sized *N. lapillus* deposited shell material (5–8 mg d$^{-1}$; Fig. 6) permitted a detailed view of short-term patterns of temporal change. The two- to three-fold increase in the rate of deposition between the first and second day of the baseline period prior to the experiment, most likely reflected a recovery from the trauma associated with obtaining estimates of wet body weight non-destructively (see methods). Although this pattern does suggest that the animals were disturbed by this procedure, it also reveals that this disturbance did not last much more than 24 h (e.g. compare day three to day two for the no-crab treatments Figs. 6a, b).

In addition, the temporary increase in the rate of deposition observed in three of the six groups following the introduction of bare stones into otherwise empty experimental cages suggests that rather subtle changes in the environment of *N. lapillus* can influence their rate of shell deposition. One interpretation of this response is that the stones somehow improved the 'quality of the habitat' as perceived by the snails. The ability to measure such changes on a daily basis would appear to provide a sensitive technique for assessing the impact of a variety of environmental stimuli on rates of shell deposition.

## Effect of perceived risk on shell form of starved snails

The exposure of *N. lapillus* to various risk stimuli in the absence of food might seem like a curious experiment to conduct because these conditions would be rather unlikely to occur in the field. These starved treatments, however, were necessary to distinguish between those morphological differences that were a byproduct of differences in

rates of growth and those that reflected a direct morphological response to risk stimuli. If I had used only fed snails in these experiments, I would not have been able to separate these effects and consequently would not have been able to determine if morphological differences in antipredatory traits resulted from different levels of feeding activity and growth rate, or from an amplification of these antipredatory traits via a direct cueing on the scents released by predatory crabs or damaged conspecific snails. The net result and hence the adaptive significance is the same, of course, whether a snail produces a thicker shell directly in response to the scent of crabs or indirectly by growing less rapidly in response to the same stimulus. By using starving snails, however, I was able to distinguish between these pathways of transduction.

The morphological responses of starved *N. lapillus* to the two risk treatments were not a straightforward function of perceived risk. Compared to controls, the scent of crabs alone resulted in significantly larger apertural teeth (Fig. 7c), a significantly thicker apertural lip (Fig. 7d) and heavier shells overall, although this last difference was not significant statistically (Fig. 7b, Table 6). Hence this form of perceived risk, which would signal predators in the vicinity but not feeding on snails, clearly resulted in adaptive morphological responses. These results parallel rather closely those reported by Appleton & Palmer (1988) for the northeastern Pacific *Nucella lamellosa* [ = *Thais lamellosa*]. They also provide convincing evidence that, although starvation alone may lead to the production of apertural teeth (Crothers, 1971), the scent of crabs clearly amplifies this response. Hence the induction of apertural teeth is at least in part a direct response to the scent of crabs.

The responses of *N. lapillus* to the scent of conspecifics being eaten by crabs, which would signal predators in the vicinity that were also consuming conspecific snails, were notably different, however. Although these stimuli should have indicated a higher immediate risk of predation, the rate of shell deposition of both shell phenotypes dropped dramatically in the 24 h following initiation of the experiment, compared both to controls and to the fish-crab treatment (Figs. 6a, b; Table 5). Similarly, both the total change in shell weight and lip thickness were either not significantly different from (protected-shore), or were significantly less than (exposed-shore), those of the no-crab treatment (Figs. 7b, d; Table 6). Only the protected-shore phenotype exhibited a significant increase in the height of apertural teeth compared to the no-crab treatment. Hence, the scent of damaged conspecifics being eaten by crabs did not appear to elicit a morphological response which would reduce the risk of predation. On the contrary, these stimuli together appeared to have a more profound effect on the level of physiological or behavioral activity (see supernormal stimulus below). The lack of response to bare stones introduced into the otherwise empty cages on day three of the experiment suggests that, in contrast to nearly all of the remaining treatments, snails in this treatment did not perceive the stones to improve the 'quality' of their living conditions.

The reduced morphological response of starved *N. lapillus* in the snail-crab treatment, compared to controls (no crab) and compared to the fish-crab treatment, is difficult to explain with the present data, although it may reflect unrealistically high stimulus concentrations (see supernormal stimulus below). If, as suggested by Appleton & Palmer (1988), the scent of conspecifics being eaten somehow induced a heightened metabolic rate which burned up the energy reserves of these starving snails more quickly and hence resulted in less shell being produced, then snails in the snail-crab treatment should have lost more body weight over the course of the experiment than those in the fish-crab treatment. No differences in body weight loss were apparent, however, between these treatments for either of the two shell phenotypes examined (Fig. 7a). Because of the duration of the experiment (94 d), the lack of differences in weight loss may be somewhat of an artifact. For example, snails in the different risk treatments may have lost weight at different rates over the early part of the experiment, but ultimately declined asymptotically to roughly the same final body weight at the end of the experiment. I cannot address this possibility with the present data.

*Effect of perceived risk on rates of feeding and growth*

Perceived risk had a dramatic effect on rates of feeding and growth in *N. lapillus* from both source populations (Figs. 8–10). The decline in feeding rate with increased risk parallels observations reported for mosquito larvae in the presence of predatory notonectid water bugs (Sih, 1980, 1984), and for sticklebacks in the presence of a simulated avian predator (Milinski & Heller, 1978). Rather remarkably, the suppression of feeding by *N. lapillus* in the snail-crab treatment was so great (Fig. 8) that they either did not gain any body weight at all (exposed-shore) or actually lost weight (protected-shore) over the 94 days of the experiment (Fig. 9) even though barnacles were available *ad libitum*.

Three observations suggest that the dramatically reduced rates of feeding by *N. lapillus* in the snail-crab treatments were a direct result of a predator-induced avoidance behavior as opposed to a generalized reduction in activity. First, when these cages were inspected or cleaned, snails of both phenotypes were almost always found about the lower margins or undersides of the barnacle-covered stones. Second, a substantial majority of the barnacles eaten in these cages was restricted to these same regions of the stones. Hence, although the snails may have been less active overall, they also appeared to restrict their movements to the regions of stones where they would have been least likely to encounter a foraging crab. Third, the size distribution of barnacles eaten was shifted towards smaller barnacles for snails of both phenotypes compared to the no-crab and fish-crab treatments (compare Figs. 8e, f to 8a–d) This patterns suggests that *N. lapillus* in the snail-crab treatment preferentially consumed prey with shorter handling times, another behavior which would reduce their exposure to foraging crabs. Because no data were recorded on the sizes of barnacles on different surfaces of the stones, however, the apparent preferential consumption of smaller barnacles may have been a product of the reduced foraging ambit of the snails. Barnacles on the lower margins and undersides of stones may have been smaller than those on the upper surfaces.

Rather curiously, although feeding activity was lower in the fish-crab treatments compared to controls (Fig. 8c, d), the distribution of barnacle mortality about the surface of stones did not suggest an obvious reduction in foraging ambit by *N. lapillus* of either shell phenotype. In addition, no differences were observed in the size distributions of barnacles eaten compared to the no-crab treatment. Needless to say, a closer examination of the behavior of these snails when exposed to such risk-related stimuli would be very illuminating.

### Effect of perceived risk on shell morphology of fed snails: comparisons among laboratory treatments

The effects of perceived risk on the shell morphology of *Nucella lapillus* in these experiments are most readily interpreted by comparing the final shell form of snails in the fish-crab and snail-crab treatments to that of snails in the no-crab treatment which served as a control for laboratory conditions. This comparison provides a measure of the effects of each treatment relative to each other. As I will discuss below, however, the shell form of snails in the laboratory controls (no-crab treatment) did differ in some cases from those of reference shells collected from the field. Because these changes in shell form were in the opposite direction in some cases from those in the two crab treatments, laboratory conditions may have heightened the differences observed among experimental groups. Nonetheless, with one exception (lip thickness of the exposed-shore phenotype), the greatest departure of shell form from that of reference snails collected from the field occurred in one of the two risk treatments (fish-crab or snail-crab).

The scent of crabs alone clearly influences shell form in an adaptive manner in *N. lapillus*. Com-pared to controls (no crab), snails of both phenotypes in the fish-crab treatment exhibited a significantly greater expression of shell traits that would reduce vulnerability to shell-breaking crabs (Hughes & Elner, 1979; Palmer, 1985a; Seed, 1978; Vermeij, 1978): apertural tooth height (+200%, +370% for exposed- and protected-shore phenotypes respectively; Fig. 11), shell weight relative to length [+16% and +4% (not significant for protected); Figs. 12a], shell weight relative to body weight (+54% and +21%; Figs. 12b), and relative lip thickness (+31% and +26%; Fig. 13). The response of exposed-shore snails provided the most convincing evidence for this. These snails grew substantially in both the no-crab and fish-crab treatments. The increase in length of more than 50% (Fig. 10a) and the approximate tripling in wet body weight (Fig. 9) transformed them from immature juveniles to the size of mature adults over the course of the experiment. Hence the differences in final shell form reflected to a very large extent differences in new shell added. In other words, these differences were not diminished very much by the similarity of original juvenile shells at the beginning of the experiment. By the same reasoning, however, the differences between the no-crab and fish-crab treatments almost certainly lead to an underestimate of the potential morphological response of the protected-shore snails. A much greater fraction of their shell at the end of the experiment was already present at the beginning because they did not grow nearly as much (Figs. 9, 10a, b).

Because of their reduced growth, the morphological differences observed between the snail-crab and no-crab treatments must also be interpreted with some caution. Even though provided with food *ad libitum*, the body weight of snails either did not change (exposed-shore) or actually decreased (protected-shore; Fig. 9), and not much new shell material was added (Fig. 10a, b). As a consequence, for example, the increased shell weight for a given wet body weight in protected-shore snails in the snail-crab treatment (Fig. 12b) was partly an artifact because they lost body weight (Fig. 9). This nearly 30% increase in relative shell weight, however, exceeded the 14%

176

loss in body weight experienced by these snails and even when scaled by shell length they exhibited an increase in shell weight and lip thickness relative to controls (Figs. 12a, 13). Here again, even though antipredatory traits of both phenotypes in this treatment developed to the same or greater extent than in the fish-crab treatment (except for apertural tooth height of exposed-shore snails, see supernormal stimulus below), these increases probably also underestimate the full impact of the scent of damaged conspecifics on shell morphology.

*Effect of perceived risk on shell morphology of fed snails: direction of change from field samples*

Although the final shell form of snails in both the fish-crab and snail-crab treatments differed from that of the controls (no crab), these differences could have resulted from either a) an amplification of antipredatory traits in the experimental groups or b) a reduced expression of antipredatory traits in the controls. For example, if the control groups produced much less well defended shells, as a product of holding these intertidal snails continuously immersed in the laboratory with superabundant food, then the greater development of antipredatory traits in the experimental groups would be an illusion. Thus, to assess the direction of change, the shells of laboratory raised snails must be compared to those collected directly from the field. Alternatively, when measuring phenotypic plasticity (see plasticity below), the range of phenotypes expressed under a particular range of conditions is the variable of interest, not the direction of change.

The shells of protected-shore snails did not change significantly when held under control conditions in the laboratory (no crab). At the end of the experiment, neither apertural teeth (Fig. 11), shell weight at a given length (Fig. 12a), nor lip thickness (Fig. 13) differed from field-collected reference shells [the decrease in shell weight at a given body weight (Fig. 12b) resulted from an increase in the size of the snail relative to the

habitable volume of the shell (see retractability below)]. Both laboratory crab treatments thus resulted in shells that were more well defended than those of protected-shore snails from the field.

In contrast, the shells of exposed-shore snails did change when held under control conditions in the laboratory. Furthermore, the direction of morphological change depended upon which reference shells were used for comparison, those collected at the beginning of the experiment (Exposed I) or those collected at the end (Exposed II). Relative to shells collected at the beginning of the experiment (Exposed I), those produced by snails in the no-crab treatment had larger apertural teeth [Fig. 11; most likely because they had nearly reached maturity by the end of the experiment (see Fig. 10a)] and were considerably thinner at the lip (Fig. 13). The shells were also heavier for a given length (Fig. 12a), although relative to body weight they did not differ from those initial reference shells (Fig. 12b). Hence, with the exception of lip thickness, both laboratory crab treatments also appeared to result in shells that were more well defended than those of exposed-shore snails from the field. Note, however, that these changes were of the same magnitude that occurred naturally at the exposed-shore site over the duration of the experiment (compare Exposed I to Exposed II reference values, Figs. 12a, b). Unfortunately, I cannot be sure whether lip thickness increased or decreased relative to initial values for two reasons: a) lip thickness was not measured for the initial reference sample (Exposed I) and the shells were subsequently destroyed while developing shell weight and body weight calibrations, and b) the initial snails used in the experiments did not span a large enough size range to allow me to extrapolate with any confidence to shells of larger size.

In conclusion, the amplification of antipredatory traits in both crab treatments do appear to represent changes in an adaptive direction; they were not an artifact of reduced expression of antipredatory traits in control snails.

## Effect of perceived risk on retractability

The changes observed in unoccupied volume of the shell, a measure of the degree to which a snail may retract into its shell, were rather intriguing and suggest another way in which snails may reduce their vulnerability to predation by crabs. Considering only snails which exhibited significant growth over the duration of the experiment (no-crab and fish-crab treatments), the fraction of the internal volume of the shell actually occupied by body tissue appeared to vary in an adaptive manner. For example, the unoccupied volume of the shell of protected-shore *N. lapillus* decreased in the absence of crabs but did not change in their presence (fish-crab, Fig. 12c, d) even though both groups grew (Fig. 9). In contrast, the unoccupied volume of the shell of exposed-shore snails decreased in the absence of crabs but increased significantly in their presence (fish-crab, Fig. 12c, d). Hence, even though body weight increased more than 150% in this latter group (Fig. 9), the snails expanded the internal volume of their shell much more than required to accommodate the increase in body size. As a consequence, they would have been able to retract further into their shell to avoid apertural probing by predatory crabs.

Of interest, these changes paralleled those observed between field-collected snails of both phenotypes. Dogwhelks from the protected shore, where the risk of crab predation was presumably higher, were able to withdraw further into their shells than those from the exposed shore (Fig. 5).

## Apertural tooth development in Nucella lapillus compared to N. lamellosa

The development of apertural teeth in fed *N. lapillus* differed in only one notable respect from that of starved snails. When provided with food, snails of the protected-shore phenotype in the snail-crab treatment developed the largest teeth by more than a factor of two compared to the fish-crab treatment (Fig. 11). When starved, no difference was observed between these two treatments. Even though snails of the protected-shore phenotype lost weight (Fig. 9), these data suggest that the availability of at least some food was essential to the development of moderate-sized apertural teeth. Note that snails in this treatment developed teeth that were at least twice as large as the largest teeth produced by any other group for this phenotype whether provided with food or not (compare Fig. 11 with Fig. 7c). Rather curiously, for the exposed-shore population, the pattern of development of apertural teeth by fed snails did not differ from that of starved snails either qualitatively or quantitatively (fish-crab > no crab $\approx$ snail-crab; Figs. 9, 11). Hence, the availability of food appeared to have no effect on the capacity of exposed-shore snails to produce teeth.

The development of apertural teeth in both starved and fed *N. lapillus* paralleled rather closely that observed for *N. lamellosa* under similar experimental conditions (Appleton & Palmer, 1988). Among starved snails of both phenotypes the largest teeth were developed in the fish-crab treatment, whereas among fed snails the largest teeth were developed in the snail-crab treatment. The one notable difference between these species occurred in fed, exposed-shore snails in the snail crab treatment. *N. lamellosa* under these conditions produced the largest apertural teeth whereas the teeth produced by *N. lapillus* under these conditions did not differ from those in the fed controls (Fig. 11).

Several observations suggest that some of the morphological responses of exposed-shore *N. lapillus* were anomalous. For example, the qualitative responses of both phenotypes in the fish-crab treatment were very similar (Table 9). In addition, in the snail-crab treatment the responses of the protected-shore phenotype was qualitatively similar to those of both phenotypes in the fish-crab treatment. Hence, where snails in the remaining three groups exhibited increases in the development of antipredatory traits compared to controls, exposed-shore snails in the snail-crab treatment exhibited no change or decreases. With the present data, however, I am not sure how to account for the seemingly anomalous response of

*Table 9.* Qualitative summary of the effects of three risk treatments on various aspects of the biology of *Nucella lapillus*. Retractability – ability of snail to withdraw into the shell (measured as the unoccupied volume of shell). Entries in each column indicate the magnitude of the change compared to controls (no crab): (0) < 10% difference compared to controls, ( − ) 10–25% less than controls, ( − − ) 25–50% less, ( − − − ) > 50% less, ( + ) 10–25% greater than controls, ( + + ) 25–50% greater ( + + + ) > 50% greater.

| Variable exhibiting response | Fish-crab | | Snail-crab | |
|---|---|---|---|---|
| | Exposed | Protected | Exposed | Protected |
| Starved snails | | | | |
| Initial daily rate of shell deposition (Fig. 6) | − | − | − − | − − − |
| Total body weight loss (Fig. 7a) | 0 | + + | 0 | + + |
| Total shell weight gain (Fig. 7b) | + | + + | − − − | + |
| Apertural tooth development Fig. 7c | + + + | + + +[†] | − − − | + + +[†] |
| Change in lip thickness (Fig. 7d) | + + + | + + + | − − | + |
| Fed snails | | | | |
| Rate of feeding (Fig. 8) | − − | − − | − − − | − − − |
| Rate of body growth (Fig. 9) | − − | − − | − − − | − − − |
| Apertural tooth development (Fig. 11) | + + + | + + +[†] | 0 | + + +[†] |
| Relative shell weight (Fig. 12a) | + | 0 | + | 0 |
| Relative retractability (Fig. 12c) | + | + | 0 | + +[§] |
| Relative lip thickness (Fig. 13) | + + | + + | + + | + + |

[†] Teeth in no-crab treatment not significantly different from zero.

[§] Snails lost weight.

this group. Perhaps, because the snails fed to crabs in the snail-crab treatment were collected from the exposed-shore site, experimental snails from this site were more sensitive to the stimuli than those from the protected-shore site.

*Phenotypic plasticity in* Nucella lapillus *compared to* N. lamellosa

In the experiments with both *N. lamellosa* and *N. lapillus*, populations exhibiting two quite different shell forms were examined for phenotypic plasticity. The patterns of variation exhibited by these species suggests that the relative plasticities of the two phenotypes differ between species. For *N. lamellosa*, the range of development of apertural teeth among experimental groups was similar for both phenotypes (Appleton & Palmer, 1988), although the range in relative shell weight among these groups was greater in snails from the protected shore (Palmer, unpublished). In

*N. lapillus*, on the other hand, exposed-shore snails exhibited a broader range of final shell form than those from the protected-shore for nearly all traits examined, regardless of whether snails were starved or fed. In addition, the protected-shore population was the only one to exhibit changes in shell form over time in the field (Figs. 4b, c). Hence, although reasons exist for believing that morphological change in one direction may be more likely than in another (Palumbi, 1984; Etter, 1988), the evidence from dogwhelks suggests no simple generalization will emerge about which populations retain a greater capacity to modify their shells ecophenotypically.

Apparent differences in phenotypic plasticity between the populations of these two species must be interpreted with caution because the experiments were started with individuals that had already spent one or more years in the field. The shells that they developed while in the field may thus have limited their ability to respond in the laboratory. To be sure that genetically based differences in phenotypic plasticity exist among populations, snails with a common history would have to be compared.

*Temporal morphological change in natural populations*

The increase in relative shell weight observed between July 10 and Oct. 14 in exposed-shore *N. lapillus* collected from the field (Fig. 4, Table 2) was unexpected. Without additional data, little can be said with confidence about the reasons for this variation. Two points, however, do seem worth mentioning. First, this temporal variation was unlikely to have been due to bias during collection: a) the snails were collected from the identical location on both occasions, and b) many of the data points for the July collection lie well outside the scatter of those for October (e.g. Fig. 4c). Second, no change of average shell weight was observed for the protected-shore population over this same time interval (Fig. 4, Table 2). Furthermore the difference in the degree of variation over time in the two field populations was consistent with their response to experimental conditions in the laboratory: exposed-shore snails appeared to be more phenotypically plastic than those from the protected shore (e.g. Figs. 11, 14). I suspect that the increase in average shell weight at the exposed-shore site may be a normal seasonal occurrence associated with a decline in the rate of growth over the course of the summer and into the fall, because shell weight tends to increase with decreasing rate of growth (Wellington & Kuris, 1983; Vermeij, 1980).

*Scaling out size differences in gastropod morphometrics*

Because of the substantial variation in shape, and the partial ability of body size to change independent of the shell, the procedures used to scale out differences in 'size' in species of gastropods whose shell varies extensively are problematical, although they are hardly unique to the morphometrics of gastropod shells. Shell length is probably the most commonly used index of size because of its convenience. Shell length, however, is a very poor predictor of wet body weight for *N. lapillus* with shells of different shape (e.g. see Fig. 4a). Snails from protected shores have a relatively longer apex than those from exposed shores, hence shell length substantially underestimates body weight. This contrasts with observations on *N. lamellosa* of the northeastern Pacific for which shell length can be an accurate predictor of body weight for populations having shells of different thickness (Palmer, 1985a). In addition, because of their different shapes, shells of protected- and exposed-shore *N. lapillus* have nearly the same weight for a given length (Fig. 4c). Hence, although one might argue that shell length is therefore a more accurate predictor of shell weight, this gives a very misleading impression about the amount of shell material committed to defense, and also about the ability of the shell to resist predation by shell breaking crabs (Currey & Hughes, 1982; Hughes & Elner, 1979), since the amount of shell per unit body weight is substantially different between these populations (Fig. 4b). When comparing differences in the commitment of resources to defense, body weight would seem to be a more relevant variable for scaling size than shell length.

For other comparisons, however, body weight may yield a misleading impression about differences in shell form. For example, body weight may either increase or decrease without any change in shell weight or dimensions. More invidiously, the fraction of the habitable volume of the shell actually occupied by animal tissue may itself vary in an adaptive manner (see retractability above). Given the potentially confounding effects of independent variation in body weight, shell length would seem to be a better measure of size for shells of similar shape.

Perhaps the search for an idealized descriptor of 'size' is unwarranted. After all, normally one is interested in the variation of one trait compared to that in another. Hence the choice of the trait by which to scale size will depend upon the question being asked. Where the energetics of defensive morphologies or the relation of defensive morphologies to life history are concerned, body weight may be the most relevant basis for comparison. On the other hand, if one is interested in the geometric distribution of shell material shell length may be a more relevant descriptor of size. Perhaps the safest tactic is to explore the variation in the trait of interest with respect to more than one descriptor of size. In this manner, potential biases associated with any single descriptor should become apparent.

*Was the scent of damaged conspecifics a supernormal stimulus?*

The dramatic inhibition of feeding and growth observed in the snail-crab treatment (Figs. 9, 10), and the concomitant effects of reduced growth on shell morphology and on estimates of phenotypic plasticity, deserve some additional comment. In the experiments described above, and in those done previously with *N. lamellosa* (Appleton & Palmer, 1988), no attempt was made to control the experimental stimuli quantitatively. The experimental groups were exposed to stimuli of arbitrary intensity to examine the presence/absence of morphological responses. If the stimulus I provided in the laboratory far exceeded any that these snails would normally encounter, this could account for two observations: a) the responses of starved dogwhelks in the snail-crab treatment were often less than in the fish-crab treatment (Fig. 7), and b) dogwhelks in the snail-crab treatment either did not grow or actually lost weight even in the presence of abundant food (Fig. 9). The number of snails eaten by crabs in the experimental aquaria, for example, were 2.9 and 3.7 snails $crab^{-1}$ $day^{-1}$ (5.8 and 7.5 snails $treatment^{-1}$ $day^{-1}$). I think it rather unlikely that individual *N. lapillus* in the field are ever exposed to the scent of conspecifics being eaten continuously at this rate by crabs for this long a period of time.

A priori, I would have expected the scent of conspecifics being eaten by crabs, which should be a more reliable predictor of increased risk to predation, to have elicited a greater morphological response than the scent of crabs alone, since the scent of crabs in the vicinity but not feeding on snails would seem to pose a lessor risk. At the very least, because both contained the same level of crab stimulus, the behavioral and morphological responses of snails in the snail-crab treatment should have been the same as those in the fish-crab treatment. Compared to the fish-crab treatments, several aspects suggest that the stimulus level in the snail-crab treatments was unnaturally high: a) among starved snails, the reduced development of apertural teeth, the smaller change in shell weight and the smaller increase in lip thickness (Fig. 7b–d), and b) among snails provided with abundant food, the lack of body growth or actual loss in body weight (Fig. 9). To verify this conjecture, snails would have to be held in the presence of crabs being fed conspecific snails at different rates.

*Adaptive behavioral and morphological variation in* Nucella lapillus

Despite the complexities of some of the patterns, two important conclusions may be drawn from the above experiments. First, both the scent of crabs and the scent of damaged conspecifics dramatically reduce the rates of feeding and growth of *N. lapillus*. Second, these stimuli can also amplify the development of several different antipredatory traits. Hence the effects of risk-related chemical cues in the environment must be considered when interpreting variation in behavior and shell morphology among natural populations of gastropods.

Although numerous gastropods flee from slow-moving predators such as starfish and other predatory gastropods (Snyder & Snyder, 1971; Vermeij, 1978, 1987), examples of escape responses to more rapidly moving predators such as crabs and fishes are rare. Nonetheless, the fact that such behaviors do occur (Geller, 1982) indicates that gastropods are capable of detecting chemical cues even from highly mobile predators. The reduced rates of feeding and growth in the fish-crab compared to the no-crab treatment reveal that, as observed for *N. lamellosa* from the northeastern Pacific (Appleton & Palmer, 1988), *N. lapillus* can also recognize and respond adaptively to the scent of predatory crabs.

The greater suppression of feeding activity of dogwhelks in the snail-crab compared to the fish-crab treatments indicates that *N. lapillus* also have an alarm response. Other marine and freshwater gastropods exhibit alarm responses to the scent of damaged conspecifics (Snyder, 1967; Atema & Stenzler, 1977; Stenzler & Atema, 1977) and these responses appear to be adaptive (Ashkenas & Atema, 1978; Hadlock, 1980). Both the reduced rates of feeding, and the tendency for snails to remain about the lower margins and undersides of stones in the snail-crab treatments, would reduce the probability that individual *N. lapillus* were encountered by foraging crabs. Hence the alarm response in *N. lapillus* also appears to be adaptive.

Perhaps the most surprising aspect of this study, and that of Appleton & Palmer (1988), is that gastropods can also respond morphologically to chemical cues released by predators and damaged conspecific snails. Such morphological responses are rare among solitary organisms (Harvell, 1986, but see Liveley, 1986). Furthermore, the morphological responses are in an adaptive direction – in the presence of these stimuli, *N. lapillus* produced heavier shells with thicker lips and more well-developed apertural teeth. All of these traits reduce the vulnerability to attack by shell-breaking crabs (Kitching *et al.*, 1966; Hughes & Elner, 1979; Palmer, 1985a; Vermeij, 1987). Because these environmental stimuli can have a significant effect on shell morphology, interpretations of morphological differences observed among natural populations must be done with caution. Such differences may result from either genetic or environmental effects, or some combination of the two (e.g. Janson, 1982; Palmer, 1985b). Thus, for example, the changes in shell morphology observed in *Littorina obtusata* (Seeley, 1986) and *N. lapillus* (Vermeij, 1982) following the introduction of *Carcinus maenas* may not reflect microevolutionary change, since such changes could equally likely have been an eco-phenotypic response to the scent of crabs.

## Acknowledgements

I thank Roger Hughes for generously agreeing to host my sabbatical visit to Wales and both he and John Davenport for assistance in numerous ways during my stay. Gary Vermeij and Lois Hammond offered useful comments on various drafts of the manuscript. This research was supported by a Natural Sciences and Engineering Research Council of Canada travel grant T7684, and by operating grant A7245 whose sustained funding I acknowledge with gratitude.

## References

Appleton, R. D. & A. R. Palmer, 1988. Water-borne stimuli released by predatory crabs and damaged prey induce more predator-resistant shells in a marine gastropod. Proc. nat Acad. Sci. U.S.A. 85: 4387–4391.

Ashkenas, L. & J. Atema, 1978. A salt marsh predator-prey relationship: attack behavior of *Carcinus maenas* (L.) and defenses of *Ilyanassa obsoletus* (Say). Biol. Bull. 155: 426.

Atema, J. & D. Stenzler, 1977. Alarm substance of the marine mud snail, *Nassarius obsoletus*: biological characterization and possible evolution. J. chem. Ecol. 3: 173–187.

Cambridge, P. G. & J. A. Kitching, 1982. Shell shape in living and fossil (Norwich Crab) *Nucella lapillus* (L.) in relation to habitat. J. Conch. Lond. 31: 31–38.

Colton, H. S., 1916. On some varieties of *Thais lapillus* in the Mount Desert Island region. A study of individual ecology Proc. Acad. nat. Sci., Philad. 68: 440–454.

Colton, H. S., 1922. Variation in the dog whelk, *Thais lapillus*. Ecology 3: 146–157.

Crothers, J. H., 1971. Further observations on the occurrence of 'teeth' in the dog-whelk *Nucella lapillus*. J. mar. biol. Ass. UK. 51: 623–639.

Crothers, J. H., 1985. Dog-whelks: An introduction to the biology of *Nucella lapillus* (L.). Field Studies 6: 291–360.

Currey, J. D. & R. N. Hughes, 1982. Strength of the dog-whelk *Nucella lapillus* and the winkle *Littorina littorea* from different habitats. J. anim. Ecol. 51: 47–56.

Etter, R. J., 1988. Asymmetrical developmental plasticity in an intertidal snail. Evolution 42: 322–334.

Geller, J. B., 1982. Chemically mediated avoidance response of a gastropod, *Tegula funebralis* (A. Adams); to a predatory crab, *Cancer antennarius* (Stimpson). J. exp. mar. Biol. Ecol. 65: 19–27.

Hadlock, R. P., 1980. Alarm response of the intertidal snail *Littorina littorea* (L.) to predation by the crab *Carcinus maenas* (L.). Biol. Bull. 159: 269–279.

Hartley, O., 1962. Analysis of variance. In: Mathematical Methods for Digital Computers, Vol. 1. Ralston, A. & H. S. Wilf (eds). Wiley, New York. pp. 221–230.

Harvell, C. D., 1986. The ecology and evolution of inducible defenses in a marine bryozoan: cues, costs, and consequences. Am. Nat. 128: 810–823.

Hughes, R. N. & R. W. Elner, 1979. Tactics of a predator, *Carcinus maenas*, and morphological responses of the prey, *Nucella lapillus*. J. anim. Ecol. 48: 65–78.

Janson, K., 1982. Genetic and environmental effects on the growth rate of *Littorina saxatilis*. Mar. Biol. 69: 73–78.

Kilburn, R. & E. Rippey, 1982. Sea Shells of Southern Africa. Macmillan South Africa, Johannesburg, 249 pp.

Kincaid, T., 1957. Local races and clines in the marine gastropod *Thais lamellosa*, a population study. Calliostoma Co., Seattle, 75 pp, 65 Plates.

Kincaid, T., 1964. Notes on *Thais (Nucella) lima* (Gmelin), a marine gastropod inhabiting areas in the North Pacific Ocean. Calliostoma Co., Seattle.

Kitching, J. A., 1976. Distribution and changes in shell form of *Thais* spp. (Gastropoda) near Bamfield, B.C. J. exp. mar. Biol. Ecol. 23: 109–126.

Kitching, J. A. & J. Lockwood, 1974. Observations on shell form and its ecological significance in Thaisid gastropods of the genus *Lepsiella* in New Zealand. Mar. Biol. 28: 131–144.

Kitching, J. A., L. Muntz & F. J. Ebling, 1966. The ecology of Lough Ine, XV. The ecological significance of shell and body forms in *Nucella*. J. anim. Ecol. 35: 113–126.

Largen, M. J., 1971. Genetic and environmental influences upon the expression of shell sculpture in the dog whelk (*Nucella lapillus*). Proc. malac. Soc. Lond. 39: 383–388.

Lively, C., 1986. Predator induced shell polymorphism in the acorn barnacle *Chthamalus anisopoma*. Evolution 40: 232–242.

Milinski, M. & R. Heller, 1978. Influence of a predator on the optimal foraging behaviour of sticklebacks (*Gasterosteus aculeatus* L.). Nature 275: 642–644.

Moore, P. G., 1985. Shell shape in living and fossil ('25ft' beach) dog-whelks, *Nucella lapillus* (L.), from the Isle of Cumbrae, Scotland. Glasg. Nat. 21: 81–91.

Palmer, A. R., 1981. De carbonate skeletons limit the rate of body growth? Nature 292: 150–152.

Palmer, A. R., 1982. Growth in marine gastropods: a non-destructive technique for independently measuring shell and body weight. Malacologia 23: 63–73.

Palmer, A. R., 1985a. Adaptive value of shell variation in *Thais* (or *Nucella*) *lamellosa*: effect of thick shells on vulnerability to and preference by crabs. Veliger 27: 349–356.

Palmer, A. R., 1985b. Quantum changes in gastropod shell morphology need not reflect speciation. Evolution 39: 699–705.

Palumbi, S. R., 1984. Tactics of acclimation: morphological changes of sponges in an unpredictable environment. Science 225: 1478–1480.

Phillips, B. F., N. A. Campbell & B. R. Wilson,, 1973. A multivariate study of geographic variation in the whelk *Dicathais*. J. exp. mar. Biol. Ecol. 11: 27–69.

Seed, R., 1978. Observations on the significance of shell shape and body form in the dogwhelk (*Nucella lapillus* (L.)) from North Wales. Nature in Wales 16: 111–122.

Seeley, R. H., 1986. Intense natural selection caused a rapid morphological transition in a living marine snail. Proc. nat. Acad. Sci., USA 83: 6897–6901.

Sih, A., 1980. Optimal behavior: Can foragers balance two conflicting demands? Science 210: 1041–1043.

Sih, A., 1984. The behavioral response race between predator and prey. Am. Nat. 123: 143–150.

Snyder, N. F. R., 1967. An alarm of aquatic gastropods to intraspecific extract. New York Agric. Experiment Sta., Ithaca, Mem. 403: 1–122.

Snyder, N. F. R. & H. A. Snyder, 1971. Defenses of the Florida apple snail, *Pomacea paludosa*. Behaviour 40: 175–215.

Sokal, R. R. & F. J. Rohlf, 1981. Biometry, W. H. Freemen and Co., San Francisco.

Spight, T. M., 1973. Ontogeny, environment, and shape of a marine snail, *Thais lamellosa* (Gmelin). J. exp. mar. Biol. Ecol. 13: 215–228.

Stenzler, D. & J. Atema, 1977. Alarm response of the marine mud snail, *Nassarius obsoletus*: specificity and behavioral priority. J. chem. Ecol. 3: 159–171.

Vermeij, G. J., 1978. Biogeography and Adaptation. Patterns of Marine Life. Harvard University Press, Cambridge. 332 pp.

Vermeij, G. J., 1980. Gastropod shell growth rate, allometry and adult size: environmental implications. In: Rhoads, D. C. & R. A. Lutz, (eds.) Skeletal Growth of Aquatic Organisms. Plenum Press, New York: 379–394.

Vermeij, G. J., 1982. Phenotypic evolution in a poorly dispersing snail after arrival of a predator. Nature 299: 349–350.

Vermeij, G. J., 1987. Evolution and Escalation. An Ecological History of Life. Princeton University Press, Princeton.

Wellington, G. M. & A. M. Kuris, 1983. Growth and shell variation in the tropical eastern Pacific intertidal gastropod genus *Purpura*: ecological and evolutionary implications. Biol. Bull. 164: 518–535.

*Hydrobiologia* **193**: 183–190, 1990.
*K. Johannesson, D. G. Raffaelli and C. J. Hannaford Ellis (eds), Progress in Littorinid and Muricid Biology.*
© 1990 *Kluwer Academic Publishers.*

# The relationship between flat periwinkle life histories and digenean infections

Gray A. Williams [1] & T. J. Brailsford [2]
*Department of Zoology, University of Bristol, Woodland Road, Bristol BS8 1UG, England, UK; Present addresses:* [1] *Department of Botany, University of Hong Kong, Pokfulam Road, Hong Kong;* [2] *Experimental Parasitology Unit, Department of Zoology, University of Nottingham, University Park, Nottingham NG7 2RD, England, UK*

*Key words: Littorina obtusata, Littorina mariae,* parasite, trematode

## Abstract

Larval digenean parasites were studied in *Littorina obtusata* (L.) and *L. mariae* Sacchi & Rastelli at Sawdern Point in West Wales. Shell parameters, ovipositor and shell colour, penis morphology and sex ratios were scored, and the influence of parasitism studied. A total of 7 species of parasites were found, although the prevalence was very low in *L. mariae*, especially in the females. The parasitic gigantism that has been described in other species of gastropod was not found in this study. Parasitic castration does occur in some infected male *L. obtusata*, resulting in severe stunting of the penis. However, this phenomena was never observed in *L. mariae*. The winkle species are congeneric and inhabit broadly similar niches on the shore; their life histories are however quite different – one being annual and the other perennial. This probably affects exposure to infection, and might explain the differing prevalence of the parasites.

## Introduction

Many aspects of the ecology of *Littorina obtusata* (L.) and *Littorina mariae* Sacchi & Rastelli are quite similar – both species reproduce by laying benthic egg capsules and live their entire life histories as mobile epiphytes on fucoid algae (Sacchi & Rastelli, 1966; Sacchi, 1967; Reimchen, 1974; Goodwin, 1975). Recent work has shown that the two species in fact inhabit different niches on sheltered shores (Williams, 1987). *L. obtusata* is found at all shore levels but reaches a peak of abundance in the mid shore on

*Ascophyllum nodosum* (L.) Le Jol. On sheltered shores the principal colour morph is *olivacea* (green) and adults reach a size of 15–17 mm. *L. obtusata* is principally a macro-algal grazer and excavates the thallus of *Ascophyllum*. In contrast *L. mariae* is found exclusively at low shore levels on *Fucus serratus* (L.) and primarily feeds by browsing on micro-epiphytes (Williams, 1987; Watson & Norton, 1987). *L. mariae* reaches an adult size of 10–12 mm and its principal colour morph is *citrina* (yellow).

The life histories of the two species are quite different (Williams, 1987). In *L. obtusata* egg mass production reaches a peak in early spring, with the eggs hatching over the following months. The winkles live for three or more years, reaching maturity in the second year when adult size is attained. Thus populations of *L. obtusata* are made up of three distinct components: an adult component, an immature component (found between the summer of one year and the summer of the next year) and a recently hatched component (found between the spring and the late summer or early winter, after which they join the immature component). This pattern has been convincingly documented for a number of sites (Guiterman, 1970; Goodwin, 1978; Williams, 1987).

*Littorina mariae* exhibits a very different life cycle. Egg masses are laid and hatch at a similar time to those of *L. obtusata* but the adult population at the time of hatching is sparse. There is a large input of newly hatched winkles which grow throughout the summer and reach adult size by the late summer or early winter of the same year. The number which reach maturity is small, due to the heavy mortality of the immature winkles. This pattern has been recorded at one study site, Sawdern Point in West Wales (Williams, 1987), the only site at which the population dynamics of *L. mariae* have been investigated.

The work described in this paper was a pilot study designed to assess the feasibility of using parasitological data as tool to investigate life history variation between the two species. It was speculated that the difference in the life histories of the winkles would be reflected in their parasite burdens, and that this might be used to confirm the observations made on the ecology of the winkles at Sawdern. The hypothesis tested was that *L. mariae* would have a reduced parasite load as compared to *L. obtusata*, due to the differences in their life histories. The importance of parasites in the ecology of winkles is well documented (Lauckner, 1987). The most commonly found parasites are larval digenea which exploit winkles as intermediate hosts. Although there is some variation in detail, a typical digenean life cycle may be described as follows. The adult parasites produce eggs in the faeces of their definitive host

(in the case of winkles they are mostly birds). These eggs hatch to produce the free-living miracidial stage which then infect the first intermediate host. This is usually some species of gastropod, such as the winkles under study. There then follows an asexually multiplicative stage which results in the production of large numbers of free living cercariae. These are shed from the host and they then seek out a second intermediate host (which might be an invertebrate; e.g. mollusc, crustacean or a fish) in which they form a metacercarial cyst. The definitive host then acquires the parasite by ingesting an infected second intermediate host. The exact details of some stages of the life cycles of most of the parasites described in this paper are not known.

What is known about the effects of parasitism in winkles is largely based on work that has been conducted on *Littorina littorea* (L.). The effects of parasitism have been cited as the cause of gigantism and castration in the winkle. Parasitism has also been suggested to affect winkle behaviour (Williams & Ellis, 1975). A number of workers have investigated digenea in *L. obtusata* and *L. mariae* (as *L. littoralis*) and James (1968) has provided a useful key for the species found. Guiterman (1970) found a similar infection incidence between the sexes in *L. obtusata* but noted a variable seasonal pattern in infection rates (probably as a result of the parasite life history and variation between the sites he studied). He suggested that infection resulted in a stimulation of growth rates in the winkles based on the fact that the infection rate was higher for the largest snails. A similar effect was noted by Goodwin (1975), with infected snails being larger than uninfected snails. Goodwin also recorded a seasonal variation in infection. In contrast to what was found by Guiterman, Goodwin noted a tendency for males to have a higher infection rate than females and this was especially true for *L. mariae*. The degree of infection was suggested to be related to the exposure of the shore; exposed shores had a lower level of infection because sheltered shores were preferred by gulls, which are the definitive hosts of many of these parasites. Both authors noted a regression of the penis and reproductive glands, the degree of which was associated with the age at which the animal was infected (i.e. before or after maturity).

The importance of parasitism on the host at the individual, population and species level is often ignored by marine ecologists (Lauckner, 1987). In this preliminary study the prevalence of parasites in two species of littorinids was examined to investigate any differences which might be associated with the difference noted in the niches of the two species, especially their life histories.

## Materials and methods

All the investigations were carried out at Sawdern Point in West Wales (Ordnance Survey Grid reference SM 888032), a sheltered rocky shore. The ecologies of the two winkle species were studied over a three year period and are described in detail elsewhere (Williams, 1987). The parasitological investigation was conducted during August, 1986.

Random collections of 200 adult winkles of each species (as denoted by thickening of the aperture lip, Goodwin, 1975) were made at the respective tidal levels occupied by the two species: mid shore for *L. obtusata* and low shore for *L. mariae*. All the collections were made at the same tidal height as designated by fixed bolts on the shore (Williams, 1987), and the collected animals were taken to the laboratory at Orielton Field Studies Centre. The subsequent work can be divided into the investigation of external and internal features of the winkles.

### External features

The wet shells of the winkles were scored for colour (using the scheme of Dautzenberg & Fischer, 1914 as revised by Reimchen, 1974). These were scored as green (*olivacea*), yellow (*citrina*) or reticulated (*reticulata*). To investigate variation in shell morphology, shell shape was scored using parameters 'a' (maximum length); 'b' (aperture length) and 'c' (maximum height), as

described by Goodwin & Fish (1977) after Colman (1932) (Fig. 1). All these measurements were made to an accuracy of 0.05 mm using vernier calipers. The blotted wet weights of the winkles were also recorded using a portable Ni-Cd balance which was accurate to 0.02 g (A & D Electronic Balances model EW-60B). Winkles were then placed in magnesium chloride solution to narcotize them before cracking the shells to examine the internal features.

### Internal features

The winkles were removed from their shells and their sex recorded. In the females the presence or absence of pigmentation on the ovipositor was scored. If the animal was male the penis was removed and measured. The length of the glandular area 'g' and tip 't' were drawn using a *camera lucida* and measured using a digitizing pad. The number of adhesive glands was also recorded. These measurements were used to estimate variations in penial size and morphology.

The digestive glands of the winkles were then dissected out and the colour, which varied from yellow through olive green to black, was recorded.

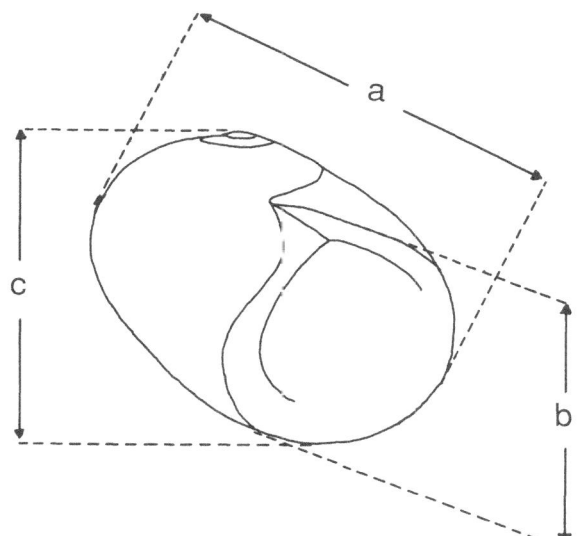

*Fig. 1.* Shell parameters measured in this study (after Goodwin & Fish, 1977).

The glands were squashed and the presence or absence of parasitic infection was noted. If the gland was infected the parasite was identified to species level using the key of James (1968). No attempt was made to assess the levels of infection for any particular snail.

## Statistical analysis

Variations in the shell morphology of the two species were compared using multivariate analysis. The data were log transformed to attain maximum separation of the groups (Janson & Sundberg, 1983). Infected and uninfected animals were compared for both species and sexes using multiple discriminant analysis and canonical variate analysis using the SPSS[X] statistical package. All the other analyses undertaken used standard statistical methods and are described below.

## Results

### Occurrence of the parasites

A total of seven species of digenea were found in *L. obtusata* and *L. mariae*. These were *Microphallus similis* (Jägerskiold, 1900), *Cercaria parvicaudata* Stunkard & Shaw, 1931; *Cercaria lebouri* Stunkard, 1932; *Cercaria buccini* Lebour, 1911; *Cryptocotyle lingua* (Creplin, 1825), *Notocotyloides petasatum* (Deslongchamps, 1824) and *Cercaria littorinae obtusatae* Lebour, 1911. The prevalence of each species is shown in Table 1. The total prevalence of all the species found was 15.5% in *L. obtusata* and 5.5% in *L. mariae*. This difference between the species was found to be highly significant according to the $z$ test for instances ($z = 3.02$, $P < 0.002$).

The sex ratio of *L. obtusata* was found to be 1 : 1, but in *L. mariae* it was 1 : 2 (males : females). A higher prevalence of parasites was found in males than in females for both species of host. In *L. obtusata* a total of 18.2% of the males were infected but only 12.9% of the females, although

*Table 1.* Prevalence of larval digenea in *L. obtusata* and *L. mariae*.

| Digenean species | Prevalence in *L. obtusata* (%) | Prevalence in *L. mariae* (%) |
|---|---|---|
| *Microphallus similis* | 11.0 | 2.5 |
| *Cercaria parvicaudata* | 2.5 | 1 |
| *Cercaria lebouri* | 0.5 | 0 |
| *Cercaria buccini* | 0.5 | 0.5 |
| *Cryptocotyle lingua* | 0.5 | 0 |
| *Notocotyloides petasatum* | 0.5 | 1 |
| *Cercaria littorinae obtusatae* | 0 | 0.5 |

this difference was not significant ($\chi^2 = 1.077$, $P = 0.25$). In *L. mariae*, however, the difference between the sexes was greater, with 11.9% of the male winkles being infected and only 2.3% of the females. In this case the difference was highly significant ($\chi^2 = 8.04$, $P < 0.005$).

The colour of the digestive gland was found to vary from black to orange, but in parasitized animals of both species there was a tendency for it to be more yellow than in normal winkles. The colour of the shell and pigmentation of the ovipositor was not found to be associated with infection and was constant between the two species. 97% of the *L. mariae* population had yellow (*citrina*) coloured shells and 98% of *L. mariae* females had unpigmented ovipositors. The principal shell colour in the *L. obtusata* population was green (*olivacea*) which accounted for 99% of the winkles examined. 99% of *L. obtusata* females had pigmented ovipositors. These interspecific differences are similar to those recorded by Goodwin & Fish (1977) and Williams (1987).

### Shell morphology

The results obtained from the shell measurements and weight of the winkles are shown in Table 2. There were some differences between the sexes of both species, similar to those described elsewhere (Goodwin & Fish, 1977; Williams, 1987). There were no obvious differences between the infected and uninfected winkles of either species. When

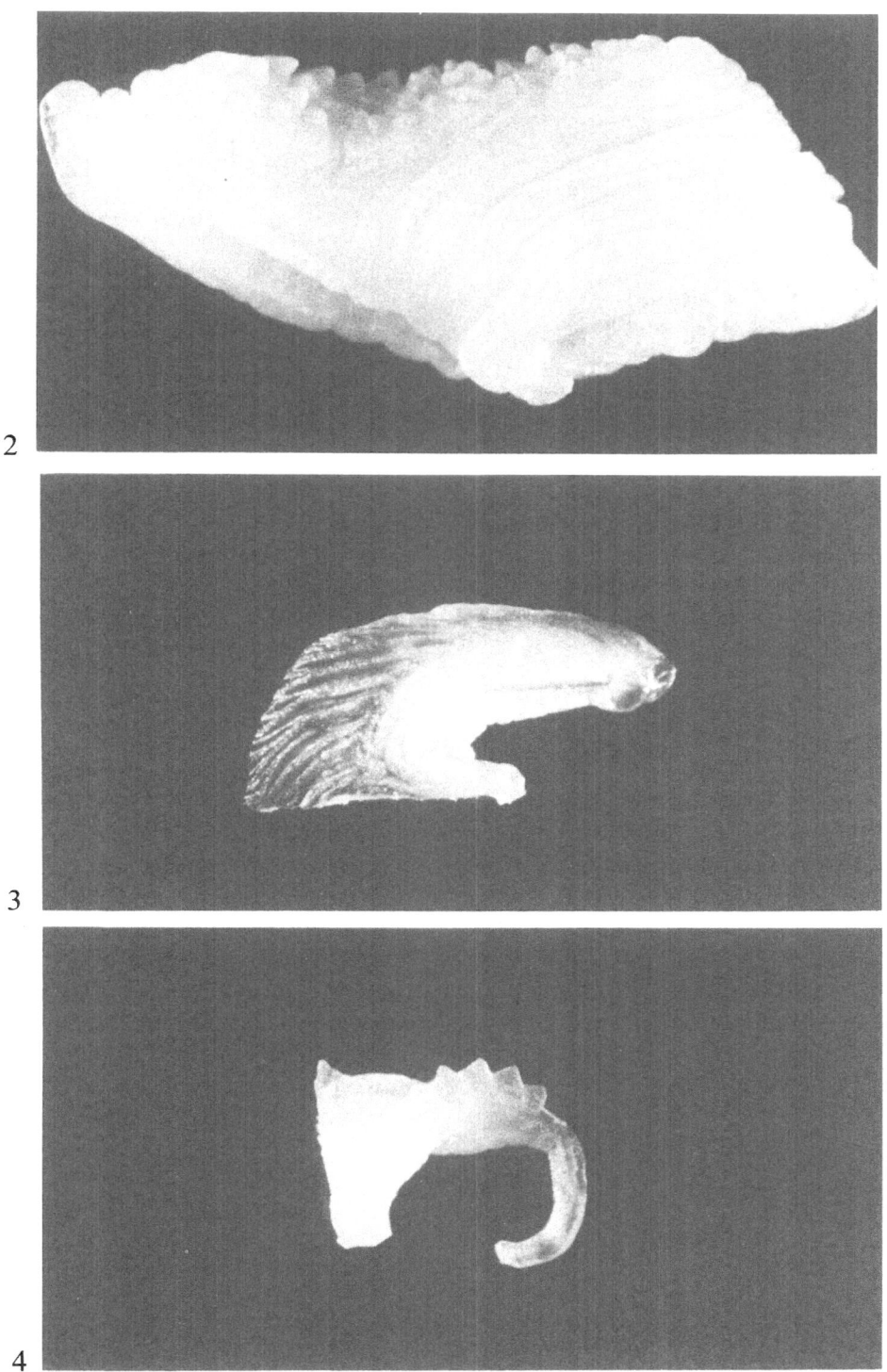

*Fig. 2.* Normal penis of *L. obtusata.*
*Fig. 3.* Stunted penis of parasitized *L. obtusata.*
*Fig. 4.* Penis of *L. mariae.* Photographs all to the same scale, for mean values see Table 3.

*Table 2.* Weight and shell measurements of *L. obtusata* and *L. mariae* (mean ± SD).

| Species | Sex | Infected | Number | Weight (g) | a (mm) | b (mm) | c (mm) |
|---------|-----|----------|--------|------------|--------|--------|--------|
| *L. obtusata* | M | No | 81 | 1.5 ± 0.2 | 16.0 ± 0.6 | 11.2 ± 0.6 | 14.9 ± 0.8 |
| *L. obtusata* | M | Yes | 18 | 1.6 ± 0.2 | 16.2 ± 0.9 | 11.4 ± 0.7 | 15.4 ± 1.0 |
| *L. obtusata* | F | No | 88 | 1.5 ± 0.2 | 16.0 ± 0.6 | 11.1 ± 0.6 | 14.8 ± 0.7 |
| *L. obtusata* | F | Yes | 13 | 1.6 ± 0.2 | 16.1 ± 1.0 | 11.1 ± 0.6 | 15.2 ± 1.2 |
| *L. mariae* | M | No | 59 | 0.3 ± 0.1 | 9.1 ± 1.0 | 7.3 ± 0.7 | 7.9 ± 0.8 |
| *L. mariae* | M | Yes | 8 | 0.4 ± 0.2 | 9.8 ± 1.3 | 7.8 ± 0.9 | 8.8 ± 1.3 |
| *L. mariae* | F | No | 130 | 0.5 ± 0.1 | 10.0 ± 1.1 | 8.0 ± 0.6 | 8.8 ± 0.9 |
| *L. mariae* | F | Yes | 3 | 0.6 ± 0.2 | 10.5 ± 1.3 | 7.8 ± 0.3 | 9.1 ± 0.8 |

the infected and uninfected individuals of each sex and species were analysed using multivariate statistics no separation of these groups was found using either raw or transformed data. Some winkles showing the classical deformities of 'gigantism' (large shells; high spire) were found, but in most cases these were not parasitized.

## Penis morphology

The measurements made of the penes of the winkles are given in Table 3. There was no difference in the number of glands present in the penes of infected and uninfected animals for either species (*L. obtusata*: Students $t = 0.62$, $P = 0.54$; *L. mariae*: Students $t = 0.13$, $P = 0.90$). However, in *L. obtusata* the total length of the penis was significantly reduced in infected animals (Students $t = 3.97$, $P = 0.0008$), but this was not the case in *L. mariae* (Students $t = 1.71$, $P = 0.11$). This is the result of severe penis

stunting which occurred in 38.9% of infected male *L. obtusata*, but was not found in uninfected winkles or in any *L. mariae* (Figs. 2–4). This effect was found to occur in 41.7% of *M. similis* infections, and was also recorded in winkles infected with *C. lebouri* and *C. buccini*.

## Discussion

In both *L. obtusata* and *L. mariae* the prevalence of digenea was much lower than that typically recorded in *L. littorea*, where prevalences of over 90% have been reported (James, 1968). It is possible that this is a reflection of the fact that at low tide *L. littorea* are found predominantly in rockpools, whereas *L. obtusata* and *L. mariae* remain upon the algae on which they feed. The exposure to miracidia is probably higher in rock pools as faecal material from the definitive hosts is likely to be more concentrated than in the open sea. In populations of *L. obtusata* from N. Wales,

*Table 3.* Penis measurements for *L. obtusata* and *L. mariae* (mean ± SD).

| Species | Infected | Number | No. of glands | t (mm) | g (mm) | Total length (mm) |
|---------|----------|--------|---------------|--------|--------|-------------------|
| *L. obtusata* | No | 81 | 36.10 ± 8.95 | 1.31 ± 0.33 | 5.21 ± 0.93 | 6.52 ± 1.05 |
| *L. obtusata* | Yes | 18 | 34.67 ± 8.88 | 1.08 ± 0.29 | 3.71 ± 1.66 | 4.79 ± 1.78 |
| *L. mariae* | No | 59 | 10.18 ± 1.96 | 1.57 ± 0.55 | 2.83 ± 0.58 | 4.39 ± 0.95 |
| *L. mariae* | Yes | 8 | 10.25 ± 1.49 | 1.27 ± 0.15 | 2.69 ± 0.65 | 3.96 ± 0.61 |

Guiterman (1970) recorded an infection incidence of 18%, which is very similar to the value found in this study. This is, however, lower than that recorded for *L. obtusata* in the White Sea, USSR, where the prevalence of parasites in the population is often above 20% and may be greater than 50% (Sergievsky, 1985). This may be due to the increased longevity of *L. obtusata* in this area which has been recorded as living for 8–10 years (A. Granovitsch, pers. comm.).

The prevalence of digenea in *L. obtusata* is much greater than in *L. mariae*. This is not due to the different tidal levels that the two species occupy as Guiterman (1970) has shown that *L. obtusata* has a greater parasite burden at low water than at mid or high water. It would therefore follow that *L. mariae* should have a higher burden than *L. obtusata* at this level, which it does not. This can however be explained by differences in exposure to infection that occur as a result of the differing life histories. *L. mariae* is an annual at this site, and thus individuals are not present on the shore for as long as *L. obtusata*. The age of the *L. mariae* used in this study was estimated at between six and eight months, whereas the *L. obtusata* were at least one or two years old. It is thus not surprising that *L. obtusata* accumulated a higher parasite burden than did *L. mariae*. This accumulation is likely to be reinforced by seasonal variations in exposure to miracidia, which is at a maximum in the summer (Guiterman, 1970). *L. mariae* were all collected in their first summer, whereas the *L. obtusata* had experienced one or more summers previously. Many authors have noticed an increase in infection with increase in body size (which is then related to age) (Robson & Williams, 1970; Hughes & Answer, 1982). Initial infection has been associated with the winkles first reproduction, taking place after the host is spent following spawning (Robson & Williams, 1971). This agrees with the results obtained in this study. *L. obtusata*, being at least one year old, will have spawned previously and would thus be more susceptible to infection than *L. mariae* which, being younger, have yet to reproduce.

The differing prevalence of parasites between the sexes is interesting, because the exposure of both sexes would be expected to be identical. The only likely explanations for this are that the male *L. mariae* are either more susceptible than the female, or else it is possible that they could have greater longevity. There is no evidence to support the latter hypothesis however, and it is likely that there is a genuine difference between the sexes in their susceptibility to the parasites. Little is known about phenomena of this type in molluscs.

Surprisingly a study of the morphometric data made it apparent that the so called 'parasitic gigantism' that is known to occur in populations of *L. obtusata* and *L. mariae* does not appear to be related to any current digenean infection, and indeed no evidence was found for any parasitic aetiology of this deformity at all. It is, however, by no means impossible that gigantism is the result of infections that occurred at an earlier time and that the burden has been subsequently lost.

Parasitic castration is a common phenomenon among invertebrates, and in the case of winkles it results in a severe stunting of the penis (Cheng, 1967). At this site, however, stunting of the penis was only found to occur in *L. obtusata*, and was never observed in *L. mariae* or uninfected *L. obtusata*. It probably can occur as a result of infection by any of the species of parasite that were found in *L. obtusata*. However, due to the small sample size, and low prevalence it is impossible to be certain about whether it can be caused by infections of *Cercaria parvicaudata* or *Cryptocotyle lingua*. Stunting of the penis only occurs in less than half of the digenean infections found in *L. obtusata*, probably because it is a developmental defect caused by infection at an early age. Animals infected after they have become sexually mature are unlikely to be affected.

The complete absence of stunted penes in *L. mariae* is interesting, and might be due to differences in the pathogenesis of the parasites due to physiological differences between the two species. This phenomenon must have an important influence upon the population dynamics of the host, as it probably renders the affected animals incapable of successful breeding, as has been shown to be the case with *L. littorea* (see Robson

& Williams, 1971). Thus in *L. obtusata* about 8 % of the adult population were effectively sterilised by digenea. However, there was obviously no such effect upon the *L. mariae* population. The results, therefore, show a significant difference in the parasite loading of *L. obtusata* and *L. mariae*. This disparity is thought to be attributable to the differences in the host winkle life histories. *L. obtusata* living for three times as long as *L. mariae* suffers greater exposure to infection and consequently has a higher prevalence of infection. The annual *L. mariae* is less heavily parasitized as a result of its shorter exposure to infectious miracidia. It is concluded that the differences in the winkles life histories are reflected in their respective parasite loads and this will influence the niche parameters and ecologies of the two species.

## Acknowledgements

This work was funded by a Natural Environment Research Council grant to G.A. W, and it would not have been possible without the support of Dr and Mrs R Crump and the staff at Orielton Field Studies Centre, to whom we are most grateful. We would also like to thank C. Little and C.J. Mapes for improving earlier drafts of the manuscript and D. Raffaelli, K. Johannesson and two anonymous referees for revising later versions.

## References

Cheng, T. C., 1967. Marine molluscs as hosts for symbioses, with a review of known parasites of commercially important species. Adv. Mar. Biol. 5: 424 pp.

Colman, J., 1932. A statistical test of the species concept in *Littorina*. Biol. Bull. 62: 223–243.

Dautzenberg, P. & H. Fischer, 1914. Etude sur le *Littorina obtusata* et ses variations. J. Conch. 62: 87–128.

Goodwin, B. J., 1975. Studies on the biology of *Littorina obtusata* and *L. mariae* (Mollusca: Gastropoda). Ph.D. thesis, University College of Wales, Aberystwyth.

Goodwin, B. J., 1978. The growth and breeding cycle of *Littorina obtusata* (Gastropoda: Prosobranchiata) from Cardigan Bay. J. moll. Stud. 44: 231–242.

Goodwin, B. J. & J. D. Fish, 1977. Inter- and intraspecific variation in *Littorina obtusata* and *L. mariae* (Gastropoda: Prosobranchiata). J. moll. Stud. 43: 241–254.

Guiterman, J. D., 1970. The population biology of *Littorina obtusata* (Gastropoda: Prosobranchiata). Ph.D. Thesis, University College of North Wales, Bangor.

Hughes, R. N. & P. Answer, 1982. Growth, spawning and trematode infection of *Littorina littorea* (L.) from an exposed shore in North Wales. J. moll. Stud. 48: 321–320.

Janson, K. & P. Sundberg, 1983. Multivariate morphometric analysis of two varieties of *Littorina saxatilis* from the Swedish west coast. Mar. Biol. 74: 49–53.

James, B. L., 1968. The distribution and keys of species in the family Littorinidae and of their digenean parasites, in the region of Dale, Pembrokeshire. Field Studies. 2: 615–650.

Lauckner, G., 1987. Ecological effects of larval trematode infestation on littoral marine invertebrate populations. Int. J. Parasit. 17: 391–398.

Lubchenco, J., 1980. Algal zonation in the New England rocky intertidal community: An experimental analysis. Ecology. 6: 333–344.

Reimchen, T. E., 1974. Studies on the biology and colour polymorphism of two sibling species of marine gastropod (*Littorina*). PhD thesis, University of Liverpool.

Robson, E. M. & I. C. Williams, 1970. Relationships of some species of Digenea with the marine prosobranch *Littorina littorea* (L.). I. The occurrence of larval Digenea in *L. littorea* on the North Yorkshire coast. J. Helminth. 44: 153–168.

Robson, E. M. & I. C. Williams, 1971. Relationships of some species of Digenea with the marine prosobranch *Littorina littorea* (L.). II. The effect of larval Digenea on the reproductive biology of *L. littorea*. J. Helminth. 45: 145–159.

Sacchi, C. F., 1967. Variabilità ed ambiente nella coppia di specie intertidali *Littorina obtusata* (L.). e *Littorina mariae* Sacchi e Rastelli (Gastropoda, Prosobranchia) a Concarneau (Bretagna meridionale). Studia ghisleriana. 3: 339–355.

Sacchi, C. F. & M. L. Rastelli, 1966. *Littorina mariae*, nov. sp.: les differences morphologiques et écologiques entre 'nains' et 'normaux' chez l''espèce' *L. obtusata* (L.) (Gastr. Prosobr.) et leur signification adaptive et évolutive. Atti Soc. ital. Sci. nat. 105: 351–370.

Sergievsky, S. O., 1985. Populational approach to the analysis of the periwinkle *Littorina obtusata* (L.) invasions with the trematode parthenitae. Helminthologia. 22: 5–14.

Watson, D. C. & T. A. Norton, 1987. The habitat and feeding preferences of *Littorina obtusata* (L.) and *L. mariae* Sacchi et Rastelli. J. exp. mar. Biol. Ecol. 112: 61–72.

Williams, G. A., 1987. Niche partitioning in *Littorina obtusata* and *L. mariae*. PhD thesis, University of Bristol.

Williams, I. C. & C. Ellis, 1975. Movements of the common periwinkle *Littorina littorea* (L.) on the Yorkshire coast in winter and the influence of infection with larval digenea. J. exp. mar. Biol. Ecol. 17: 47–58.

*Hydrobiologia* **193**: 191–198, 1990.
*K. Johannesson, D. G. Raffaelli and C. J. Hannaford Ellis (eds), Progress in Littorinid and Muricid Biology.*
© 1990 *Kluwer Academic Publishers.*

# Feeding behaviour in *Littorina littorea*: a study of the effects of ingestive conditioning and previous dietary history on food preference and rates of consumption

D.W. Imrie[1], C.R. McCrohan[2] & S.J. Hawkins[3]
[1] *Department of Environmental Biology,* [2] *Department of Physiological Sciences, University of Manchester, Manchester M13 9PL, England, UK;* [3] *Department of Marine Biology, University of Liverpool, Port Erin, Isle of Man, UK*

*Key words:* feeding, preference, learning, phagostimulant, intertidal

## Abstract

Feeding responses of the generalist herbivore, *Littorina littorea* (L.), to the perceived 'taste' of macroalgae were assessed with respect to the effects of recent dietary intake and to overlapping versus non-overlapping distributions of winkles and algae. The extent of grazing on artificial substrates impregnated with crude algal extracts was used as a measure of rate of response to the odour of preferred algae, and of feeding preference among less preferred algae, in a variety of designs. Adult *L. littorea* collected from a site where a range of algae were present showed preference among extracts of fucoids, whereas adults from a nearby site showed no such preference. Juvenile *L. littorea* of two weight cohorts collected from the former site responded faster to *Porphyra umbilicalis* extract-containing substrate than similar-sized animals from the latter site. Juveniles, fed either *Porphyra*, *Ulva lactuca*, or starved for two weeks in the laboratory, responded similarly to *Ulva* versus *Porphyra* extracts in a dose-dependent manner across a range of concentrations, although the *Porphyra*-maintained group consumed more of each, and the starved group less over seven days. Juveniles maintained on a mixed diet of *Ulva* and *Porphyra* consumed more *Porphyra* extract and less *Ulva* extract over the same period. These results are discussed in relation to the possible role of ingestive conditioning and previous dietary history in determining the occurrence and extent of chemically-mediated feeding preference in *L. littorea*.

## Introduction

Among the order Gastropoda, the underlying mechanisms of feeding behaviour of opistho-branchs and pulmonates have been intensively studied. The simplicity of their nervous systems and their large and individually identifiable neurones have led to the widespread use of these groups in neuroethological studies, for which their stereotyped, rhythmic and dependable feeding behaviour is particularly suitable (Benjamin, 1983). The factors which interact to determine the feeding behaviour of prosobranchs are by comparison poorly understood, despite the fact that the literature concerning their diets is extensive (e.g. Steneck & Watling, 1982). This is especially true of herbivorous intertidal prosobranchs, whose influence as consumers is such that they often determine the composition of the communities in which they occur (Hawkins & Hartnoll, 1983; Petraitis, 1987). This influence is believed to be a product of the grazers' ability to recognise potential prey items as well as their assimilation capability. Studies of feeding in herbivorous intertidal prosobranchs have tended to concentrate on the latter factor (Steneck & Watling, 1982; Padilla, 1984, 1985), whilst assuming that those factors which determine preference at the pre-ingestive stages of feeding – the so-called kairomonic influence exerted by odours of prey items (Thomas, 1982) – are less relevant due to the apparent superabundance of food. This argument, first put forward by Bovbjerg (1965, 1968) to explain the inability of lymnaeid snails to detect algae, has however been challenged in the case of selective grazers such as *Aplysia* spp. (Audesirk, 1975).

Foraging efficiency may be enhanced by preference for an abundant, nutritious food source. For relatively nonselective grazers, though, sampling new food items (which could at some future point constitute the only available source of food in a fluctuating environment) may also be adaptive. The herbivorous prosobranch *Littorina littorea* (L.), for example, is willing to ingest a wide range of macrophytes (Watson & Norton, 1985), even though microalgae probably constitute the bulk of this species' normal diet.

The tenets of optimal foraging theory (as defined by Hughes 1980) suggest that factors such as experience and memory may influence the response to a given prey item. The concepts of switching of food preference and olfactory-based search images formed by consumers have been discussed in relation to the feeding behaviour of opisthobranchs (Hall *et al.*, 1982, 1984), pulmonates (Croll & Chase, 1980) and predatory prosobranchs (Williams *et al.*, 1983); there is evidence that recent dietary experience may alter the overall feeding response in these groups.

This paper presents observations and experiments which collectively explore the possibility of plasticity in the feeding behaviour of *L. littorea*, and specifically the effects of previous dietary experience and ingestive conditioning on the food preference of this generalist grazer.

## Materials and methods

### Preparation of algal extracts

Crude extracts of a range of seaweeds collected from sites in North Wales were prepared in a manner similar to that used in the determination of chlorophyll content of algae (Parsons *et al.*, 1984). 1 g (tissue-damped wet weight) of each alga was finely chopped and then ground with mortar and pestle, using carborundum powder and 1 ml filtered seawater, for a few minutes. The resultant slurry was diluted with filtered seawater to 10 ml volume, then homogenised for ten minutes. The homogenate was centrifuged at 1000 rpm for ten minutes and the supernatant collected. This was the fraction used in all experiments involving the use of artificial feeding substrates.

Extracts of the ephemeral red alga, *Porphyra umbilicalis*, the ephemeral green alga, *Ulva lactuca*, and the fucoids, *Fucus serratus*, *Fucus vesiculosus* and *Pelvetia canaliculata*, were prepared using the above method. *Porphyra* and *Ulva* extracts were diluted with filtered seawater to produce solutions containing 13%, 25%, 50%, and 100% of the original extract concentration.

Filtered seawater was used as a blank control in all experiments.

### Assessment of effect of collection site on responses to algal extracts

Adult (> 20 mm) and juvenile (< 10 mm) *L. littorea* were collected from two adjacent sites at Rhos-on-sea, Gwynedd. The first site ('breakwater') consisted of a large boulder mound upon which grew several common algal species, notably fucoids, green algae and the foliose red alga, *Porphyra umbilicalis*. The second site ('musselbed'), less than 20 m distant but separated from the first by sand and a freshwater runoff, was a flat rocky outcrop virtually concealed by the barnacle, *Semibalanus balanoides* and the mussel, *Mytilus edulis* (L.). No macrophytes were present at this site, due presumably to competition for space.

Adults were allowed to acclimate in static seawater tanks at 13 °C for 21 days, fed *Ulva lactuca* for the first seven days. Two groups of animals (shell length 22 mm) were then presented with a choice of three fucoid extracts and a blank control, each of which was incorporated into one of four circles (diameter 50 mm) cut into a 0.25 mm thick layer of a cellulose/alginate mixture, coated on glass plates. The cellulose base acted as a carrying medium for the extracts, which were retained within the cellulose/alginate matrix after air-drying. Scoring the substrate prevents the spread of extracts beyond the defined test areas. Neither cellulose nor alginate act as a phagostimulant for *L. littorea*.

Two plates were placed vertically in each corner of two cuboid glass tanks, measuring 300 mm by 200 mm by 220 mm, filled with filtered seawater; 16 adult *L. littorea* from each sample site were then introduced and allowed to move freely within this area. Feeding behaviour left readily apparent rasp marks on the cellulose substrate. The rank order of grazing on the treatments was determined for each plate after 24 h. The degree of preference between treatments was assessed using the non-parametric procedure devised by Meddis (1984). Details of this experimental design are presented elsewhere (Imrie *et al.*, 1989).

Juvenile *L. littorea* from the above sites were arranged according to weight into two cohorts of $0.4 \pm 0.1$ g and $1.0 \pm 0.3$ g live weight, containing 100 animals each, per site. Each cohort was starved for a week, then placed within a rectangular area at the centre of a static tank and allowed to move at will. Each tank contained two cellulose-coated glass plates containing a range of concentrations of *Porphyra* extract applied, in $5 \times 10^{-6}$ l aliquots, to seventy-five discrete 1 cm$^2$ areas per plate. These loci were demarcated by scoring the substrate as in the previous experiment. Each extract dosage was introduced to 15

loci per plate; there were a total of 30 loci per concentration per tank. Allocation of a given concentration of *Porphyra* extract to squares within the matrix of possible locations was by Latin square design, to avoid biases incurred through localized grazing intensities. The number of squares grazed (defined arbitrarily as $> 75\%$ removal of a square) was judged by eye at 24 h intervals for seven days. No measurement was made at 144 h of the smaller cohorts' grazing.

*Assessment of effect of recent dietary experience on responses to algal extracts*

Juvenile *L. littorea* (shell length 7–8 mm) were collected from rockpools devoid of macrophytes in the upper eulittoral from Rhosneigr, on Anglesey, Gwynedd, where they occurred in large numbers. 1200 winkles were divided into four groups of 300 and subdivided into lots of 15. Each subgroup was placed in a mesh bag containing 0.3 g (tissue-damped wet weight) of seaweed, except for one group (referred to hereafter as group **S**) which was starved for two weeks. Of the other groups, one (group **P**) was fed *P. umbilicalis* fronds; the second (group **U**), *U. lactuca* fronds; and the third (group **U + P**), approximately 0.15 g of each alga. Each subgroup was independently suspended in a tidal flow system operating on a 6 h immersion: 6 h emersion regime. A light regime of 8 h light: 16 h dark was imposed throughout. Ambient temperature was 13–14 °C. After 14 days all seaweed remnants were removed from the mesh bags and the animals left in the tidal system for a further 24 h to evacuate gut contents.

Pairs of subgroups were pooled, and each of the ten resultant replicate factions containing 30 animals from each pretreatment group was sealed in a clean mesh bag containing a glass plate prepared as for the previous experiment. In half of the plates, 1 cm$^2$ squares scored in the cellulose substrates were treated with concentrations of *Porphyra* extract arranged in Latin square matrices as described above; the other half differed only in the use of *Ulva* extract instead of *Porphyra* extract.

Experiments were conducted in a constant temperature room at 13 °C, on a 8 h light: 16 h dark cycle. The plates were arranged horizontally in a series of static seawater tanks. The number of treatment sites grazed ($>75\%$ removal) was estimated by eye after seven days continuous immersion.

### Assessment of effect of maintenance diet on long-term survivorship

Preparatory to an unrelated experiment, several hundred juvenile *L. littorea* from Rhosneigr were maintained on intact *Ulva*, *Porphyra*, or both seaweeds, or starved as in the previous experiment. Animals were grouped in tens and suspended with their food source in mesh bags in a tidal system on a 6 h immersed: 6 h emersed cycle, and a 12 h light: 12 h dark cycle. The numbers of animals alive or dead were counted after 60 days.

## Results

### Effect of collection site on algal extract preference

When presented with a choice of three wrack species' extracts, adult animals from the breakwater site show statistically significant preferences (nonspecific test, $P < 0.05$; *post hoc* specific test, *P. canaliculata* = *F. serratus* > *F. vesiculosus* = control: $P < 0.05$). Animals from the musselbed site, however, would not have shown statistically significant preference even if we had adopted an hypothesis, formulated *a priori*, predicting the same rank order of preference as expressed by breakwater animals ($P > 0.05$). Preference rankings obtained from the two groups are similar, however (specific test: respective ranks of each treatment are the same between collection sites; $P < 0.001$), indicating that the degree of preference is reduced in the latter group but the relative preference rankings are the same.

### Effect of collection site on response rate to a preferred algal extract

Juvenile *L. littorea* of both cohorts from the breakwater site were found to have consumed more of the total substrate available than similar-sized juveniles from the musselbed site, at any given time (Fig. 1). Differences in cumulative frequency of each treatment's grazed sites became significant within 48 h for the 1.0 g cohort (G-test of association, null hypothesis: ratio of grazed to ungrazed squares is homogeneous, within treatment concentrations, between groups of animals from different sample sites; 100%, 50%, 25% & 13%, $P < 0.001$; control, $P < 0.01$). Among the 0.4 g cohort, the grazing rates of juveniles from

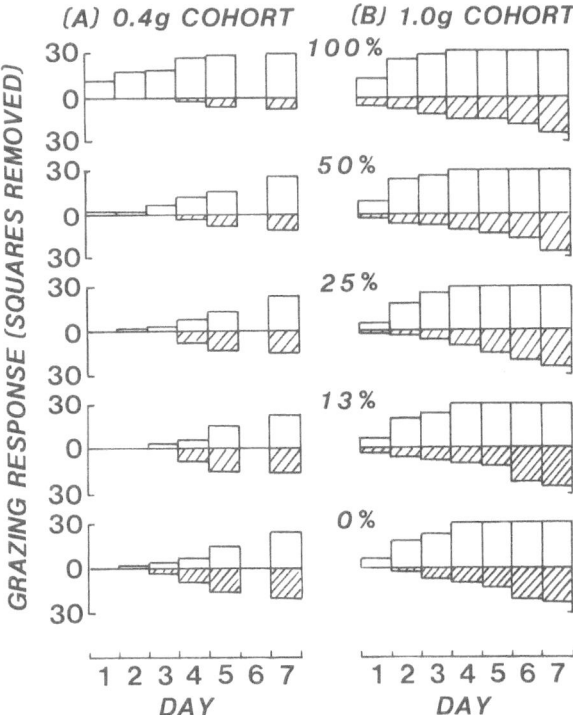

*Fig. 1.* Consumption rates of a range of *P. umbilicalis* extract concentrations impregnated into artificial substrates, by juvenile *L. littorea* in two size cohorts, from two sample sites. The percentage figures refer to the concentration of an extract applied to squares with respect to the original extract strength (1 g seaweed equivalent per 10 ml). The shaded areas indicate removal of squares by animals from the musselbed site; the unshaded areas indicate removal by animals from the breakwater site.

the two collection sites on squares impregnated with the highest *Porphyra* extract's concentration differed significantly after 24 h (G-test, $P < 0.001$) but grazing on the next highest treatment concentration did not differ significantly until day 7.

*Effect of recent dietary experience on resposes to algal extracts*

The numbers of substrate squares removed by each maintainance diet group, of both types and every concentration of extract, are listed in Table 1. Due to the replication within treatment and pretreatment groups, it was possible to fit Model I regression lines to the grazing response data (Fig. 2). Grazing response (transformed by square root to normalise frequency data) showed positive regression on treatment concentration ($P < 0.05$) in group **U + P**, **U** and **S**. Group **P**'s response was consistently high and did not vary greatly with treatment dose. Analysis of covariance showed that responses to both *Porphyra* and *Ulva* extract concentration matrices contrasted mainly in that the elevations of the dose-response

*Table 1.* Removal of treatment squares by juvenile *L. littorea* maintained on different diets. Abbreviations: UPU = group **U + P**, presented with *Ulva* extract matrix; UPP = group **U + P**, *Porphyra* extract; PU = group **P**, *Ulva* extract; PP = group **P**, *Porphyra* extract; UU = group **U**, *Ulva* extract; UP = group **U**, *Porphyra* extract; SU = group **S**, *Ulva* extract; SP = group **S**, *Porphyra* extract.

| % Conc^n | No. of squares removed | | | | | | | |
|---|---|---|---|---|---|---|---|---|
| | UPU | UPP | PU | PP | UU | UP | SU | SP |
| 100 | 6 | 15 | 15 | 15 | 8 | 13 | 10 | 8 |
| | 9 | 15 | 7 | 13 | 9 | 11 | 8 | 9 |
| | 6 | 14 | 12 | 12 | 15 | 15 | 8 | 12 |
| | 4 | 13 | 11 | 9 | 9 | 11 | 9 | 5 |
| | 5 | 15 | 13 | 14 | 9 | 13 | 10 | 11 |
| 50 | 5 | 15 | 15 | 15 | 8 | 9 | 8 | 8 |
| | 5 | 15 | 8 | 11 | 9 | 10 | 6 | 8 |
| | 5 | 12 | 12 | 11 | 8 | 14 | 7 | 12 |
| | 3 | 11 | 10 | 8 | 7 | 12 | 7 | 6 |
| | 5 | 15 | 12 | 14 | 9 | 10 | 8 | 10 |
| 25 | 2 | 15 | 15 | 15 | 5 | 7 | 7 | 7 |
| | 6 | 15 | 7 | 10 | 9 | 6 | 6 | 7 |
| | 5 | 9 | 11 | 10 | 8 | 8 | 7 | 13 |
| | 2 | 8 | 10 | 10 | 7 | 8 | 7 | 4 |
| | 5 | 13 | 13 | 12 | 6 | 7 | 8 | 7 |
| 13 | 2 | 11 | 15 | 15 | 4 | 5 | 7 | 6 |
| | 8 | 15 | 7 | 8 | 8 | 6 | 4 | 6 |
| | 3 | 9 | 13 | 9 | 7 | 6 | 7 | 10 |
| | 3 | 8 | 11 | 6 | 4 | 7 | 6 | 3 |
| | 4 | 8 | 13 | 12 | 4 | 6 | 7 | 6 |
| 0 | 1 | 10 | 14 | 14 | 4 | 4 | 5 | 7 |
| | 3 | 13 | 6 | 7 | 5 | 5 | 3 | 5 |
| | 3 | 8 | 10 | 10 | 6 | 5 | 5 | 10 |
| | 1 | 8 | 9 | 7 | 4 | 6 | 4 | 2 |
| | 2 | 7 | 11 | 10 | 7 | 5 | 5 | 5 |

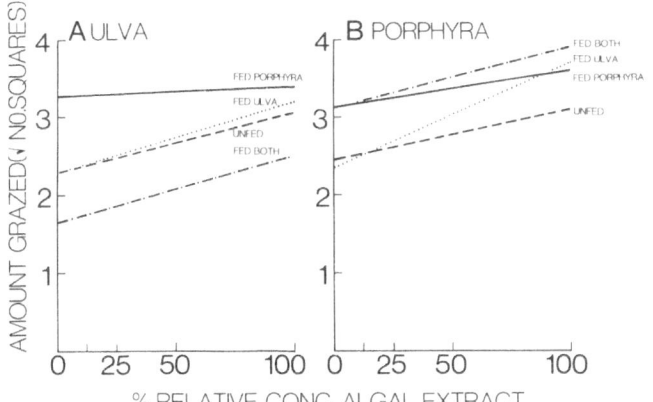

*Fig. 2.* Model I regression lines fitted to square-root-transformed data from Table 1. Fig. 2A shows the grazing levels of groups of juvenile *L. littorea*, maintained for 14 days on a variety of dietary combinations, when presented with a range of concentrations of *Ulva lactuca* extract (suspended in 1 cm² squares of artificial substrate) for seven days. Fig. 2B shows the grazing levels of identical groups presented with *Porphyra umbilicalis* extract over the same period.

lines differed between pretreatment groups ($P < 0.01$). The slopes of the regression lines were also significantly different due to the high levels of response shown by the *Porphyra*-prefed group, regardless of treatment type or concentration. The regression coefficients of the transformed dose-response curves of animals maintained on the other diets (*Ulva/Porphyra*, *Ulva* and starved) were similar ($P > 0.05$); the functional response of their regression lines differed in elevation only.

Analysis of covariance within pretreatment groups revealed a large discrepancy in elevations

*Table 2.* Survivorship of juvenile *L. littorea* maintained on different diets for 60 days. See text for dietary abbreviations.

| Diet | Alive | Dead | Total |
|------|-------|------|-------|
| **P** | 30 | 60 | 90 |
| **U + P** | 26 | 84 | 110 |
| **U** | 6 | 154 | 160 |
| **S** | 3 | 147 | 150 |
| Total | 65 | 345 | 510 |

of the regression lines fitted to the transformed dose-response data, reflecting different grazing levels on *Ulva* versus *Porphyra* concentration matrices within group **U + P** ($P < 0.001$) and a smaller discrepancy within group **U** ($P < 0.05$). The regression coefficients of the fitted lines did not vary significantly within groups between treatments. Figure 2 shows that the difference in grazing level in the winkles fed a mixed diet has produced a rearrangement of the ranked order of responses of maintenance diet groups between *Porphyra* and *Ulva* concentration matrices. Group **U + P** grazed more on *Porphyra* matrices and less on *Ulva* matrices than any other group over the seven day period.

*Mortalities of juveniles maintained on different diets*

Survivorship levels of animals in the various dietary groups are shown in Table 2. Ratios of alive versus dead juvenile *L. littorea* after 60 d are heterogeneous among the groups provided different food sources (G-test of homogeneity:

*Table 3.* Results of simultaneous test procedure to identify homogeneous groupings within the survivorship data from Table 2. A significantly large Gh value indicates that the survivorship levels of the compared maintainance diet groups are heterogeneous. Abbreviations: *** = $P < 0.001$; ns = not significant. See text for dietary abbreviations.

| Grouping | Gh value | df | Significance |
|----------|----------|-----|--------------|
| **U & S** | 0.430 | 3 | ns |
| **U, S & U + P** | 40.050 | 3 | *** |
| **U + P & P** | 2.301 | 3 | ns |

$P < 0.001$). A simultaneous test procedure showed that proportionately more animals fed on diets containing *Porphyra* survived than those whose diet lacked *Porphyra* (Table 3).

**Discussion**

*L. littorea* has been shown to be able to discriminate between extracts of a wide range of algae (Bertness *et al.*, 1983; Imrie *et al.*, 1989), including many which are common on British shores. The ranked order of grazing on these extracts, when they are incorporated into artificial substrates of the type used in this study, correspond closely with ranked preferences between intact algae, measured in terms of arrestive effect and removal by grazing (Watson & Norton, 1985). The means by which *L. littorea* detects the presence of phagomodulatory agents in algae is not known.

The only obvious difference between the Rhos-on-sea sample sites was the presence either of an encrusting layer of barnacles and mussels, or of macroalgae. It is, therefore, tempting to ascribe the observed disparity of grazing rates on *Porphyra* extract concentration matrices to the effects of ingestive conditioning, acting in such a way as to reinforce an inherent inclination in juveniles to respond to this highly palatable extract. Likewise, the presence of a preference hierarchy among non-preferred fucoid algal extracts, expressed by adult *L. littorea* from the breakwater site but not by those from the musselbed site, could be a consequence of enhanced discrimination resulting from previous dietary experience. Instances of long-term aversive conditioning to unpalatable food items are to be found in behavioural studies involving terrestrial and freshwater gastropods (see Audesirk & Audesirk, 1986 for review). In the absence of any knowledge as to the actual diets of experimental animals prior to collection, however, no conclusive evidence can be produced to either prove or disprove either hypothesis; we wish merely to draw attention to this anomaly.

The results of the experiment in which juvenile *L. littorea* were presented with controlled diets are complex, and caution must be exerted in their interpretation. We have identified three important trends in the data, however. Firstly, maintenance diet influences the response to a given algal extract. Secondly, superimposed on this trend, the rank order of maintenance diet groups' feeding rates remain constant regardless of which extract is presented – with the exception of group U + P, which of all the groups responds to *Porphyra* extract most and to *Ulva* extract least over the seven day period, regardless of dosage. Thirdly, the response of the *Ulva*-maintained group to *Porphyra* extract is more concentration-specific than any other.

Clearly, the rank order of grazing on the extracts is not a simple function of maintenance diet; there is no simple relationship between quality of diet (including the concommitant effects of satiation level) and feeding response. Similarly, we must discard ingestive conditioning to the odour of a single prey species such as found in the land snail *Achatina fulica* (Croll & Chase, 1980) or the nudibranch *Aeolidia papillosa* (Hall *et al.*, 1982) as the sole determinant of the differences found. It appears that only simultaneous experience of *Porphyra* and *Ulva* fronds affects the relative responses to the extracts of these two species, such that preference between these algal extracts is enhanced. Previous assays to determine the chemically-mediated basis of food preference in *L. littorea* established that adult winkles, removed from a site where both of these palatable algae occurred, displayed strong preference for *Porphyra* extract (Imrie *et al.*, 1989). It is likely that the individuals used in that study had previously encountered one or both of these algae. It is impossible to devise controls for these potentially confounding factors, without rearing *L. littorea* in the laboratory; this has to our knowledge never been attempted.

The *Ulva*-prefed group's feeding response to a range of *Porphyra* extracts is harder to explain. It is possible that *Porphyra* contains a similar phagostimulant to *Ulva*'s; previous experience of this phagostimulant in *Ulva* fronds may predispose *L. littorea* to respond to it, if it occurs in higher concentrations in *Porphyra* extract.

The survivorship data, although tentative, indicate that the presence of *Porphyra* in the maintenance diet might increase the percentage survival of juvenile *L. littorea*, and suggests that *Porphyra* is inherently a superior source of nutrition to *Ulva*. In the absence of any other source of food, therefore, those individuals which, when presented with a choice between *Ulva* and *Porphyra*, consume at least some of the latter, may be more likely to survive over a two month period than those which consume only *Ulva*. Given the similar responses of 'naive' animals to *Porphyra* versus *Ulva* extracts, the response of those which have encountered both algae could be considered adaptive. Note that this differential response is in relation to algal extracts in isolation and need not necessarily entail switching *sensu* Murdoch (1969). The relative importance of edibility versus attractiveness of food items to *L. littorea* is considered elsewhere (Imrie *et al.*, 1989).

To summarize, the observations and experiments described in this paper suggest that *L. littorea* is flexible in its behavioural response to potential food items, in ways which could under certain circumstances lead to the increased likelihood of survival under suboptimal conditions. Much more detailed and controlled work in laboratory and field conditions is required to satisfactorily test the validity of this hypothesis.

## Acknowledgement

This work was funded by the SERC, UK.

## References

Audesirk, G. J. & T. E. Audesirk, 1986. Behaviour of gastropod molluscs. In A. O. D. Willows (ed.), The Mollusca, vol. 8: Neurobiology and Behaviour, Part 1. Academic Press, Lond.: 1–94.

Audesirk, T. E., 1975. Chemoreception in *Aplysia californica*. I. Behavioural localization of distance chemoreceptors used in food-finding. Behav. Biol. 15: 45–55.

Benjamin, P. R., 1983. Gastropod feeding: Behavioural and neural analysis of a complex multicomponent system. In A. Roberts and B. L. Roberts (eds), Neural Control of

Rhythmic Movements. Cambridge University Press, Lond.: 159–193.

Bertness, M. D., P. O. Yund & A. F. Brown, 1983. Snail grazing and the abundance of algal crusts on a sheltered New England rocky beach. J. exp. mar. Biol. Ecol. 71: 147–164.

Bovbjerg, R. V., 1965. Feeding and dispersal in the snail *Stagnicola reflexa* (Basomatophora: Lymnaeidae). Malacologia 2: 199–207.

Bovbjerg, R. V., 1968. Responses to food in Lymnaeid snails. Physiol. Zool. 41: 412–423.

Croll, R. P. & R. Chase, 1980. Plasticity of olfactory orientation to foods in the snail *Achatina fulica*. J. comp. Physiol. 136: 267–277.

Hall, S. J., C. D. Todd & A. D. Gordon, 1982. The influence of ingestive conditioning on the prey species selection in *Aeolidia papillosa* (Mollusca: Nudibranchia). J. anim. Ecol. 51: 907–921.

Hall, S. J., C. D. Todd & A. D. Gordon, 1984. Prey-species selection by the anemone predator *Aeolidia papillosa* (L.): the influence of ingestive conditioning and previous dietary history, and a test for switching behaviour. J. exp. mar. Biol. Ecol. 82: 11–33.

Hawkins, S. J. & R. G. Hartnoll, 1983. Grazing of intertidal algae by marine invertebrates. Oceanogr. mar. Biol. ann. Rev. 21: 195–202.

Hughes, R. N., 1980. Optimal foraging theory in the marine context. Oceanogr. mar. Biol. ann. Rev. 18: 423–481.

Imrie, D. W., S. J. Hawkins & C. R. McCrohan, 1989. The olfactory-gustatory basis of food preference in the herbivorous prosobranch *Littorina littorea* (L.). J. moll. Stud. 55: 217–225.

Meddis, R., 1984. Statistics using Ranks: a Unified Approach. Blackwell, N.Y., 449 pp.

Murdoch, W. W., 1969. Switching in general predators: experiments on predator specificity and stability of prey populations. Ecol. Monogr. 39: 335–354.

Padilla, D. K., 1984. The importance of form: differences in competitive ability, resistance to consumers and environmental stress in an assemblage of coralline algae. J. exp. mar. Biol. Ecol. 79: 105–127.

Padilla, D. K., 1985. Structural resistance of algae to herbivores. Mar. Biol. 90: 103–109.

Parsons, T. R., Y. Maita & C. M. Lalli, 1984. A Manual of Chemical and Biological Methods for Seawater Analysis. Pergamon Press, Oxford, 173 pp.

Petraitis, P. S., 1987. Factors organizing rocky intertidal communities of New England: herbivory and predation in sheltered bays. J. exp. mar. Biol. Ecol. 109: 117–136.

Steneck, R. S. & L. Watling, 1982. Feeding capabilities and limitations of herbivorous molluscs: a functional group approach. Mar. Biol. 68: 299–319.

Thomas, J. D., 1982. Chemical ecology of the snail hosts of schistasomiasis: snail-snail and snail-plant interactions. Malacologia 22: 81–91.

Watson, D. C. & T. A. Norton, 1985. Dietary preferences of the common periwinkle, Littorina littorea (L.). J. exp. mar. Biol. Ecol. 88: 193–211.

Williams, L. G., D. Rittscoff, B. Brown & M. R. Carriker, 1983. Chemotaxis of oyster drills Urosalpinx cineria to competing prey odors. Bull. mar. biol. Lab. Woods Hole 164: 536–548.

*Hydrobiologia* **193**: 199–215, 1990.
*K. Johannesson, D. G. Raffaelli and C. J. Hannaford Ellis (eds), Progress in Littorinid and Muricid Biology.*
© 1990 *Kluwer Academic Publishers.*

# Shell microstructure and mineralogy of the Littorinidae: ecological and evolutionary significance

John D. Taylor & David G. Reid
*Department of Zoology, British Museum (Natural History), Cromwell Road, London SW7 5BD, England, UK*

*Key words:* shell dissolution, calcite, aragonite, shell structure, latitudinal variation

### Abstract

An examination of the shell microstructure and mineralogy of species from 30 of the 32 genera and subgenera of the gastropod family Littorinidae shows that most species have a shell consisting of layers of aragonitic crossed-lamellar structure, with minor variations in some taxa. However, *Pellilitorina, Risellopsis* and most species of *Littorina* have partly or entirely calcitic shells. In *Pellilitorina* the shell is made entirely of calcitic crossed-foliated structure, while in the other two genera there is only an outer calcitic layer of irregular-prismatic structure. A cladistic analysis shows that the calcitic layers have been independently evolved in at least three clades. The calcite is found only in the outermost layers of the shell and in species inhabiting cooler waters of both northern and southern hemispheres. Calcium carbonate is more soluble in cold than warm water and, of the two polymorphs, calcite is about 35% less soluble than aragonite. We suggest that calcitic shell layers are an adaptation of high latitude littorinids to resist shell dissolution.

### Introduction

Calcite and aragonite are the two mineral polymorphs of calcium carbonate normally used by molluscs in the construction of their shells. Of the two forms aragonite is much the most widespread, with calcite being found in relatively few taxa. Characters of the microstructure, mineralogy and disposition of shell layers have been extensively used in phylogenetic analysis, particularly in the Bivalvia (Taylor *et al.*, 1969, 1973; Waller, 1978; Uozumi & Suzuki, 1981; Shimamoto, 1986) and Archaeogastropoda (MacClintock, 1967; Batten, 1975). Other gastropods have been little studied, mainly because in most of the Caenogastropoda the shell microstructure is relatively uniform, consisting of several layers of aragonitic crossed-lamellar structure (Bøggild, 1930; Bandel, 1979; Taylor, unpubl.). However, some species of a few families, namely the Littorinidae, Muricidae and Buccinidae, are known to possess an additional calcitic layer (Bøggild, 1930; Lowenstam, 1954; Petitjean, 1965; Togo, 1974).

In a general review of the distribution of calcite and aragonite in molluscan shells, Lowenstam (1954) used the distribution of calcite in the Littorinidae, along with examples in bivalves, as instances of the control of shell mineralogy by environmental temperature. He presented evidence that littorinid species living at higher latitudes tend to have a higher proportion of a calcite in their shells. Out of the thirteen littorinid species examined, the four which contained both calcite and aragonite lived at high latitudes, and of the

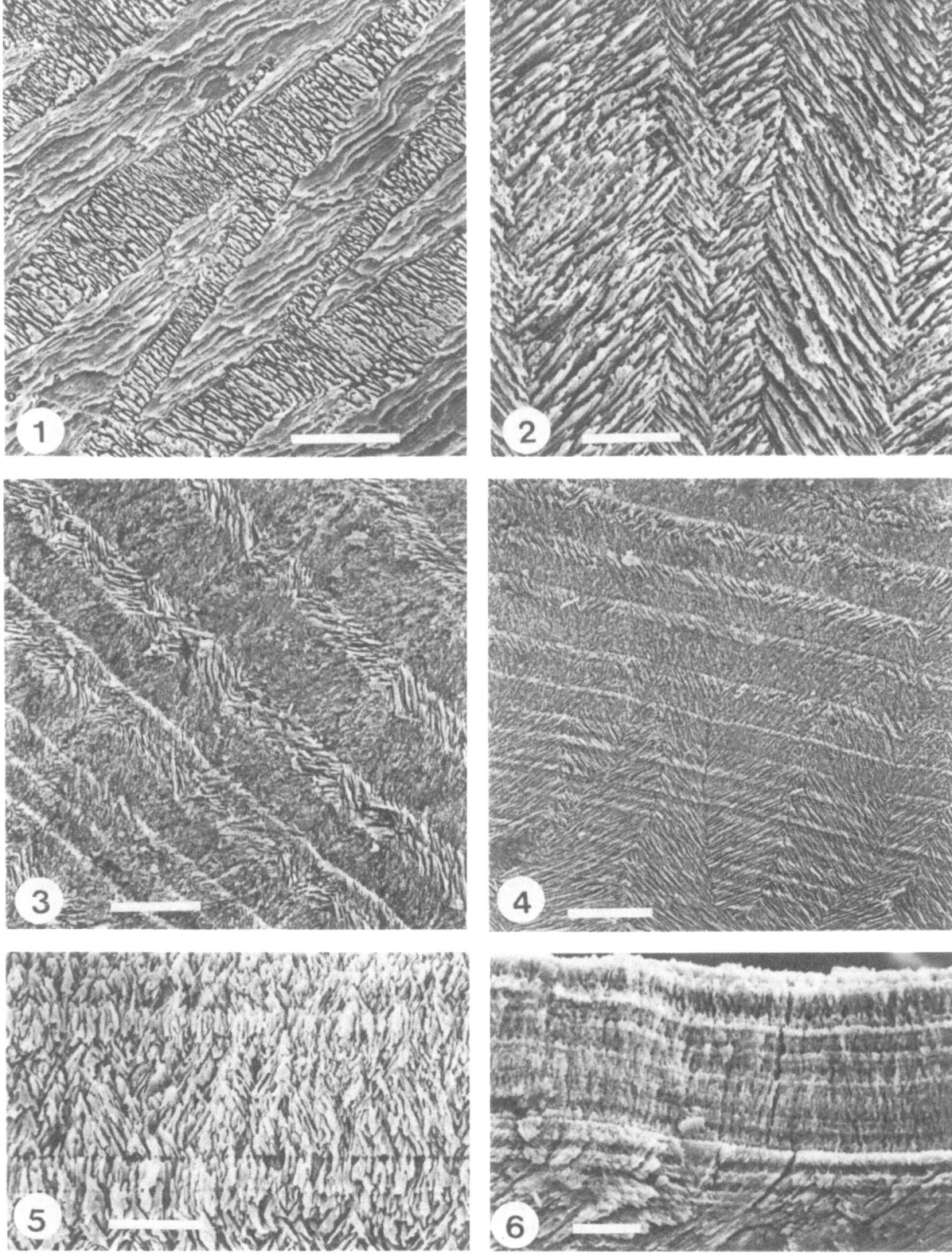

nine species with aragonitic shells, three showed trace amounts of calcite at the northernmost edges of their ranges. Subsequent studies have shown that the distribution of calcite in bivalve shells has a strong phylogenetic control and only in *Mytilus* has a temperature effect been clearly demonstrated (Dodd, 1963, 1964; Kennedy *et al.*, 1969; Carter, 1980).

Lowenstam (1954) grouped all the littorinid species that he examined into the one genus *Littorina* and did not consider the phylogenetic distribution of the calcite. The family Littorinidae comprises about 172 living species and a recent phylogenetic analysis of the family (Reid, 1989) has demonstrated a number of major clades within the family which have separate evolutionary and biogeographic histories. The distribution of calcite among these clades was unclear. Therefore, in order to establish the distribution of shell minerals and shell microstructure types among the Littorinidae, we examined species from 30 of the 32 genera and subgenera of the family (classification after Reid, 1989, and this volume). For comparison we examined the shell of the terrestrial snail *Pomatias elegans* Müller. This species was taken as a representative of the Pomatiasidae (superfamily Littorinoidae) which has been used as an outgroup in a phylogenetic analysis of the Littorinidae (Reid, 1989).

The objectives of this study were to explore the use of shell mineralogy and microstructure as phylogenetic characters and to examine the possible adaptive or environmental significance of variations in shell mineralogy among the Littorinidae. Apart from the work of Lowenstam (1954) only a few other species of littorinids have been examined, either for shell mineralogy or microstructure (Bøggild, 1930; Arnaud & Bandel, 1978; Bandel, 1979; Kobayashi *et al.*, 1983), and there has been no previous systematic study of littorinid shell structures.

## Methods

The microstructure of the gastropods was investigated at optical level using acetate peels (method after Taylor *et al.*, 1969; Kennish *et al.*, 1980) taken from ground and etched sections of shells which had been vacuum-embedded in resin blocks. The same etched shell sections, together with pieces of shell growth surfaces, were used for scanning electron microscopy.

Shell mineralogy was determined by X-ray diffraction of powder scrapings or pieces of individual shell layers from selected species.

## Results

A number of distinct shell microstructures are found in the Littorinidae. Most of these are found in other gastropods and some have been previously described in some detail. The extensive literature and confusing terminology associated

*Fig. 1. Pomatias elegans*, section of crossed-lamellar layer showing alternating lamellae. Scale bar = 10 μm.

*Fig. 2. Pomatias elegans*, crossed-lamellar layer showing different orientation of crystals in adjacent lamellae. Scale bar = 6 μm.

*Fig. 3. Littorina keenae*, outermost part of the outer layer showing banded appearance of fine crossed-lamellar structure with bands of coarser structure. Scale bar = 10 μm.

*Fig. 4. Littorina keenae*, showing how banded structure of Fig. 3 passes transitionally into normal crossed-lamellar structure. Scale bar = 10 μm.

*Fig. 5. Pomatias elegans*, outer spherulitic-prismatic structure. Scale bar = 2.5 μm.

*Fig. 6. Lacuna vincta*, outer spherulitic-prismatic layer and crossed-lamellar layer. Scale bar = 5 μm.

202

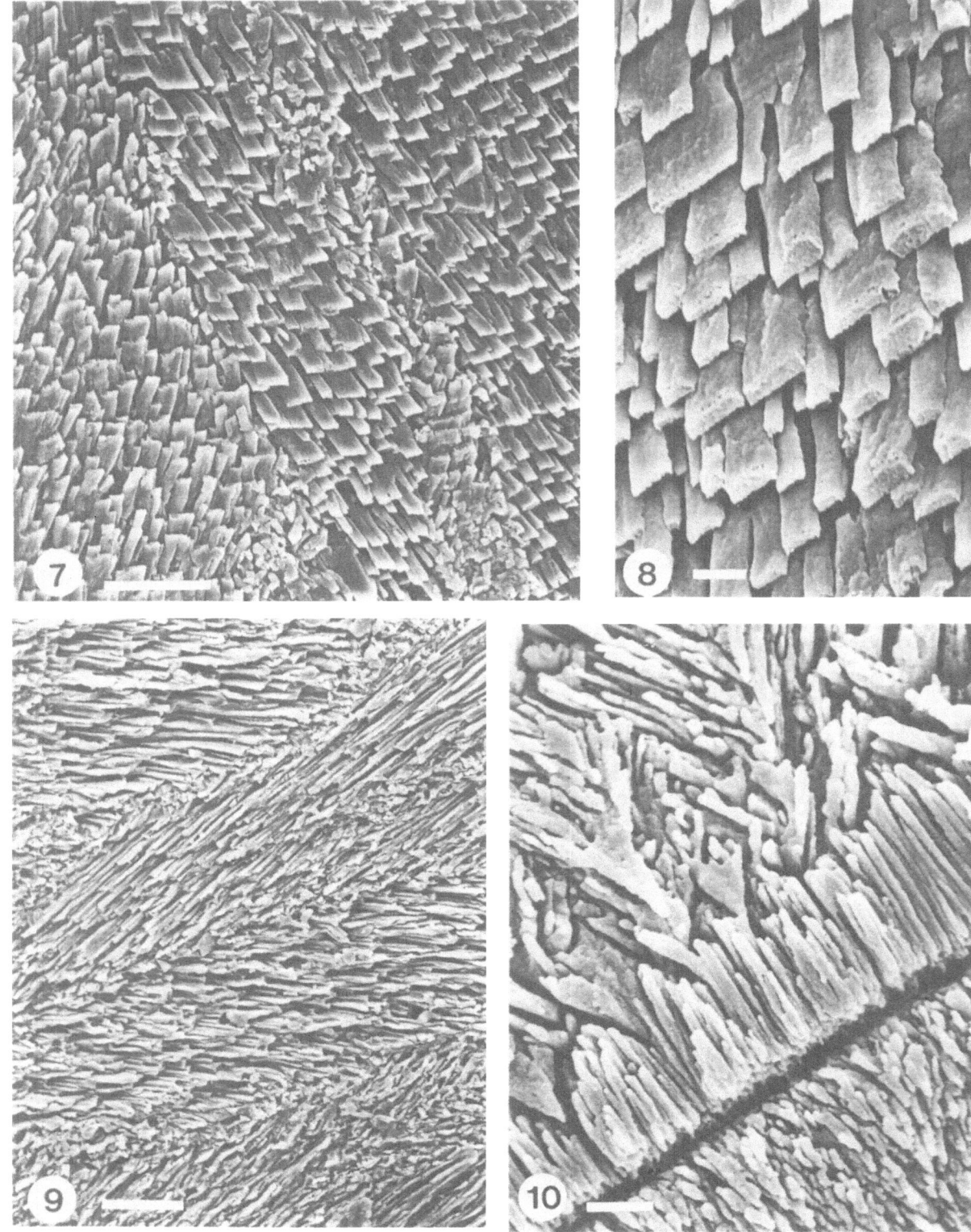

with the description and definition of shell micro-structures have recently been reviewed by Carter & Clark (1985). They have attempted to define and illustrate most of the structures and their variations found in the Mollusca. Consequently, only brief descriptions of the microstructures found within the Littorinidae are given below.

*Crossed-lamellar structure*

This is one of the most common and well known of the molluscan structures (Taylor *et al.*, 1969; Carter & Clark, 1985; Watabe, 1988) and consists of a lamellate 'plywood-like' fabric. Within adjacent first order lamellae the second order lamellae are composed of long lath-like crystallites, which are inclined at a high angle in opposing directions (Figs 1, 2). The height axes of the primary lamellae are aligned normal to the secretory surface. In those littorinids with an outermost crossed-lamellar layer the long axes of the primary lamellae are usually arranged concentrically, parallel to the shell margin. In the second crossed-lamellar layer, the long axes are arranged radially and normal to the shell margin. There is considerable variation between taxa in the size, thickness and branching of the first order lamellae and in the angle of inclination of second order lamellae. Crossed-lamellar structure sometimes arises from initial spherulitic growth (Fig. 22) or sheets of aragonitic needle-like crystals (Fig. 10).

A distinctive variety of crossed-lamellar structure is found in some littorinids (mainly species of *Nodilittorina*). The outermost part of the outer shell layer appears in optical sections to be homogeneous, with prominent growth increments. However, at higher magnifications it is seen to consist of very fine crossed-lamellar structure in which the aragonitic crystallites are very thin and the boundaries between first order lamellae indistinct. Alternating with this fine structure are narrow bands of coarser crossed-lamellar structure (Fig. 3), but in bands too narrow to allow the development of the usual first order lamellae. The fine grained structure passes transitionally into normal crossed-lamellar structure (Fig. 4). The banded structure probably represents a tidal secretory regime, as has been shown for the calcitic layer of *Littorina littorea* (L.) (Ekaratne & Crisp, 1982).

*Crossed-foliated structure*

This structure is similar to crossed-lamellar structure but made of calcite. Elongate calcitic crystals are arranged into blocks, with the laths in adjacent blocks being aligned mainly in two opposing directions (Figs 7-9). However, the structure is less ordered than crossed-lamellar structure, the primary lamellae are more irregular and do not extend through the thickness of the shell layer. Additionally, the laths are orientated in more than two directions. This structure was first named by MacClintock (1967) from limpets, but Carter & Clark (1985) considered it to be merely calcitic crossed-lamellar structure.

*Irregular-prismatic/grained structure*

In littorinids this structure is always calcitic. It consists of prisms with very irregular boundaries

*Fig. 7. Pellilitorina setosa*, inner growth surface of crossed-foliated structure, showing domains of calcitic crystals with different orientations. Scale bar = 5 μm.

*Fig. 8. Pellilitorina setosa*, inner growth surface showing calcitic crystals outcropping at the surface. Scale bar = 1 μm.

*Fig. 9. Pellilitorina setosa*, etched section of crossed-foliated layer showing differing orientations of calcitic crystals. Scale bar = 5 μm.

*Fig. 10 Pomatias elegans*, section showing junction of the spherulitic-prismatic layer (lower right) and crossed-lamellar layer, with initial part of the crossed-lamellar layer consisting of a sheet of short aragonitic needles. Scale bar = 1 μm.

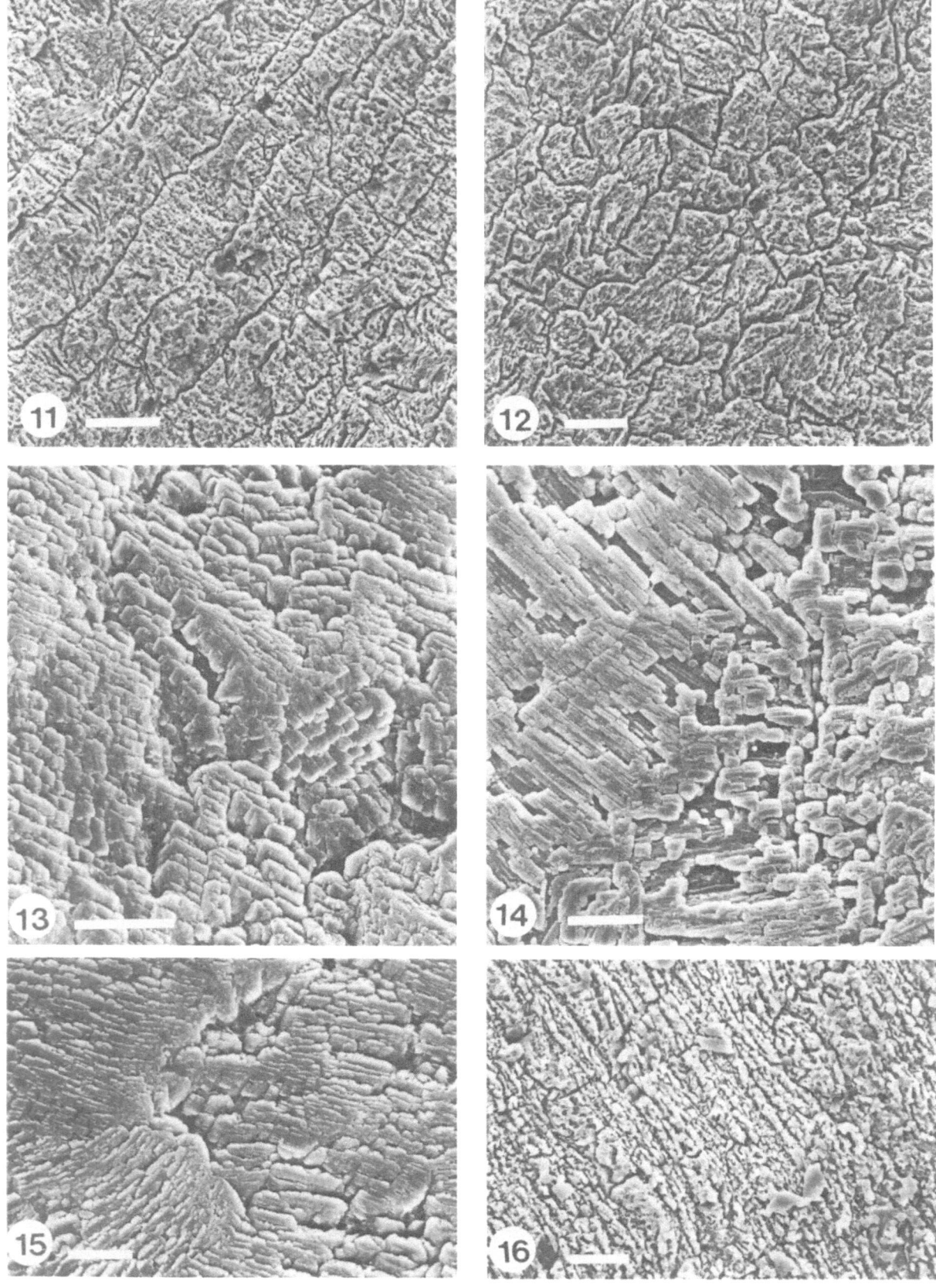

in both vertical and horizontal sections (Figs 11, 12, 17, 18). Frequently, in the outermost part of the shell layer the prisms are elongate and the boundaries more distinct, but in the inner parts the microstructure consists of irregular 'grains' (Fig. 23) and corresponds to what Bøggild (1930) called the *grained structure*. Both the prisms and the grains have an irregularly banded substructure (Fig. 12). In places the prism and grain boundaries are indistinct and the structure consists of irregular layers of calcitic crystallites producing a laminated appearance in section (Fig. 16). The inner growth surface of the irregular-prismatic structure consists of domains of crystal growth, with each domain consisting of irregular, laminar and dendritic growths of small (1–2 μm) calcitic crystallites (Figs 13–15). The growth orientation of the crystallites varies in adjacent domains. The boundaries of the domains seen at the surface correspond to the boundaries of the prisms and grains seen in section. The irregular laminar structure seen within these primary units corresponds to the overgrowing layers of calcitic crystallites. A similar structure is found in the outer layer of the bivalve *Chama pellucida* Broderip (Taylor & Kennedy, 1969) and in the outer layer of some muricid gastropods (Petitjean, 1965).

### Spherulitic-prismatic structure

This is an aragonitic structure in which fine fibrous crystallites radiate in three dimensions from an initial spherulite or common points of origin. The structure forms the outer layer of *Pomatias*, *Lacuna* and *Laevilitorina* (Figs 5, 6, 21). Bandel (1979) has illustrated the structure from other caenogastropods.

### Lath-type fibrous-prismatic structure

This is an aragonitic prismatic structure consisting of elongate lath-like crystals which are inclined at a shallow angle (15°) to the depositional surface of the shell (Figs 19, 20). The structure is found in the third layer of *Pomatias*.

### Intersected crossed-acicular structure

This is similar to crossed-lamellar structure, but the primary lamellae are not developed. Lath-like crystallites are aligned in two prominent dip directions, but seen in section the laths often intersect in a 'cross-stitch' pattern. This structure is found in the inner layer of *Pomatias*.

### Distribution of microstructures

Each gastropod shell consists of several distinct layers which may differ in microstructure and mineralogy. The distribution of the major microstructures among the genera and subgenera of Littorinidae and in *Pomatias* is shown in Table 1. Additionally, the main types of microstructural

*Figs 11–16. Littorina littorea*, irregular-prismatic layer.

*Fig. 11.* Section showing irregular prism boundaries. Scale bar = 10 μm.

*Fig. 12.* Section showing irregular 'grained' structure. Scale bar = 10 μm.

*Fig. 13.* Growth surface of prismatic structure showing stepped calcitic crystallites. Scale bar = 5 μm.

*Fig. 14.* Growth surface showing lamellar dendrites of calcitic crystallites. Scale bar = 3 μm.

*Fig. 15.* Growth surface showing domains of calcitic growth, with differing orientation in adjacent domains. Scale bar = 3 μm.

*Fig. 16.* Section showing laminar growth common in the prismatic structure. Scale bar = 5 μm.

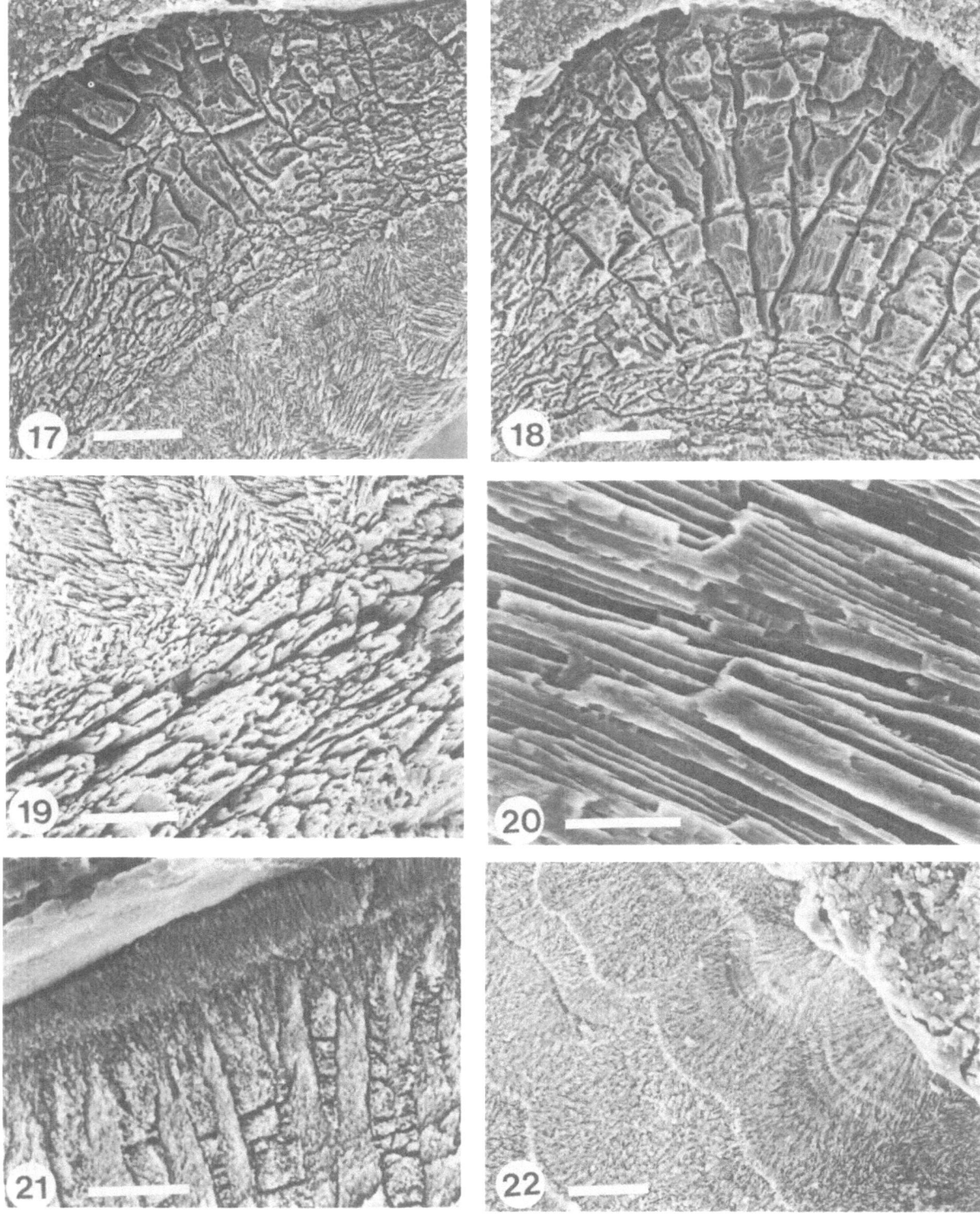

combination are illustrated diagrammatically in Fig. 24.

In the outgroup (*Pomatias elegans*) the shell is entirely aragonitic and consists of four layers, with firstly, a thin outer spherulitic-prismatic layer, followed by a crossed-lamellar layer, a fibrous-prismatic layer and an inner crossed-acicular layer (Figs 1, 2, 5, 10, 19, 20).

In *Lacuna* and *Laevilitorina* there is a thin outer spherulitic-prismatic layer, followed by two successive crossed-lamellar layers. In seven other genera and their component subgenera (*Cremnoconchus, Bembicium, Tectarius, Littoraria, Peasiella, Cenchritis, Melarhaphe* and *Mainwaringia*) the shell consists of three layers of crossed-lamellar structure. In the outermost layer, the primary lamellae are arranged concentrically in relation to the shell aperture, and radially in the second layer. The orientation in the third layer, which is found only in the upper whorls, is less clear and reflects repeated adjustments of the secreting surface during shell growth.

In most *Nodilittorina* species (*N. interrupta* (C.B. Adams) is an exception) together with *Littorina striata* King & Broderip and *L. keenae* Rosewater, the outermost part of the outer shell layer consists of very fine crossed-lamellar structure, with prominent growth increments (Fig. 3) which passes transitionally within the layer into normal crossed-lamellar structure (Fig. 4). Inwards from this are two further crossed-lamellar layers (Fig. 25).

The Antarctic species *Pellilitorina setosa* (E.A. Smith) has a shell consisting of only a single layer of calcitic crossed-foliated structure (Figs 7, 8,

9). Arnaud & Bandel (1978) reported this species as having an aragonitic crossed-lamellar microstructure.

A calcitic layer is also found in the small New Zealand species *Risellopsis varia* (Hutton) which has a three-layered shell. The outermost layer is calcitic, and formed from irregular-prismatic structure. This is coarsely crystalline beneath the keel-like spiral ribs (Figs 17, 18), but irregularly granular elsewhere. The two inner layers are both formed of crossed-lamellar structure.

Most species of the northern Pacific and northern Atlantic genus *Littorina* (excepting *L. striata* and *L. keenae*) have an outer calcitic layer consisting of irregular-prismatic structure. In many species the layer is extremely thick and makes up most of the shell material (Fig. 26) and the two inner crossed-lamellar layers are correspondingly extremely thin. The innermost crossed-lamellar layer frequently appears in sections to have sublayers. These probably mark adjustments made during growth in the orientation of the secreting surface high in the spire.

In summary, the mineral calcite is found only in *Pellilitorina, Risellopsis* and most species of *Littorina*. If the distribution of calcite is superimposed on a cladogram of relationships for the family Littorinidae which has been derived from a large set of anatomical characters (Reid, 1989) (order in Table 1 derived from the cladogram), it is seen that calcite has been independently developed in three distinct clades. *Risellopsis* and *Littorina* are unrelated, but both have apparently similar calcitic microstructures in the form of the irregular-prismatic structure.

*Fig. 17. Risellopsis varia*, showing outer irregular-prismatic structure with crossed-lamellar layer below. Scale bar = 20 μm.

*Fig. 18. Risellopsis varia*, detail of irregular-prismatic structure forming spiral rib. Scale bar = 20 μm.

*Fig. 19. Pomatias elegans*, fibrous-prismatic layer (lower) and crossed-lamellar layer (upper). Scale bar = 5 μm.

*Fig. 20. Pomatias elegans*, fractured section of fibrous-prismatic layer, showing elongate, aragonitic prisms. Scale bar = 5 μm.

*Fig. 21. Laevilitorina bennetti*, thin spherulitic-prismatic layer beneath periostracum followed by crossed-lamellar layer. Scale bar = 5 μm.

*Fig. 22. Bembicium auratum*, initial spherulitic structures in outer part of crossed-lamellar layer. Scale bar = 10 μm.

208

*Table 1.* The distribution of calcite, aragonite and the major shell microstructures among the Littorinidae and *Pomatias elegans* (Pomatiasidae). Shell layers listed sequentially from the outside to the inside of the shell. Abbreviations: C = calcite, A = aragonite, sp = spherulitic-prismatic structure, ip = irregular-prismatic structure, fp = fibrous-prismatic structure, cf = crossed-foliated structure, cl = crossed-lamellar structure, fcl = fine crossed-lamellar/crossed lamellar, ca = crossed-acicular structure, details in text.

| Species | Mineralogy | Microstructures |
|---|---|---|
| *Pellilitorina setosa* (E. A. Smith, 1875) S. Orkney Is | C | cf |
| *Lacuna (Lacuna) pallidula* (da Costa, 1778) Kimmeridge, Dorset | A | sp/cl/cl |
| *Lacuna (Epheria) vincta* (Montagu, 1803) Oban, Scotland | A | sp/cl/cl |
| *Cremnoconchus syhadrensis* (Blanford, 1863) W. Ghats, India | A | cl/cl/cl |
| *Bembicium auratum* (Quoy & Gaimard, 1834) Magnetic I., Qld. | A | cl/cl |
| *Bembicium melanostoma* (Gmelin, 1791) S.E. Australia | A | cl/cl/cl |
| *Risellopsis varia* (Hutton, 1873) Takapuna, New Zealand | C + A | ip/cl/cl |
| *Laevilitorina (Pellilacunella) bennetti* (Preston, 1916) Palmer Arch. | A | sp/cl/cl |
| *Laevilitorina (Macquariella) antarctica* (von Martens, 1885) Melchior Arch. | A | sp/cl/cl |
| *Laevilitorina (Laevilitorina) caliginosa* (Gould, 1849) Macquarie I. | A | sp/cl/cl |
| *Melarhaphe neritoides* (Linnaeus, 1758) Rhodes, & Farr, Scotland | A | cl/cl/cl |
| *Peasiella roepstorffiana* (Nevill, 1885) Orpheus I., Qld. | A | cl/cl/cl |
| *Cenchritis muricatus* (Linnaeus, 1758) Florida | A | cl/cl/cl |
| *Tectarius (Tectarius) grandinatus* (Gmelin, 1791) Polynesia | A | cl/cl/cl |
| *T. (Echininus) spinulosus* (Philippi, 1847) Singapore | A | cl/cl/cl |
| *T. (Tectininus) antonii* (Philippi, 1846) Cuba | A | cl/cl/cl |
| *T. (Echininiopsis) viviparus* (Rosewater, 1982) Guam | A | cl/cl/cl |
| *Littoraria (Protolittoraria) pintado* (Wood, 1828) Hawaii | A | cl/cl/cl |
| *L. (Palustorina) sulculosa* (Philippi, 1846) Broome, W. Australia | A | cl/cl/cl |
| *L. (Littoraria) irrorata* (Say, 1822) Piney Point, Maryland | A | cl/cl/ |
| *L. (Lamellilitorina) albicans* (Metcalfe, 1852) Sarawak | A | cl/cl/cl |
| *L. (Littorinopsis) scabra* (Linnaeus, 1758) Mahé, Seychelles | A | cl/cl/cl |
| *L. (Bulimilittorina) aberrans* (Philippi, 1846) Punta Morales, Costa Rica | A | cl/cl/cl |
| *Nodilittorina (Fossarilittorina) meleagris* (Potiez & Michaud, 1838) St Thomas | A | fcl/cl/cl |
| *N. (Fossarilittorina) modesta* (Philippi, 1846) Costa Rica | A | fcl/cl/cl |
| *N. (Echinolittorina) africana* (Philippi, 1847) Uvongo Beach, Natal | A | fcl/cl/cl |
| *N. (Echinolittorina) interrupta* (C. B. Adams, 1847) Cahuita, Costa Rica | A | cl/cl/cl |
| *N. (Nodilittorina) acutispira* (E. A. Smith, 1892) Port Jackson, NSW | A | fcl/cl/cl |
| *N. (Nodilittorina) pyramidalis* (Quoy & Gaimard, 1833) Merimbula, NSW | A | fcl/cl/cl |
| *N. (Nodilittorina) unifasciata* (Rray, 1826) Westernport Bay, NSW | A | fcl/cl/cl |
| *Littorina (Liralittorina) striata* King & Broderip, 1832 Madeira | A | fcl/cl/cl |
| *L. (Planilittorina) keenae* Rosewater, 1978 San Diego, California | A | fcl/cl/cl |
| *L. (Littorina) brevicula* Philippi, 1844 Hong Kong | C + A | ip/cl/cl |
| *L. (Littorina) kurila* Middendorff, 1848 Valdez, Alaska | C + A | ip/cl/cl |
| *L. (Littorina) littorea* (Linnaeus, 1758) several localities | C + A | ip/cl/cl |
| *L. (Littorina) mandshurica* Schrenck, 1851 Hokkaido | C + A | ip/cl/cl |
| *L. (Littorina) scutulata* Gould, 1849 Victoria, British Columbia | C + A | ip/cl/cl |
| *L. (Littorina) squalida* Broderip & Sowerby, 1829 Japan | C + A | ip/cl/cl |
| *L. (Neritrema) aleutica* Dall, 1872 Adak I., Aleutian Is | C + A | ip/cl/cl |
| *L. (Neritrema) mariae* Sacchi & Rastelli, 1966 Brighton, Sussex | C + A | ip/cl/cl |
| *L. (Neritrema) nigrolineata* Gray, 1839 Sandy Haven, Wales | C + A | ip/cl/cl |
| *L. (Neritrema) obtusata* (Linnaeus, 1758) Westgate, Kent | C + A | ip/cl/cl |
| *L. (Neritrema) saxatilis* (Olivi, 1792) Anglesey, Wales | C + A | ip/cl/cl |
| *L. (Neritrema) sitkana* Philippi, 1846 N.W. USA | C + A | ip/cl/cl |
| *L. (Neritrema) subrotundata* (Carpenter, 1864) Westport, Washington | C + A | ip/cl/cl |
| *Mainwaringia leithii* E. A. Smith, 1876 Bombay | A | cl/cl/cl |
| *Mainwaringia rhizophila* Reid, 1986 Santubong, Sarawak & Hong Kong | A | cl/cl/cl |
| *Pomatias elegans* Müller, 1774 Oxford, England | A | sp/cl/fp/ca |

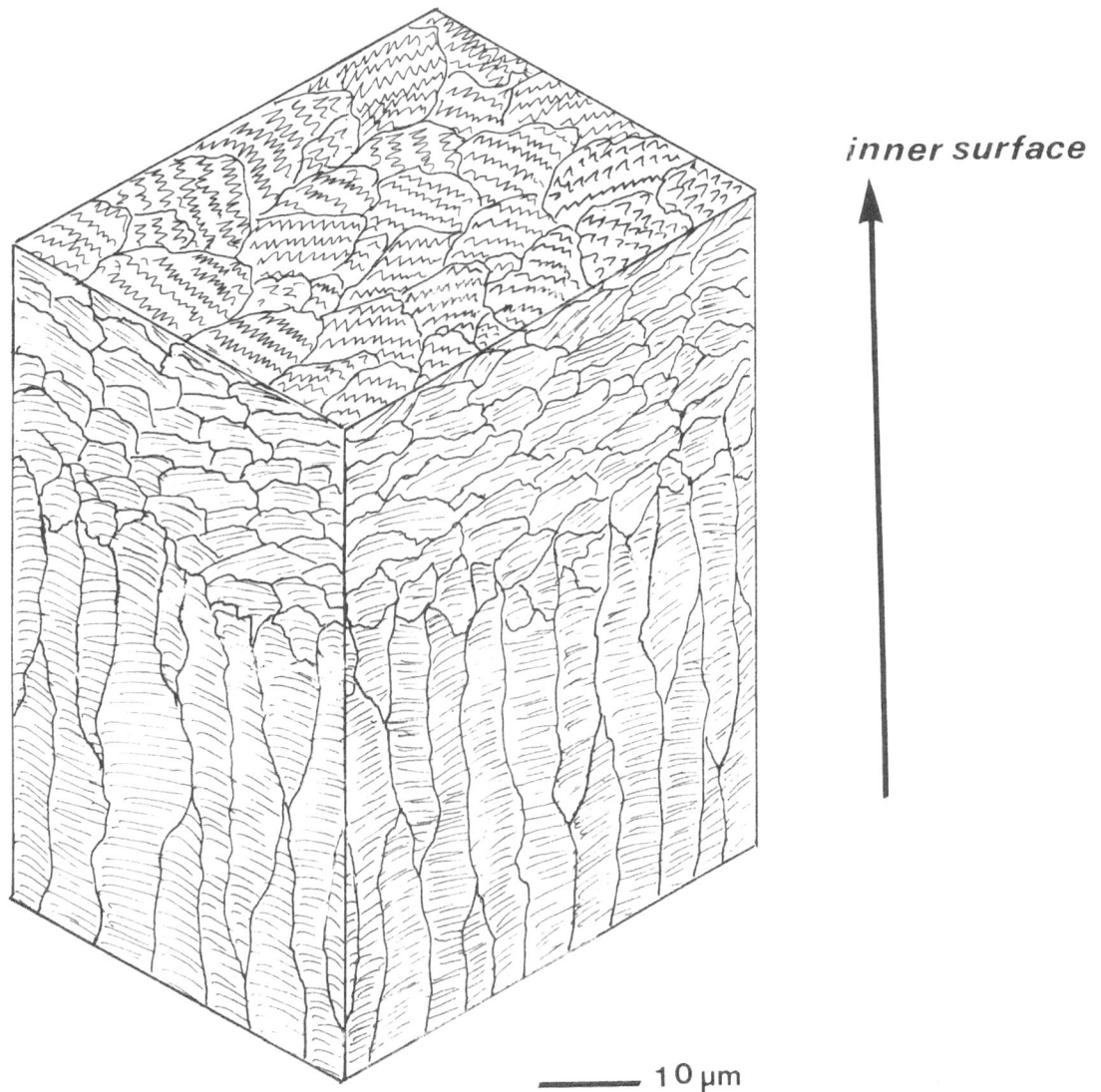

*inner surface*

_____ 10 μm

*Fig. 23.* Diagrammatic representation of the calcitic, irregular-prismatic structure.

However, the calcitic fabric of crossed-foliated structure found only in *Pellilitorina* is distinct.

The more detailed distribution of calcitic layers among species within the genus *Littorina* is shown in Fig. 27. *Littorina striata* and *L. keenae* are the two least derived species of the genus and have an outer shell layer of the fine crossed-lamellar structure with conspicuous growth banding, very similar to that seen in the sister-group *Nodilittorina*. All other species in the genus have an outer calcitic layer and two crossed-lamellar layers. The inclu-

sion of *Mainwaringia* within the clade *Littorina* is uncertain (Reid, this volume). The two species of *Mainwaringia* have an aragonitic shell with three crossed-lamellar layers similar to that of *Littoraria*.

Lowenstam (1954) found trace amounts of calcite in some normally aragonitic species, collected at the northernmost edge of their distributional range. We were unable to confirm the presence of calcite in *Littoraria irrorata* (Say) (*Littorina nebulosa* of Lowenstam, 1954) collected from

210

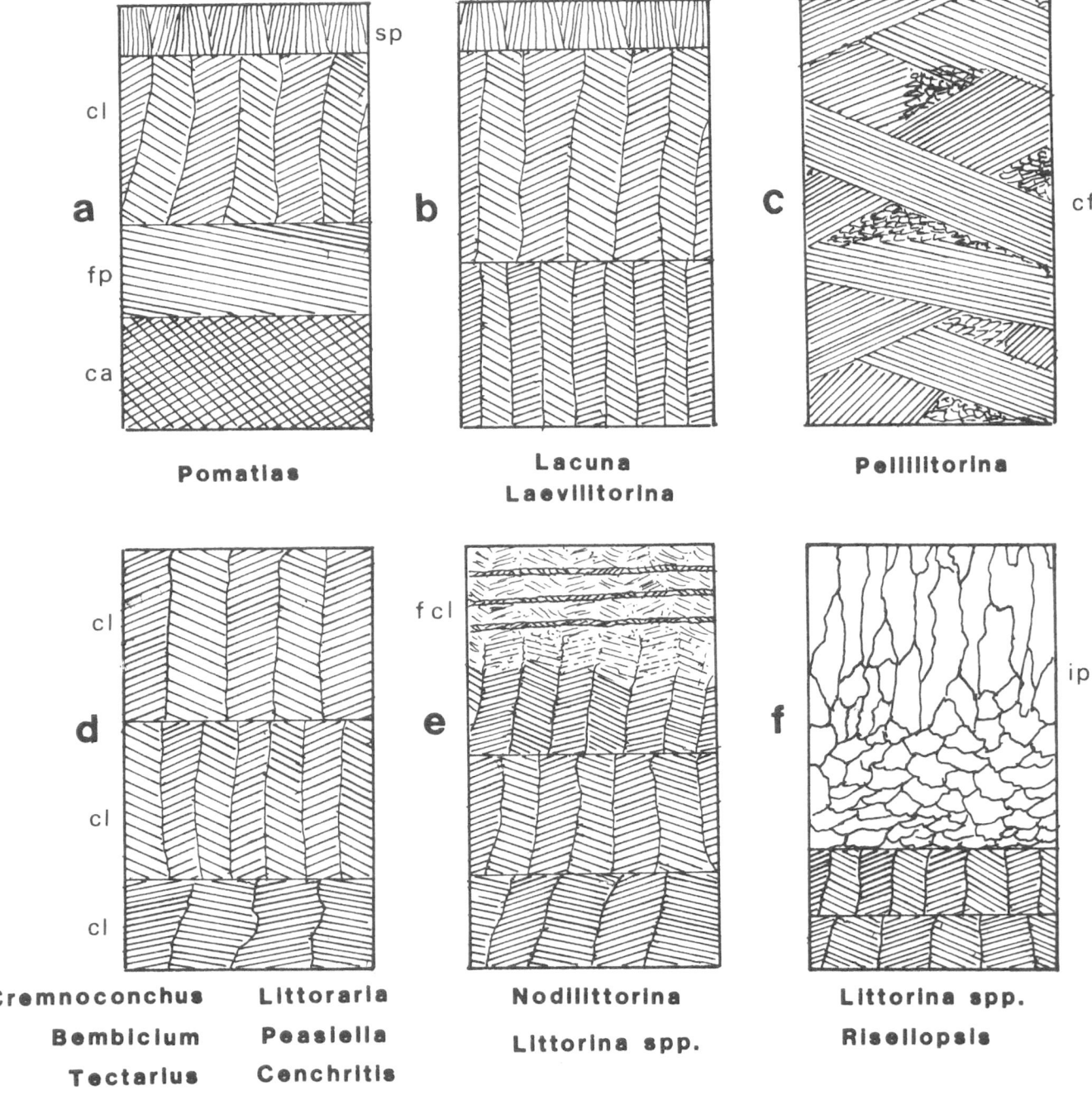

Fig. 24. Diagrammatic summary of the main combinations of shell microstructures found in the Littorinidae.
sp = spherulitic-prismatic structure; fp = fibrous-prismatic structure; ip = irregular-prismatic structure; cl = crossed-lamellar structure; cf = crossed-foliated structure; ca = crossed-acicular structure; fcl = fine crossed-lamellar/crossed lamellar.

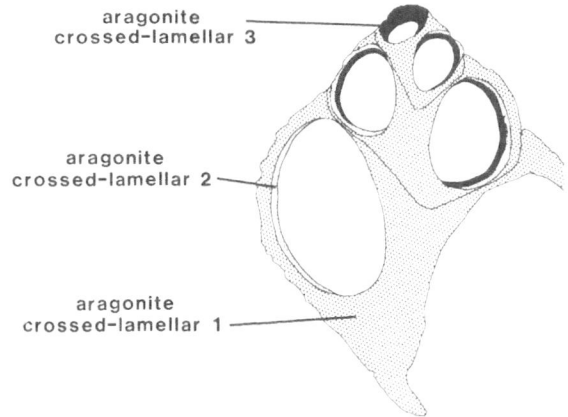

*Fig. 25.* Section of the shell of *Littorina striata* showing the disposition of the shell layers.

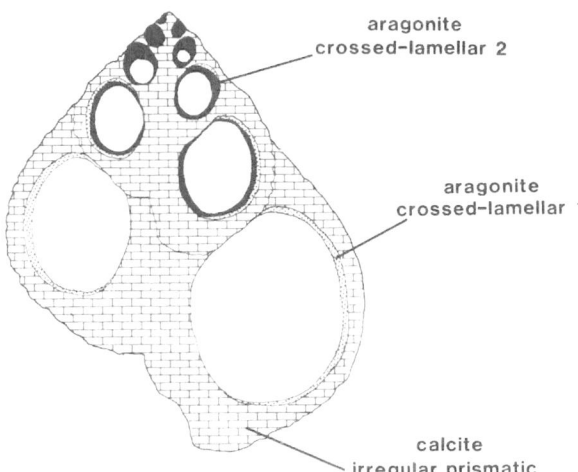

*Fig. 26.* Section of the shell of *Littorina littorea* showing the disposition of the shell layers.

Maryland. However, shells are often extensively bored and the records of calcite may possibly have resulted from contamination in the boreholes. We found no evidence of the mixed calcite and aragonite layers reported by Mutvei *et al.* (1985) for some *Haliotis* species.

**Discussion and conclusions**

The results of this study show that compared with other families of caenogastropods there is within

the Littorinidae a surprising amount of variation in shell microstructure and mineralogy.

The outgroup *Pomatias elegans* has a shell of four layers and only two of these, the spherulitic-prismatic layer and the crossed-lamellar layer, are present in the Littorinidae. The spherulitic-prismatic layer is found only in the genera *Lacuna* and *Laevilitorina* and may be a plesiomorphic character. Apart from those taxa having calcitic layers (discussed below), most of the other littorinid genera have shells constructed from up to

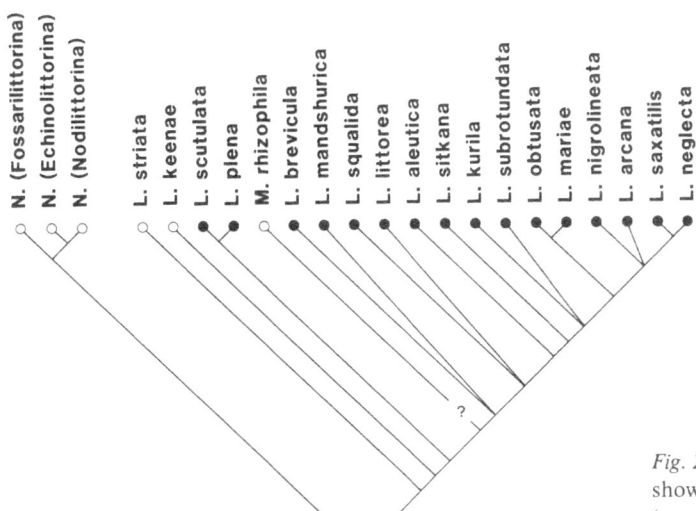

*Fig. 27.* Cladogram of relationships among *Littorina* species showing species with calcitic (solid circles) and aragonitic (open circles) outer layers. For details of characters and construction of cladogram see Reid (this volume).

three layers of crossed-lamellar structure. This is a common pattern found in many other caenogastropod families (Bøggild, 1930; Bandel, 1979; Kobayashi *et al.*, 1983).

A variation on this is found in most *Nodilittorina* species, *Littorina striata* and *L. keenae*, where the outer part of the outer shell layer consists of a very fine crossed-lamellar structure with prominent growth increments. *Nodilittorina* is the proposed sister-group of *Littorina* (Reid, 1989) and this is supported by the presence of this structure in the two least-derived species of *Littorina*.

The most interesting feature of the shell structure of the littorinids is the distribution of calcite. Most gastropod and bivalve shells are made entirely of aragonite and calcite is found in relatively few taxa (Bøggild, 1930; Taylor, *et al.*, 1969). The phylogenetic analysis shows that calcitic layers have been independently derived as two different microstructures in three different clades, *Pellilitorina*, *Risellopsis* and *Littorina*. The calcitic layer of irregular-prismatic structure is similar in the last two genera, but in *Pellilitorina* it is of the distinctive crossed-foliated structure. If the inclusion of *Mainwaringia* in the clade *Littorina* is accepted, the lack of a calcitic layer in the former is anomalous; either the calcitic layer has been lost in *Mainwaringia* or it has been developed independently in the *scutulata-plena* and *brevicula-neglecta* clades (Fig. 27). If *Mainwaringia* is not related to *Littorina*, the calcitic layer is an unreversed synapomorphy within the latter. The position of *Mainwaringia* in the cladogram is discussed by Reid (1990; this volume).

A feature of the taxa with calcitic layers is that they are all found at temperate to high latitudes; *Pellilitorina* in Antarctic seas (Arnaud & Bandel, 1978), *Risellopsis* in New Zealand and *Littorina* in the northern Pacific and northern Atlantic (Reid, 1989). Not all high latitude littorinids (e.g. *Lacuna* and *Laevilitorina*) have developed calcite. However, the shells of tropical littorinids (including *Mainwaringia*) are all entirely aragonitic. The lowest latitudinal occurrence of a calcitic species is *Littorina brevicula* Philippi, which ranges from Hokkaido to as far south as Hong Kong.

Another feature of the distribution of calcite is

that it always occurs as the outermost shell layer and in *Pellilitorina* forms the whole shell thickness. In *Littorina* species the calcitic outer layer is usually very thick and forms the bulk of the shell material (Fig. 26). The layer occupies an homologous position to the fine crossed-lamellar/crossed-lamellar layer of the sister-group *Nodilittorina*, and also *Littorina striata* and *L. keenae* (Fig. 25). In *Risellopsis*, the calcitic layer is in an homologous position to the spherulitic-prismatic layer of *Lacuna*, but the homology of the layer in *Pellilitorina* is uncertain.

This leads to the question of the possible functional significance, if any, of the calcitic layers. Coddington (1988) has recently discussed cladistic tests of adaptational hypotheses and concludes (p. 3) that '... an adaptation is an apomorphic function promoted by natural selection, as compared with the plesiomorphic function'. To propose an adaptive function, a character must first be shown to be apomorphic at a particular level of a cladogram (derived from other, non correlated, characters), and the function compared with that of the plesiomorphic state in the appropriate outgroup. The first criterion is fulfilled by the apomorphic development of calcitic layers in the three clades of Littorinidae.

Most discussion concerning the distribution and functional significance of calcitic shell structures has been restricted to the Bivalvia, where a number of families have outer calcitic shell layers (review by Carter, 1980). Any hypothesis to explain the distribution of calcite in the Littorinidae has to account for its occurrence both in outer shell layers and in species from high latitudes of northern and southern hemispheres. Recent work on the biomechanical properties of shells has highlighted the functional significance of many microstructural features of molluscan shells (Currey, 1988), and it is likely that the selection of calcite over the more usual aragonite concerns its physical or chemical behaviour.

The most obvious possibility is that calcitic layers somehow enhance the strength of shells. There have been no direct mechanical tests of littorinid irregular-prismatic structure, but from mechanical tests on other microstructures there is no evidence that calcitic structures are stronger than those made of aragonite (Taylor & Layman, 1972; Currey, 1980, 1988). In fact the evidence suggests the reverse. Additionally, Currey (1988) has shown how the juxtaposition of shell layers with different microstructural fabrics and orientations can enhance the mechanical strength of shells. This possibility has not been tested for littorinid shells. The problem with any biomechanical hypothesis is that there is good evidence that the most highly armoured and strongest shells are found in the tropics (Vermeij, 1978), where predation intensity is much greater than at higher latitudes. Mechanical tests in the family Thaididae, for example, have shown that the shells of tropical species are considerably stronger than their counterparts from temperate seas (Vermeij & Currey, 1980).

The relative hardness of the two minerals might be important in relation to resistance to abrasion, particularly as calcite occurs as an outer layer. However, calcite is significantly softer than aragonite (Carter, 1980). Another proposition concerns the relative densities of the minerals, because calcite is about $0.14$ g/cm$^3$ less dense than aragonite. Carter (1980) considered that the advantage derived from the density differences could account for the largely calcitic shells of the swimming Pectinidae. However, such a property would not be as obviously advantageous in benthic gastropods and would not show a latitudinal effect.

Another possibility is that the secretion of calcite is a consequence of influences of environmental temperature on the physiological controls of shell mineralogy, with calcite being selectively precipitated at lower temperatures (Lowenstam, 1954; Dodd, 1964; Kennedy et al., 1969; Carter, 1980). If this were true, then it would be expected that the secretion of calcite would be widespread in molluscs from high latitudes and deep water. In fact, it is limited to rather few taxonomic groups.

The most attractive hypothesis to explain the high latitude occurrence of calcitic shell layers relates to the relative solubilities of calcite and aragonite in seawater. There are considerable differences in the solubilities of the two minerals,

aragonite being about $35\%$ more soluble than calcite in warm ($25\ ^{\circ}$C) and cold ($5\ ^{\circ}$C) seawater of normal salinity of $35\%_0$ (Morse, 1983; Mucci, 1983). Both minerals are more soluble in cold than warm water. Under normal conditions in the open sea, shell dissolution is not a problem because seawater is usually supersaturated with respect to calcite. Nevertheless, there are marked latitudinal differences in the degree of saturation, attributable to the increased solubility of $CO_2$ at lower temperatures. Surface ocean water in the tropics is supersaturated with respect to calcite by a factor of 5.5 and to aragonite by a factor of 2.5. In seas at higher latitudes the factors are 3.7 for calcite and 1.7 for aragonite (Broeker et al., 1979; Mackensie et al., 1983).

The latter observations apply to water in the open ocean, but in seawater over the intertidal zone there are large diurnal differences in the $CO_2$ content of the seawater (Daniel & Boyden, 1975). The build-up of dissolved $CO_2$ concentrations during the night results from plant respiration in the absence of photosynthesis. Night-time inshore seawater frequently becomes undersaturated with respect to calcite, to a degree sufficient to cause the dissolution of calcium carbonate, even in tropical seas (Schmalz & Swanson, 1969; Trudgill, 1976). Intertidal shelled animals are particularly exposed to these dissolution effects, which, because of the increased solubility of $CO_2$ at low temperatures, will be more pronounced at high latitudes. Alexandersson (1975, 1978, 1979) has described rapid post-mortem etching, dissolution and disintegration of shells in shallow water over wide areas of the Baltic and North Seas. This is probably a general phenomenon at high latitudes.

We suggest that because of its significantly lower solubility, calcite will be selected preferentially over aragonite; shells built with calcite in the outermost layers will be considerably more resistant to dissolution by seawater undersaturated with respect to calcium carbonate. This hypothesis could be tested by an experiment comparing the dissolution rates of calcitic and aragonitic littorinid shells in seawater at various levels of calcium carbonate saturation. In addition to the mineralogy, different microstructures may be effective in resisting dissolution. Walter & Morse (1984) have shown that finely crystalline microstructures dissolve more rapidly than coarsely crystalline microstructures which have less reactive surfaces. It is possible that the coarsely grained irregular-prismatic structure may be more resistant than the finely crystalline, crossed-lamellar structure.

The change from an aragonitic to a calcitic outer shell layer could be a fairly simple evolutionary step. There is evidence that $Mg^{2+}$ concentrations such as occur in normal seawater inhibit calcite but not aragonite precipitation (Mackensie *et al.*, 1983; Meyer, 1984). In order to precipitate calcite the cells of the outer mantle surface must selectively reduce the $Mg^{2+}$ concentration in the extrapallial fluid from which the shell crystals are precipitated (Lorens & Bender, 1980).

The hypothesis presented here accounts for both the occurrence of calcite in the outer layers of shells and also the restriction of calcite to taxa at higher latitudes. At a more general level, a range of common intertidal animals on north Atlantic shores have shells with either calcitic outer layers or shells made entirely of calcite; examples are the dogwhelk *Nucella lapillus* (Linnaeus), the mussel *Mytilus edulis* (Linnaeus), patelloidean limpets and barnacles. Calcitic shells are generally less common in sublittoral habitats. It is thus likely that the factors influencing the evolution of shell mineralogy in the Littorinidae are affecting other prominent members of the intertidal fauna.

## Acknowledgements

We are very grateful to Dr Gordon Cressey for the X-ray determinations and to Dr Paul Henderson for advice concerning calcium carbonate solubility.

## References

Alexandersson, E. T., 1975. Etch patterns on calcareous sediment grains: petrographic evidence of marine dissolution of carbonate minerals. Science 189: 47–48.

Alexandersson, E. T., 1978. Destructive diagenesis of carbonate sediments in the eastern Skagerrak, North Sea. Geology 6: 324–327.

Alexandersson, E. T., 1979. Marine maceration of skeletal carbonates in the Skagerrak, North Sea. Sedimentology 26: 845–852.

Arnaud, P. M. & K. Bandel, 1978. Comments on six species of marine Antarctic Littorinacea (Mollusca, Gastropoda). Tethys 8: 213–230.

Bandel, K., 1979. Ubergange von einfacheren Strukturtypen zur Kreuslamellen-struktur bei Gastropodenschalen. Biomineralization 10: 9–38.

Batten, R. L., 1975. The Scissurellidae – are they neotenously derived Fissurellids? (Archeogastropoda). Am. Mus. Novit. 2567: 1–37.

Bøggild, O. B., 1930. The shell structure of the mollusks. K. danske Vidensk. Selsk. Skr. Roekke 9: 233–326.

Broeker, W. S., T. Takahashi, H. J. Simpson & T. H. Peng, 1979. Fate of fossil fuel carbon dioxide and the global carbon cycle. Science 206: 409–418.

Carter, J. G., 1980. Environmental and biological controls of bivalve shell mineralogy and microstructure. In D. C. Rhoads & R. A. Lutz (eds). Skeletal Growth of Aquatic Organisms. Plenum Press, New York: 69–113.

Carter, J. G.& G. R. Clark, 1985. Classification and phylogenetic significance of molluscan shell structures. In T. W. Broadhead (ed.), Mollusks, Notes for a Short Course. University of Tennessee Department of Geological Sciences Studies in Geology 13: 50–71.

Coddington, J. A., 1988. Cladistic tests of adaptational hypotheses. Cladistics 4: 3–22.

Currey, J. D., 1980. Mechanical properties of mollusc shell. Symp. Soc. exp. Biol. 34: 75–97.

Currey, J. D., 1988. Shell form and strength. In E. R. Trueman & M. R. Clarke (eds), The Mollusca, 11: 183–210.

Daniel, M. J. & C. R. Boyden, 1975. Diurnal variations in physico-chemical conditions within intertidal rockpools. Field Studies 4: 161–176.

Dodd, J. R., 1963. Palaeoecological implications of shell mineralogy in two pelecypod species. J. Geol. 71: 1–11.

Dodd, J. R., 1964. Environmentally controlled variation in the shell structure of a pelecypod species. J. Paleont. 38: 1065–1071.

Ekaratne, S. U. K. & D. J. Crisp, 1982. Tidal micro-growth bands in intertidal gastropod shells, with an evaluation of band-dating techniques. Proc. r. Soc. B. 214: 305–323.

Kennedy, W. J., J. D. Taylor & A. Hall, 1969. Environmental and biological controls on bivalve shell mineralogy. Biol. Rev. 44: 499–530.

Kennish, M. J., R. A. Lutz & D. C. Rhoads, 1980. Preparation of acetate peels and fractured sections for observations of growth patterns within the bivalve shell. In D. C. Rhoads & R. A. Lutz (eds), Skeletal Growth of Aquatic Organisms. Plenum Press, New York: 597–606.

Kobayashi, I., K. Mano, F. Isogai & O. Masae, 1983. Biomineral formation of gastropods in comparison with that of pelecypods. In P. Westbroek & E. W. DeJong (eds), Biomineralization and Biological Metal Accumulation. Reidel, Dordrecht, Netherlands: 261–266.

Lorens, R. B. & M. L. Bender, 1980. The impact of solution chemistry on *Mytilus edulis* calcite and aragonite. Geochim. Cosmochem. Acta 44: 1265–1278.

Lowenstam, H. A., 1954. Factors affecting the aragonite: calcite ratios in carbonate-secreting organisms. J. Geol. 62: 284–322.

MacClintock, C., 1967. Shell structure of patelloid and bellerophontoid gastropods. Bull. Peabody Mus. nat. Hist. 22: 1–140.

Mackensie, F. T., W. D. Bischoff, F. C. Bishop, M. Loijens, J. Schoonmaker & R. Wollast, 1983. Magnesian calcites: low temperature occurrence, solubility and solid state behaviour. Reviews in Mineralogy 11: 97–144.

Meyer, H. J., 1984. The influence of impurities on the growth rate of calcite. J. Crystal Growth 66: 639–646.

Morse, J. W., 1983. The kinetics of calcium carbonate dissolution and precipitation. Reviews in Mineralogy 11: 227–264.

Mucci, A., 1983. The solubility of calcite and aragonite in seawater at various salinities, temperatures, and one atmosphere total pressure. Am. J. Sci. 283: 780–799.

Mutvei, H., Y. Dauphin & J.-P. Cuif, 1985. Observations sur l'organisation de la couche externe du test des *Haliotis* (Gastropoda): un cas exceptionnel de variabilité minéralogique et microstructurale. Bull. Mus. natn. Hist. nat., Paris, 4 sér. 7: 73–91.

Petitjean, M., 1965. Structure microscopique, nature minéralogique et composition chimique de la coquille des Muricidés (Gastèropodes Prosobranches). Importance systématique de ces caractères. Thèse à la Faculté des Sciences de l'Université de Paris, 131 pp.

Reid, D. G., 1989. The comparative morphology, phylogeny and evolution of the gastropod family Littorinidae. Phil. Trans. r. Soc., Lond. Ser. B 324: 1–110.

Schmalz, R. F. & F. J. Swanson, 1969. Diurnal variations in the carbonate saturation of seawater. J. sedim. Petrol. 39: 255–267.

Shimamoto, M., 1986. Shell microstructure of the Veneridae (Bivalvia) and its phylogenetic implications. Sci. Rep. Tôhoku Univ. Ser. 2. Geology 56: 1–39.

Taylor, J. D. & W. J. Kennedy, 1969. The shell structure and mineralogy of *Chama pellucida*. Veliger 11: 391–398.

Taylor, J. D. & M. Layman, 1972. The mechanical properties of bivalve (Mollusca) shell structures. Palaeontology 15: 73–87.

Taylor, J. D., W. J. Kennedy & A. Hall, 1969. The shell structure and mineralogy of the Bivalvia. I. Introduction. Nuculacea-Trigonacea. Bull. Brit. Mus. nat. Hist. Suppl. 3: 1–125.

Taylor, J. D., W. J. Kennedy & A. Hall, 1973. The shell structure and mineralogy of the Bivalvia. II. Lucinacea-Clavagellacea. Conclusions. Bull. Br. Mus. nat. Hist. Zool. 22: 253–294.

Togo, Y., 1974. Shell structure and growth of protoconch and teleoconch in *Neptunea* (Gastropoda). Chishitsugaku Zasshi 80: 369–380.

Trudgill, S. T., 1976. The marine erosion of limestones on Aldabra Atoll, Indian Ocean. Z. Geomorph. Suppl. 26: 164–200.

Uozumi, S. & S. Suzuki, 1981. The evolution of shell structures in the Bivalvia. In T. Habe & M. Omori (eds), Studies of Molluscan Paleobiology. Niigata University, Niigata, Japan: 63–77.

Vermeij, G. J., 1978. Biogeography and Adaptation. Harvard University Press, Cambridge, Massachusetts, 332 pp.

Vermeij, G. J. & J. D. Currey, 1980. Geographical variation in the strength of thaidid snail shells. Biol. Bull. 158: 383–389.

Waller, T. R., 1978. Morphology, morphoclines, and a new classification of the Pteriomorphia (Mollusca: Bivalvia). Phil. Trans. r. Soc., Lond. Ser. B 284: 345–365.

Walter, L. M. & J. W. Morse, 1984. Reactive surface area of skeletal carbonates during dissolution: effect of grain size. J. sedim. Petrol. 54: 1081–1090.

Watabe, N., 1988. Shell structure. In E. R. Trueman & M. R. Clarke (eds), The Mollusca, 11: 69–104.

Hydrobiologia **193**: 217–221, 1990.
*K. Johannesson, D. G. Raffaelli and C. J. Hannaford Ellis (eds), Progress in Littorinid and Muricid Biology.*
© 1990 *Kluwer Academic Publishers.*

# Differences in shell properties between morphs of *Littoraria pallescens*

Laurence M. Cook
*Department of Environmental Biology, University of Manchester, Manchester M13 9PL, England, UK*

*Key words:* polymorphism, stabilizing selection, Littorinidae, mangrove snail

## Abstract

Differences between the shells of the dark and the yellow morphs of the leaf-inhabiting mangrove snail *Littoraria pallescens* (Philippi) have been examined. In the samples studied yellows are less heavy than darks, which agrees with earlier findings that they are thinner and more easily broken. In both morphs there is a decrease in variance of shell shape between small-shelled and large-shelled individuals, and in both size classes yellows are less variable in shell shape than darks. The interpretation of these differences in relation to stabilizing selection is discussed. It is concluded that they are not simply consequences of the normal growth process but indicate the direct action of natural selection.

## Introduction

*Littoraria pallescens* (Philippi) is a very widespread species on mangroves in the Indo-Pacific region. Like other members of the genus, there is a pelagic larval phase in the life cycle. The young snails settle and grow on mangrove trees and spend the rest of their lives above the water level. A number of mangrove tree species are colonized, species of *Rhizophora* being commonly used. The snails are almost always on the leaf surfaces, rather than on bark, typically from high tide level to 3 m above on the seaward side of the trees. Bark dwelling members of the same genus usually occur on the same trees. In common with many other leaf inhabiting snails but unlike the bark dwellers, *L. pallescens* exhibits visible polymorphism. The taxonomy and biology of the genus are discussed by Reid (1986).

The difference in physical properties between shells of different colours in *Littoraria pallescens*

has been examined as part of a study of the factors which may contribute to maintenance of their polymorphism. *L. pallescens* has three main colour classes, which have been called yellow, orange and dark (Cook, 1983, 1986). Reid (1986, 1987) refers to the orange phenotype, which is always the least common, as pink. Yellow and orange have shells of those colours, sometimes with the addition of one or two thin brown bands. The dark category consists of shells of an olive, brown colour with pale flecks. These probably have the yellow or orange ground colour covered by a series of bands which are completely or almost completely fused to one another (Freeman, 1986), so that, as in helicid snails such as *Cepaea*, the phenotype is determined by ground colour and banding loci (Murray, 1975; Cain, 1988). The genetic basis of the morph differences has not, however, been established in *Littoraria*. The dark shells are stronger, thicker and heavier than yellow ones (Cook *et al.*, 1985; Cook &

Freeman, 1986). This may be a direct consequence of the presence of the banding regions or an adjustment through additional thickening. As well as any other effect it may have, the difference in visual appearance between morphs therefore would also affect the chance of surviving damage from attack or mechanical stress.

If *L. pallescens* is attacked by predators, and the colour morphs differ on average in shell strength, then it may be expected that variance in those shell parameters which affect susceptibility to predation will also differ. Mollusc shells are classical material for the study of selective elimination of less fit variants. Weldon (1901) and Di Cesnola (1907) showed that variance of shell dimensions declines between younger and older age groups in two species of land snail. Berry & Crothers (1968) showed that populations of the dogwhelk, *Nucella lapillus* (L.), showed reduction in variance of shell shape between young and old cohorts collected from exposed shores but little or no reduction in those from sheltered shores. By analogy, it may be expected that in *L. pallescens*, yellows should exhibit smaller variance in shape, and a greater reduction with age, than darks. The present investigation was designed to confirm the difference in shell properties observed previously (Cook *et al.*, 1985), and to test this prediction.

## Materials and methods

Thirty-nine samples, amounting to a total of over 2500 individuals, were collected from leaves of mangroves of the genus *Rhizophora* at sites on the north coast of Papua New Guinea between the Huon Peninsula and Lasanga island in Morobe Province. Details are given by Cook (1986). The animals were removed from the shells, which were dried and stored for further examination. In the present investigation, material from 8 of these samples was used. Dark was always the commonest morph. The yellow shells were therefore taken from each of the 8 samples, and then the same number of darks showing a similar range of size. A total of 346 individuals of each morph was selected in this way, equal numbers of each morph

coming from each site in order to control for any variation in shell parameters which may occur between sites. The height and breadth of each shell was measured to the nearest 0.1 mm, using vernier callipers, and the shell was weighed on a digital balance to the nearest 10 mg.

## Results

For the dark morph the shell height had a range in the sample from 4 mm to 27 mm, with a mean and standard deviation of 15.1 $\pm$ 3.9. The range in the yellow sample was 4 mm to 22 mm, with a mean standard deviation of 10.5 $\pm$ 3.7. The mean for darks is significantly greater than for yellows ($t = 7.02, P < 0.001$). The median values are similar, however, being in the 12 mm class for darks and the 11 mm class for yellows.

The relation of shell height and breadth to weight is given by the equations in Table 1. There is a linear relation of the logarithm of the linear dimensions to the logarithm of weight, with slopes close to, but a little below, one-third. The gradient given is the major axis, as best representing the relationship between dimensions when there is no dependence of one variable upon another. It cannot be tested for significance, but using the regression coefficients or the reduced major axis (the geometric mean of the regression coefficients), tests may be carried out (Ricker, 1973). For all four equations the regression coefficients are significantly less than one-third, and the reduced major axis is significantly less in three of the four cases ($P < 0.01$, the exception is for height on

*Table 1.* Relation of inear dimensions to weight in samples of dark and yellow shells, each consisting of 346 individuals. The slope given is the major axis; $r$ is the correlation coefficient.

| (a) log height on log weight | | |
|---|---|---|
| Dark: | $y = 0.29x + 4.11$, | $r = 0.87$ |
| Yellow: | $y = 0.31x + 4.09$, | $r = 0.96$ |
| | | |
| (b) log breadth on log weight | | |
| Dark: | $y = 0.28x + 3.63$ | $r = 0.94$ |
| Yellow: | $y = 0.29x + 3.61$, | $r = 0.96$ |

weight in darks). Shells therefore get relatively thicker as they increase in height or breadth. The difference between slopes for the yellow and dark morphs is not significant.

Relative robustness was examined by calculating log (height × breadth/weight$^{2/3}$) for each individual in each colour class. For the darks, the mean and standard error were 7.50 ± 0.012, while for yellows they were 7.56 ± 0.010. The mean for yellows is significantly larger than for darks ($t = 3.25$, $P < 0.01$), showing that, weight for weight, yellows have the larger external dimensions. This test is therefore consistent with the earlier finding that yellows were thinner and less strong.

In order to examine change in variance with size, the samples of each morph have been ranked in order of increasing weight, and then divided into smaller-sized and larger-sized groups of 173 individuals each. The major axes shown in Table 1 have been used to represent the relation of height and breadth to weight for each morph. The minimum deviation of each individual from the major axis in each dimension was then calculated. This is given by $(y - ax - c)/(1 + a^2)^{1/2}$, where $y$ is the log height or log breadth, $x$ is log weight and $a$ and $c$ are the slope and intercept respectively (Cook, 1979). The values of $a$ and $c$ are given in Table 1. The shortest distance from point to line is the square root of the sum of squares of the deviations in the two defined dimensions. Mean squared deviations are then calculated for the large-shelled and small-shelled halves of each distribution. The results are given in Table 2.

The yellow morph shows a smaller mean squared deviation than the dark for both the small-shelled and the large-shelled sections of the distributions. In both morphs, the mean squared deviation declines with increase in size, to a greater extent in yellows than in darks. Tested as variance ratios, the four ratios shown in Table 2 are significant ($P < 0.01$). The thinner-shelled morph has the smaller variation overall, and the variance ratios indicate a greater reduction in variation with increase in size, than in the thicker-shelled, dark morph.

*Table 2.* Mean squared deviation of observations from major axis for height on weight and breadth on weight. The data are divided by weight into two equally sized samples of 173 individuals each for the dark and the yellow morph. For method of calculation of mean squared deviation, see text.

| Morph | Weight | | Ratio |
|---|---|---|---|
| | Lighter half | Heavier half | |
| Dark | 0.0392 | 0.0182 | 2.16 |
| Yellow | 0.0215 | 0.0073 | 2.94 |
| Ratio | 1.82 | 2.48 | |

## Discussion

In this study, the yellow morph has been shown to be less heavy for given external dimensions than the dark. It also shows smaller variability in shell shape than does the dark morph and a greater reduction in variability with increase in size. The result is compatible with stabilizing selection acting upon individuals as they grow. Of course, it does not follow that the explanation is correct; an alternative explanation could perhaps be that dark individuals experience haphazard secondary thickening of the shell which increases their variability, although in itself this does not explain why variability should decrease with size. Selection appears to operate through visual predation (Hughes & Mather, 1986; Reid, 1987), however, and may do so through thermoregulatory differences. It is not surprising if it imposes stabilizing selection.

The question of what kind of change in variance should occur with increase in size has been discussed by Manly (1985). He points out that decrease in variance can only be interpreted with certainty as a consequence of stabilizing selection if a control series without selection does not show a similar decline in variance. The best situation would be one where individuals can be followed throughout their lives to ensure that the index used is invariant with age. This is undoubtedly true, but the alternative approach of comparing change in variance in samples living under different selective conditions can be very informative. Thus, Berry & Crothers (1968) found that the variance of the character they studied in *Nucella lapillus* decreased with size at a rate which was related to the degree of exposure of the shore on which they lived. In another example, quoted by Manly, Parker (1971) subjected some samples of small salmonid fish to predation by other salmon species, while there was no predation in other samples. Variation decreased more in the samples exposed to predation. The approach used here, of comparing reduction in variation in two classes known to differ in physical properties of the shell, also goes some way towards meeting Manly's objection.

Manly (1985) also suggests that the experimental procedure may lead to a decrease in the coefficient of variation. The argument concerns the relation of one linear measurement, $y$, to another, $x$, which is used as an indication of age. If these are related as $y = a\,x$, then their standard deviations should be $S_y \simeq a\ S_x$ and the coefficients of variation will be $\dfrac{S_y}{y} \simeq \dfrac{S_x}{x}$. *If the x* variable is divided into very small class intervals then the estimated standard deviation for each of them will remain approximately constant, and the coefficient of variation of $y$ will vary inversely with $x$. This criticism applies only under the particular conditions specified by Manly, however, which do not usually hold in experimental studies. It seems much more likely that the coefficient of variation (or the standard deviation of the logarithms of the variable) should remain constant or increase with increase in size in the absence of modifying factors. For most organisms, the starting point for increase in linear dimensions is an exponential increase in volume. Starting with unit volume, the volume at time $t$ is $W_t = e^{rt}$, where $r$ is a growth constant greater than 1. Pielou (1977) shows the variance at time $t$ to be $e^{rt}(e^{rt} - 1)$. Consequently, the coefficient of variation is $e^{-\frac{1}{2}rt}(e^{rt} - 1)^{\frac{1}{2}}$, which rapidly converges on unity (or some other number depending on the units of initial volume) with increase in $t$. Accidental variation superimposed on the growth process will increase, rather than reduce, the variation (Cook, 1979).

In organisms such as those molluscs which grow more or less exponentially and do not have a fixed adult size, the final distribution of size for a cohort depends on the interaction of growth and mortality, and it seems reasonable to expect that reduction in variability indicates selective elimination. In endotherms, and many molluscs, there is a fixed adult size, and the growth pattern is often sigmoidal. In that case, variability may decrease as a consequence of the processes determining adult size. When this occurs, it must often indicate constraints built into the normal growth process by selection operating through long periods of evolution to optimise survival. Foote & Cowie (1988) compared variability in early and late whorls in adult shells of the land snail *Theba pisana* (Müller), and showed that variance declines with growth. Since the sample consists of adults collected when they were living the result is not due to selective elimination, and they conclude that developmental homeostasis is involved. As with other types of homeostasis, developmental homeostasis reduces direct dependence on the environment and is a substitute for direct selective tuning exerted by the environment from generation to generation. We do not know the details of the growth pattern in *L. pallescens*. The evidence presented here, however, shows that variance declines with growth, and the differences between the two morphs are most readily interpreted as a direct effect of natural selection.

## Acknowledgements

The material was collected on a visit to Papua New Guinea with a party from Oakham School in 1984, supported by the Royal Society and with the permission of the Morobe Provincial Government. I am grateful to Mr. J.H.K. Lindsay for making the measurements, to Dr. Tim Gunn for organization and the Professor R.J. Berry for comments on the typescript.

# References

Berry, R. J. & J. H. Crothers, 1968. Stabilizing selection in the dogwhelk (*Nucella lapillus*). J. Zool., Lond. 155: 5–17.

Cain, A. J., 1988. The scoring of polymorphic colour and pattern variation and its genetic basis in molluscan shells. Malacologia 28: 1–15.

Cook, L. M., 1979. Variation in the Madeiran lizard *Lacerta dugesii*. J. Zool., Lond. 187: 327–340.

Cook, L. M., 1983. Polymorphism in a mangrove snail in Papua New Guinea. Biol. J. linn. Soc. 20: 167–173.

Cook, L. M., 1986. Site selection in a polymorphic mangrove snail. Biol. J. linn. Soc. 29: 101–113.

Cook, L. M. & P. M. Freeman, 1986. Heating properties of morphs of the mangrove snail *Littoraria pallescens*. Biol. J. linn. Soc. 29: 295–300.

Cook, L. M., J. D. Currey & V. H. Sarsam, 1985. Differences in morphology in relation to microhabitat in littorinid species from a mangrove in Papua New Guinea (Mollusca: Gastropoda). J. Zool., Lond. 206: 297–310.

Di Cesnola, A. P., 1907. A first study of natural selection in *Helix arbustorum*. Biometrika 5: 387–399.

Foote, M. & R. H. Cowie, 1988. Developmental buffering as a mechanism for stasis: evidence from the pulmonate *Theba pisana*. Evolution, Lancaster, Pa. 42: 396–399.

Freeman, P. M., 1986. Thermal and physical properties of the shell of the Indo-Pacific mangrove snail *Littorina pallescens* and associated species. M.Sc. thesis, University of Manchester. 131 pp.

Hughes, J. M. & P. B. Mather, 1986. Evidence for predation as a factor in determining shell colour frequencies in a mangrove snail *Littorina* sp. (Prosobranchia: Littorinidae). Evolution, Lancaster, Pa. 40: 68–77.

Manly, B. F. J., 1985. The statistics of natural selection. Chapman & Hall, London. 484 pp.

Murray, J. J., 1975. The genetics of the Mollusca. In R. C. King (ed), Handbook of genetics, vol. 3. Plenum, N.Y., 3–31.

Parker, R. P., 1971. Size selective predation among juvenile salmonid fishes in a British Columbia inlet. J. Fish. Res. Bd Can. 28: 1503–1510.

Pielou, E. C., 1977. Mathematical ecology. Wiley, N.Y., 385 pp.

Reid, D. G., 1986. The littorinid molluscs of mangrove forests in the Indo-Pacific region. British Museum (N.H.), London, 227 pp.

Reid, D. G., 1987. Natural selection for apostasy and crypsis acting on the shell colour polymorphism of a mangrove snail, *Littoraria filosa* (Sowerby) (Gastropoda: Littorinidae). Biol. J. linn. Soc. 30: 1–24.

Ricker, W. E., 1973. Linear regressions in fishery research. J. Fish Res. Bd Can. 30: 409–434.

Weldon, W. F. R., 1901. A first study of natural selection in *Clausilia laminata* (Montagu). Biometrika 1: 109–124.

*Hydrobiologia* **193**: 223–231, 1990.
*K. Johannesson, D. G. Raffaelli and C. J. Hannaford Ellis (eds), Progress in Littorinid and Muricid Biology.*
© 1990 *Kluwer Academic Publishers.*

# Adaptive and non-adaptive variation in two species of rough periwinkle (*Littorina*) on British shores

J. Grahame, P.J. Mill & A.C. Brown
*Department of Pure & Applied Biology, University of Leeds, Leeds LS2 9JT, England, UK*

*Key words: Littorina saxatilis, Littorina arcana*, shell morphology, intraspecific variation

## Abstract

Rough periwinkles are notoriously variable in shell characters. There are many reports of substantial local variation on single shores which not only make identification difficult but also may be difficult to understand in terms of likely selective pressures. We show that despite local variation in southern Britain there is evidence of a broader scale of change which is likely to be explicable in adaptive terms. At the same time, along an extreme environmental gradient on a single shore in south Wales we show that there are changes in morphology which are related to avoidance of water loss.

## Introduction

Many workers have reported considerable variation in populations of *Littorina saxatilis* (Olivi) on the same shore or on nearby shores. Janson (1983) showed that over about 2 km of Swedish coast with alternating exposed cliffs and sheltered boulder bays there was a corresponding alternation in morphs between those with a small, thin shell and a large aperture (exposed morph) and those with a larger, thicker shell with a smaller aperture (sheltered morph). Janson & Sundberg (1983) showed that populations only 60 m apart (though from different habitats) on these shores differed in shell morphology and penial gland number, while Janson & Ward (1984) showed that there was allozyme heterogeneity between what they called subpopulations, sometimes only metres apart. At one locus most of the differentiation was habitat-related and might therefore be indicative of selective processes.

Several authors have suggested adaptive reasons for the variation in the morphology of the shell and soft parts. Thus Brandwood (1985) and Johannesson (1986) have shown that the shape and strength of the shell depends upon the risk of crab predation, dislodgement by wave action, and the relative cost of shell construction. Hart & Begon (1982) found that populations from cliff crevice and boulder field populations on an exposed shore in Wales differed in shell and body characters; those from cliff crevices having relatively heavier brood weights but lighter shells. Moreover these crevice animals had larger and heavier individual embryos, suggesting greater parental investment per juvenile. Grahame & Mill (1986) showed that along a gradient of wave exposure in Wales both *L. saxatilis* and *L. arcana* Hannaford Ellis showed increasing foot area and aperture size with increasing exposure. We suggested that, since *L. arcana* had the greater foot area at any one site on the shore, and since it reached the most exposed rocks, where *L. saxatilis* was not found, tolerance to wave action might be an important ecological difference between the two species where they are sympatric. We also noted that on shores where *L. arcana* is absent, *L. saxatilis* appeared to achieve a larger aperture size than did animals in these Welsh populations where both species co-occur.

On a single rock slope at St Ann's Head, Dyfed, Wales, we have observed that the shells of *L. saxatilis* become more pointed with higher spires, and the apertures are relatively smaller, as one progresses up the slope away from mean high water springs. *L. arcana* is confined to the lowest portion of the slope and the shore below.

In this paper we suggest that, while local variation is often important, there is also a broader geographical scale of variation in *L. saxatilis* on wave-exposed shores. This is manifested in two shell shapes which are repeated over hundreds of kilometres, depending on whether *L. arcana* is found on the shore or not.

## Methods

### Field locations and species identifications

Collections of *Littorina* were made at locations from shores along the south, southwest and west coasts of Britain, between Kent and Gwynedd (Fig. 1). All the locations were exposed to the

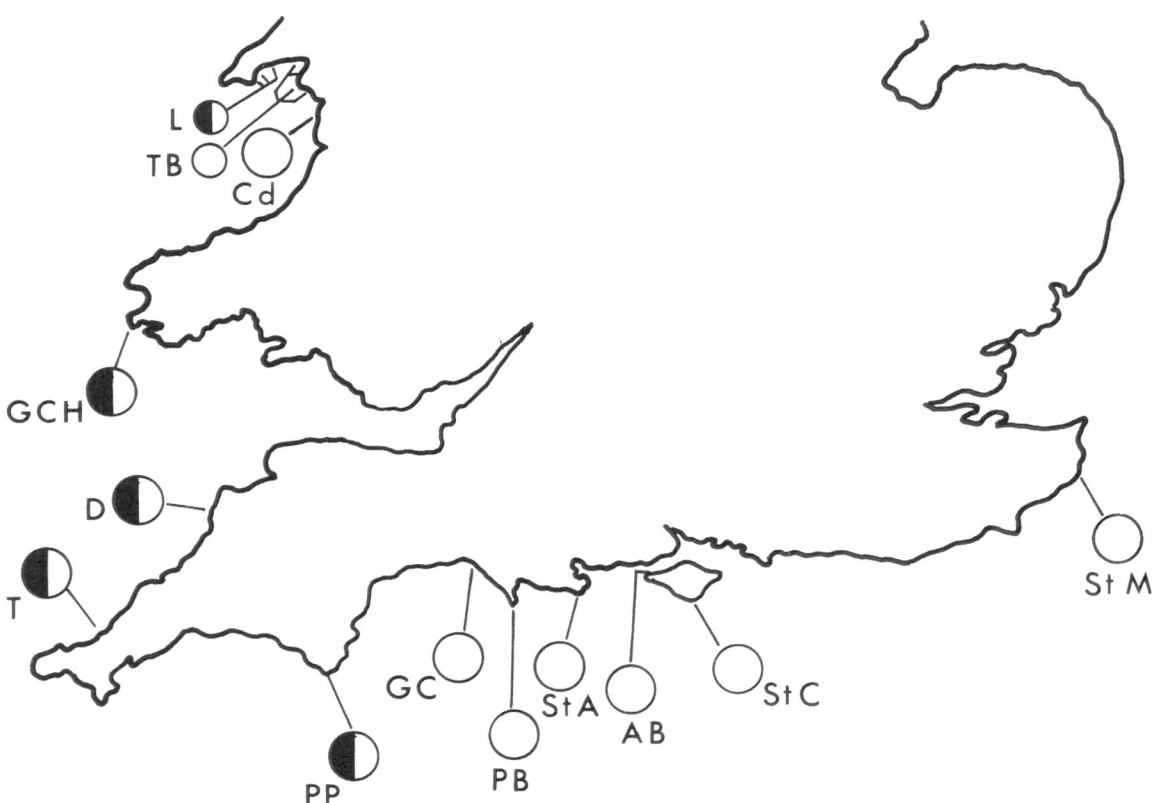

*Fig. 1.* Locations of collections used in this study. Shores on which *L. saxatilis* occurs without *L. arcana* are indicated with open circles, those on which both species occur together with half solid circles. The St Ann's Head site (see text) is very near the Great Castle Head site. Abbreviations (anticlockwise): L, Lleyn Peninsula; TB, Tremadoc Bay; Cd, Cae-du; GCH, Great Castle Head; D, Duckpool; T, Trevaunance; PP, Prawle Point; GC, Golden Cap; PB, Portland Bill; StA, St Alban's Head; AB, Alum Bay; StC, St Catherine's Head; StM, St Margaret's at Cliffe. The circles for the first two of these are smaller than the rest reflecting the use of these samples for brood size analysis alone, and not shell shape (see text).

open sea and were chosen so as to be as similar as possible to one another, showing maximum exposure for that section of the coast. Therefore we are dealing here with the shape of rough periwinkles under conditions of considerable (though unquantified) exposure to wave action.

Details of site locations and sample sizes are given in Tables 1 and 2. At each site about 200 specimens were taken; from these collections animals were removed and determined as to species before measurement, the aim being to obtain a sample of measured shells of between 20 and 30. The Alum Bay sample is unusually small due to a combination of scarcity of animals and an unfavourable tide. At Prawle Point shells from two sites about 350 m apart have been used in the analysis; these samples were pooled after initial analysis showed no difference between them. Similarly at Trevaunance shells from two sites 100 m apart have been used but in this case the shells are different and have been kept as two groups in the analysis. At other locations collections from single restricted sites have been analysed. Bodies were extracted from the shells after boiling, or in some cases after deep-freezing and thawing, and determined for sex and species. The only wholly reliable criterion for initially distinguishing *L. saxatilis* from *L. arcana* is among females, where the presence of a brood pouch indicates *L. saxatilis* and a jelly gland indicates *L. arcana* (Hannaford Ellis, 1979). At Prawle Point and the two Cornish locations, where the two species occur together, only females have been used – for further details, see Mill & Grahame (this volume). The remaining location of co-occurrence is Great Castle Head where, after experience using the female reproductive tract character as an initial criterion, the shells were found to be sufficiently distinct with respect to colour and sculpture to separate the species, and hence we have included males from this site in the analysis. Similarly we have included males and females at the sites where *L. saxatilis* occurs alone. Throughout, great care was taken to use only those shells which were attributable unambiguously to one species or the other, and to avoid selecting shells on the basis of size or shape.

Table 1. Site and sample size data for shells used in shape analysis.

| Site and Ordnance Survey Grid reference | Numbers in analysis | |
| --- | --- | --- |
| | *L. saxatilis* | *L. arcana* |
| Cae-du, Wales (SH 565059) | 27 | absent |
| Great Castle Head (Westdale Bay), Wales (SM 797056) | 33 | 59 |
| Duckpool, Cornwall (SS 199114) | 21 | present, not measured |
| Trevaunance, Cornwall (SW 725519) | | |
| cliff | 26 | 15 |
| boulder | 7 | absent |
| Prawle Point, south Devon (SX 775351) | 29 | 28 |
| Golden Cap, south Devon (SY 407918) | 30 | absent |
| Portland Bill, Dorset (SY 675683) | 19 | absent |
| St Alban's Head, Dorset (SY 959754) | 21 | absent |
| Alum Bay, Isle of Wight (SZ 304851) | 13 | absent |
| St Catherine's Point, Isle of Wight (SZ 496753) | 31 | absent |
| St Margaret's at Cliffe, Kent (TR 368444) | 28 | absent |

*Measurement*

Measurements of shell dimensions and opercular area (Fig. 2) were made using a digitizer tablet and microcomputer (Grahame & Mill, 1986). In addition, for animals from some of the sites, foot areas were measured as described by Grahame & Mill (1986). For the site at Cae-du and for further sites in Cardigan Bay we have obtained limited shell measurements and have determined weights of the full brood pouch of *L. saxatilis*. Brood

Table 2. Site and sample size data for shells used for brood size determinations of *L. saxatilis* (all are in Wales).

| Site and Ordnance Survey Grid reference | Numbers of *L. saxatilis* in analysis | *L. arcana* |
| --- | --- | --- |
| Penrhyn Du (SH 323266) | 30 | present |
| Llanbedrog (SH 334315) | 30 | present |
| Pen-y-chain (SH 435353) | 25 | present |
| Graig Ddu (SH 522374) | 30 | absent |
| Llandanwg (SH 568291) | 31 | absent |
| Shell Island (SH 552265) | 31 | absent |
| Llwyngwril (SH 595108) | 30 | absent |
| Cae-du (SH 565059) | 30 | absent |

pouch weights were determined by removing the entire brood pouch and contents, drying at 80 °C for 12 hours (determined to give constant weight), and weighing to the nearest 0.01 mg.

Values for apical angle (Fig. 2) were all normally distributed and have not been transformed before analysis. In the case of the other dimensions the data were skewed and hence these measurements have been transformed to logarithms to base 10 before analysis. All values used in analyses of brood weight and of water loss have also been transformed to logarithms.

Data have been analysed using canonical variate analysis (Reyment *et al.*, 1984) in the Statistical Analysis System (SAS) package (SAS Institute Inc., 1985). Foot area has been shown to vary with degree of exposure (Grahame & Mill, 1986); it is also likely to be influenced by shell shape, and this may be reflected in shell dimensions. Thus we have used the STEPWISE proce-

dure in SAS to identify those shell dimensions which are the best predictors of foot area, relying on forward selection and the maximum $R^2$ methods of selection of predictor variables. The problem of selection of predictor variables has been discussed by Sokal & Rohlf (1981) and we have followed their practise of using the 5% significance criterion to add a variable to the model and 10% for that variable to remain in the model. As would be expected from the discussion in Sokal & Rohlf (1981), the FORWARD and MAXR options in STEPWISE identified the same models.

*Desiccation*

Two collections of *L. saxatilis* were made – the first from about mean high water spring tide level ('low' level) and the second from the limit of their vertical extent upward (about 7 m from mean high water springs, 'high' level) on a rock face which slopes upwards towards the sea on St Ann's Head, Dyfed (SM 809028). Mean high water spring level was determined by direct observation of an appropriate tide. In each case about 100 animals were collected. In the laboratory the animals were placed in 500 ml beakers with sea water-dampened paper towel, covered, and left overnight. The next day they were removed, blotted dry, weighed and placed in the cells of plastic trays divided into individually numbered compartments. Two such sets were made up, one each from the high and low sites, and each set was sealed in a large plastic box with about 500 g of anhydrous calcium chloride. After three hours the animals were removed, reweighed, and replaced in the desiccators. The experiment was continued for a further three hours and terminated after the third set of weights was obtained. The animals were then killed in boiling water. The shells were dissolved in 10% nitric acid in order to recover all the body tissue, which was dried at 80 °C for 12 hours and weighed to the nearest 0.1 mg. Such brief acid treatment has been used previously by Grahame (1977) and has no detectable effect on weight determinations. The opercula were kept

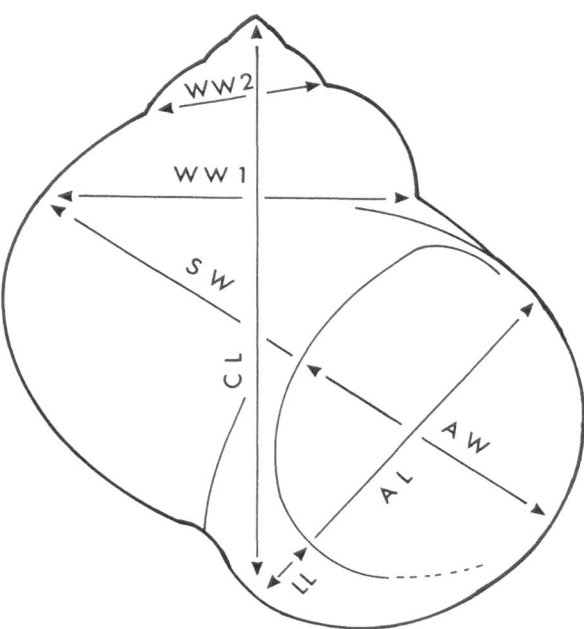

*Fig. 2.* Camera lucida drawing of a shell showing the dimensions measured. Abbreviations: CL, columellar length; SW, shell width; AW, aperture width; WW1, whorl width 1; WW2, whorl width 2; AL, aperture length; LL, lip length. The apical angle is defined as the angle subtended by lines joining the extent of whorl width 1 to the apex of the shell.

for measurement of area and circumference using the digitizer.

Since species could not be determined reliably before the experiment, the low level sample included both *L. arcana* and *L. saxatilis*, while the high level sample was entirely *L. saxatilis*. The data for weight loss under desiccation have been analysed using analysis of covariance.

## Results

### Shell shape

The scatter diagram of a total of 387 shells from the sites described on the first three canonical variates is shown in Fig. 3. There are differences between the samples represented in the figure. This was shown by using the Mahalanobis dis-

tance and its associated *F* statistic (Reyment *et al.*, 1984). In this application the degrees of freedom for *F* are the number of variables measured (nine) and the quantity (observations – groups – variables + 1). With 387 shells from sixteen samples (groups) and nine variables, the latter is 363. As there are sixteen samples, there are 120 comparisons, which causes difficulties in the interpretation of significance in multiple tests. A conservative solution to this problem is to divide the desired level of probability by the number of tests (comparisons) to obtain a new level of probability against which the computed values of *F* may be tested. Thus the 5% level should be tested using the probability level $P = 0.0004$ (i.e. 0.05/120). Using this criterion, samples from two pairs of shores were found not to be significantly different in shell shape, namely *L. saxatilis* at Golden Cap and St Margaret's at

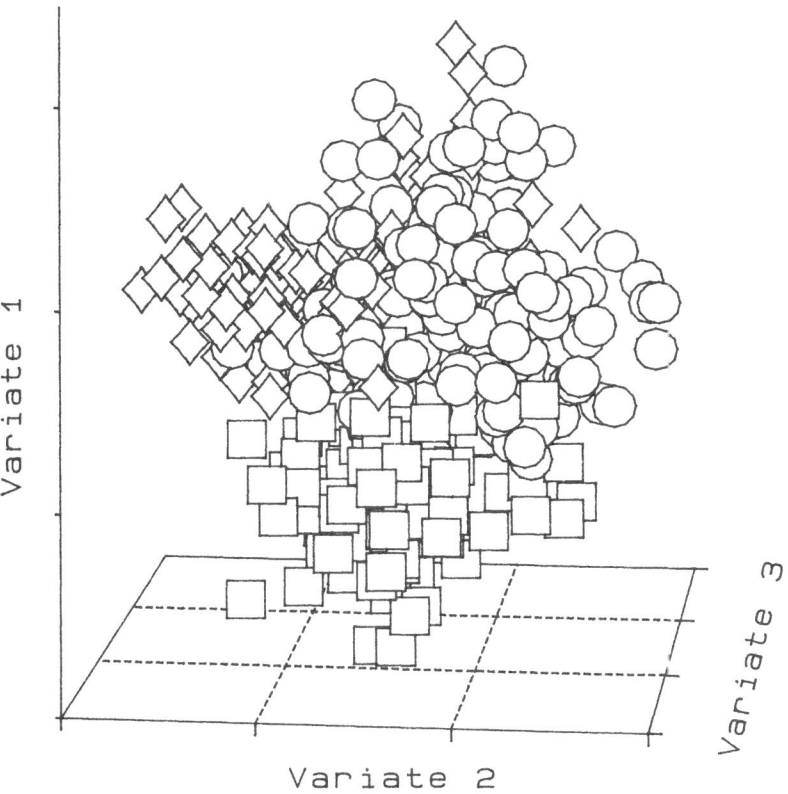

*Fig. 3*. Scatter diagram of individual shells from three groups plotted on the first three canonical variates in a canonical variate analysis (see text). ○ *L. saxatilis*, allopatric; ◇ *L. saxatilis*, sympatric; ⊏ *L. arcana*.

228

Cliffe ($F = 2.451$, $P = 0.0102$) and at Portland Bill and St Catherine's Point ($F = 2.271$, $P = 0.0174$).

To see if the results were influenced by size alone we repeated the analysis after transforming to remove the effect of size (Reist, 1984). We have used a geometric mean approach, namely

$$\log X_1 - (\Sigma \log X_i^P)/P$$

where $X$ is the measure of the variable and $P$ the number of variables involved in the transformation. Again, the measure of apical angle was excluded from this transformation. This procedure resulted in no alteration in either the plot of the shells on the canonical variates or in the relative values of the coefficients forming the standardized canonical vectors.

We consider that there is local between-site variation which is in accord with those workers reporting differences between sites on the same shore. However, it is also clear from Fig. 3 that *L. arcana* separates almost perfectly from *L. saxatilis* on canonical variate 1, while on canonical variate 2 the shells of *L. saxatilis* show a good degree of separation into those which come from shores without *L. arcana* (we refer to these as 'allopatric') and those with *L. arcana* ('sympatric').

The vectors allow identification of the dimensions most important in producing the separations indicated in Fig. 3. For the separation of the species on variate 1, the four most important dimensions are operculum area ($-1.47$), aperture length ($-1.41$), lip length ($1.46$) and whorl width 1 ($2.65$). For the separation of *L. saxatilis* on variate 2, the dimensions are columellar length ($-10.16$), apical angle ($-1.16$), shell width ($2.45$) and whorl width 1 ($5.50$). In terms of the shape of the shell, these dimensions express the relative size of the aperture and operculum, and the globosity and jugosity of the shells. A more detailed account and appraisal of much of these data is given in Grahame & Mill (1989).

For a limited number of sites (St Margaret's at Cliffe, Portland Bill, Golden Cap and Great Castle Head) measurements of foot area were

*Table 3.* Regression statistics for foot area on shell variables. The partial regression coefficient and its significance level are shown, together with the contribution of $R^2$ and the standard partial regression coefficient (SPRC).

| Variable | b | P | R² | SPRC | n |
|---|---|---|---|---|---|
| *Littorina saxatilis* | | | | | |
| aperture length | 1.033 | 0.0001 | 0.895 | 0.542 | 99 |
| operculum area | 0.418 | 0.0010 | 0.011 | 0.418 | 99 |
| *Littorina arcana* | | | | | |
| columella length | 1.881 | 0.0001 | 0.920 | 0.959 | 26 |

made. The regression analysis identified the dimensions aperture length and, to a much lesser extent, operculum area as predictor dimensions in the case of *L. saxatilis*. Only columellar length was identified in the case of *L. arcana*, for which the sample size is very small (Table 3).

For eight sites in north Wales (Fig. 1), four in north Cardigan Bay and four on the south side of the Lleyn Peninsula, we have data for brood weight and the shell dimensions described above. We have used the brood weight data in a multiple regression which identifies aperture width (and to a lesser extent operculum area) as predictors of brood weight (Table 4). However, because shell measurements were made by ACB in this case, while JG made all those in the preceding canonical variate analysis, we have not included the latter data set in the shell shape analysis.

Thus, of the four dimensions identified as being of importance in separating the species (Fig. 3), two (aperture length and operculum area) are also of importance in predicting foot area (Table 3). However, the dimensions most important in separating the forms of *L. saxatilis* are not the same as the most important in predicting brood size

*Table 4.* Regression statistics for brood weight in *Littorina saxatilis* on shell variables. The statistics shown are as in Table 3.

| Variable | b | P | R² | SPRC | n |
|---|---|---|---|---|---|
| aperture width | 2.158 | 0.0001 | 0.644 | 0.580 | 237 |
| operculum area | 0.477 | 0.0005 | 0.018 | 0.260 | 237 |

(aperture width and operculum area, Table 4), although one of them (columella length) is important in predicting foot area in *L. arcana*.

### Water loss

The numbers in the experiment for which data were obtained were 37 high site animals and 55 low site animals. All the females (25) in the former were *L. saxatilis* and it is therefore assumed that this was the only species at this site (except for *Melaraphe neritoides* (L.), which was not included in the experiment). Among the low site animals, 22 were mature females; 13 (59%) were *L. saxatilis* and 9 (41%) were *L. arcana*. By eye there is no apparent difference in shape between the species at this site (this has not yet been analysed) and this experiment therefore investigates water loss from animals of a high shore shape adopted by *L. saxatilis* and those of a lower shore shape adopted by both species. The possibility of differences in water loss between *L. saxatilis* and *L. arcana* of the same shape remains to be investigated.

Figure 4 shows the relationship between water loss during the second period of desiccation and dry body weight at the high and low sites. The regression slope for the low site animals is not significantly different from that for the high site ones (probability of different slopes, $P = 0.9017$) but the intercepts are different (probability of $intercept_1 = intercept_2$, $P = 0.0041$). At a dry body weight of 10 mg, the regressions allow estimates of a loss of 0.56 mg of water over the three hours at the high site, and 0.79 mg at the low site, i.e. 5.6% and 7.9% of dry body weight respectively.

Figure 5 shows the relationship between water loss over the same time period and operculum perimeter. This measure was chosen since much of the water is presumably lost from a drying animal past the seal made between operculum and shell. In this case the slopes of the regression lines appear to differ by eye, although the null hypothesis that $slope_1 = slope_2$ was not rejected ($P = 0.1552$). Since the lines actually cross, interpretation of the intercepts is difficult. However, it

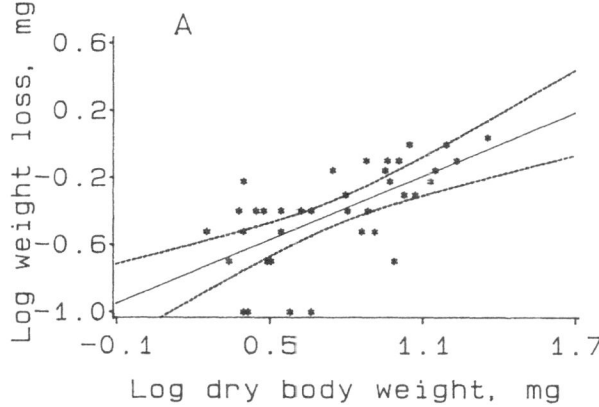

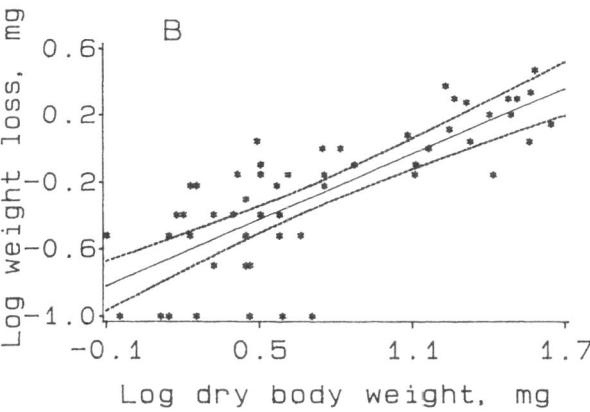

*Fig. 4.* Water loss as a function of dry body weight at the high shore site (Fig. 4a) and low shore site (Fig. 4b), St Ann's Head. The least squares regression lines and 95% confidence limits are shown. Sample sizes: high shore 37; low shore, 55.

appears that the smaller animals from both sites lose similar amounts of water past the opercular seal, but that the larger animals from the high site may lose less water in this way. Nevertheless the crossing of the lines suggests that some other factor, perhaps behavioural, is operating to reduce water loss in larger animals from the high shore site.

By calculating the perimeter each operculum would have if it were a perfect circle and dividing this into the observed perimeter value, we have obtained ratios expressing the departure of the

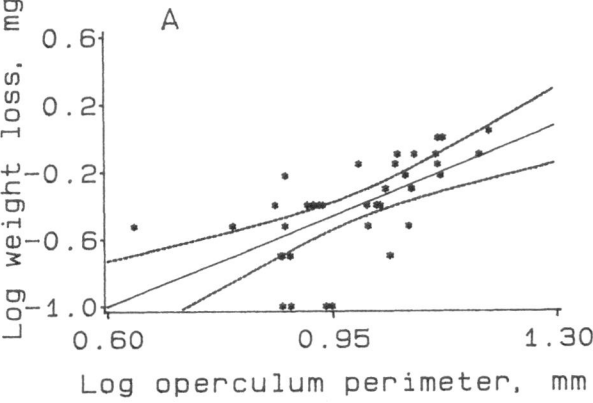

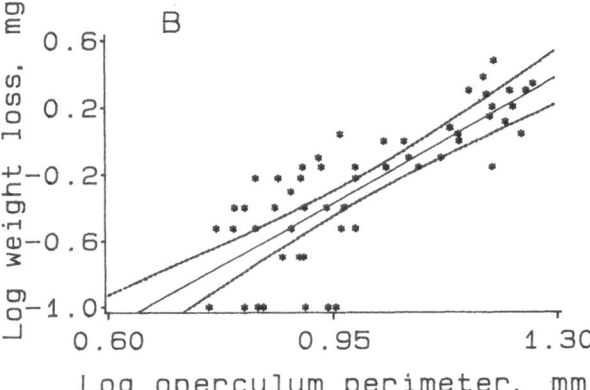

*Fig. 5.* Water loss as a function of operculum perimeter at the high shore site (Fig. 5a) and low shore site (Fig. 5b), St Ann's Head. The least squares regression lines and 95% confidence limits are shown. Sample sizes: high shore 37; low shore, 55.

operculum from circularity. This criterion has been adopted because the plane with minimum perimeter for a given area is a circle; thus the smaller the ratio the more nearly does the operculum approach a perfect circle, i.e. a minimal perimeter. The average ratios were 1.018 for high site animals, and 1.038 for low site ones. This difference is significant at the level $P = 0.0655$ (Mann-Whitney U-Wilcoxon rank sum W test). This indicates a slight increase of circularity of the operculum with height on the slope, with consequent reduction of the operculum perimeter. We suggest that this is so despite the slight increase in

the likelihood of Type I error inherent in accepting a probability level of 0.0655 rather than the conventional level of 0.05.

## Discussion

There is evidence that, despite variation between nearby shores, and sometimes even samples, there is in southern Britain a larger scale pattern of variation in *L. saxatilis*. Canonical variate analysis has identified operculum area and aperture length as important dimensions in separating *L. saxatilis* from *L. arcana* (see also Grahame & Mill, 1989), while multiple regression analysis has suggested that these dimensions are closely related to foot area. This indicates that a soft part character of importance in separating the species is foot area, a finding in accord with Grahame & Mill (1986). Thus, the shell shape differences between the species discussed here may be adaptive. The shell dimensions indicated as important in separation of groups of *L. saxatilis* – columellar length, apical angle, shell width and whorl width 1 – have not been convincingly identified with the other soft part dimension for which data are available – the brood weight. We therefore do not have good evidence of the shape differences between *L. saxatilis* on exposed promontories being adaptive.

All these dimensions are of course highly intercorrelated but the multiple regression analysis has identified those shell dimensions most involved in predicting the soft part dimensions. The analysis reported above is incomplete in the sense that, as yet, we lack data for all these dimensions taken from the same set of animals and this investigation is now in hand.

We have found that on a single and unusual rock slope (unusual in that it slopes upwards towards the sea) at St Ann's Head in south-west Wales there is a change in shape of rough periwinkles up the slope away from the level of mean high water springs. This is associated with a reduction in water loss per gram dry body weight. The amounts of water involved are very small, and it remains to be shown that this is of importance to the animals. To what extent the shape of the shells may be responsible for the slight reduction in water loss with increased exposure to desiccation is arguable. There is evidence of an increase in the circularity of the operculum high on the slope, with concommitant reduction in the perimeter of the operculum, but this difference was not statistically significant at the $P = 0.05$ level. Nevertheless we suggest that there is an adaptive change in shape of *L. saxatilis* on the St Ann's Head rock slope.

Appleton & Palmer (1988) and Palmer (this volume) have demonstrated a phenotypic response in two species of thaid gastropods to chemical stimuli from crab predators feeding on prey. The experimental thaids increased shell thickness during growth when exposed to the stimuli. In *L. saxatilis*, Smith (1981) has reported variation in shape of pre-emergence juveniles removed from the brood pouch, indicating genetically controlled variation in shell shape, and Janson (1982) has shown both environmental and genetic effects on growth rate. A phenotypic response may be involved in the change in shape of *L. saxatilis* on the St Ann's Head slope, induced perhaps by growth under conditions of desiccation. However, preliminary work indicates that there are differences in the shape of pre-emergence juvenile shells from the high and low sites, suggesting that there must be a genetic component as well; this is being investigated further. Johannesson & Johannesson (this volume) also discuss phenotypic differences in shape among rough periwinkles.

We conclude that much of the variation in shape evident in rough periwinkles may be adaptive, but that only in the case of foot area and its related shell dimensions were we able to demonstrate this.

## Acknowledgements

We are grateful to the University of Leeds for support for this work from the research fund of the University, and to undergraduate students for their interest in the St Ann's Head animals.

## References

Appleton, R. D. & R. A. Palmer, 1988. Water-borne stimuli released by predatory crabs and damaged prey induce more predator-resistant shells in a marine gastropod. Proc. Natn. Acad. Sci. U.S.A. 85: 4387–4391.

Brandwood, A., 1985. The effects of environment upon shell construction and strength in the rough periwinkle *Littorina rudis* Maton (Mollusca:Gastropoda). J. Zool., Lond. 206: 551–565.

Grahame, J., 1977. Reproductive effort and r- and K-selection in two species of *Lacuna* (Gastropoda: Prosobranchia). Mar. Biol. 40: 217–224.

Grahame, J. & P. J. Mill, 1986. Relative size of the foot of two species of *Littorina* on a rocky shore in Wales. J. Zool., Lond. 208: 229–236.

Grahame, J. & P. J. Mill, 1989. Shell shape variation in *Littorina saxatilis* (Olivi) and *L. arcana* Hannaford Ellis; a case of character displacement? J. mar. biol. Ass. U.K., in press.

Hannaford Ellis, C. J., 1979. Morphology of the oviparous rough winkle, *Littorina arcana* Hannaford Ellis, 1978, with notes on the taxonomy of the *L. saxatilis* species-complex (Prosobranchia:Littorinidae). J. Conch. 30: 43–56.

Hart, A. & M. Begon, 1982. The status of general reproductive-strategy theories, illustrated in winkles. Oecologia (Berlin) 52: 37–42.

Janson, K., 1982. Genetic and environmental effects on the growth rate of *Littorina saxatilis*. Mar. Biol. 69: 73–78.

Janson, K., 1983. Selection and migration in two distinct phenotypes of *Littorina saxatilis* in Sweden. Oecologia (Berlin) 59: 58–61.

Janson, K. & P. Sundberg, 1983. Multivariate morphometric analysis of two varieties of *Littorina saxatilis* from the Swedish west coast. Mar. Biol. 74: 49–53.

Janson, K. & R. D. Ward, 1984. Microgeographic variation in allozyme and shell characters in *Littorina saxatilis* Olivi (Prosobranchia: Littorinidae). Biol. J. linn. Soc. 22: 289–307.

Johannesson, B., 1986. Shell morphology of *Littorina saxatilis* Olivi: the relative importance of physical factors and predation. J. exp. mar. Biol. Ecol. 102: 183–195.

Reist, J. D., 1984. An empirical evaluation of several univariate methods that adjust for size variation in morphometric data. J. Fish. Res. Bd Can. 63: 1429–1439.

Reyment, R. A., R. E. Blackith & N. A. Campbell, 1984. Multivariate morphometrics, 2nd Edition. Academic Press, London, 233 pp.

SAS Institute Inc., 1985. SAS User's Guide: Statistics, Version 5 Edition. North Carolina: Cary, 956 pp.

Smith, J. E., 1981. The natural history and taxonomy of shell variation in the periwinkles *Littorina saxatilis* and *Littorina rudis*. J. mar. biol. Ass. U.K. 61: 215–241.

Sokal, R. S. & F. J. Rohlf, 1981. Biometry, 2nd edition. W.H. Freeman and Company, San Francisco, 859 pp.

Hydrobiologia **193**: 233–240, 1990.
K. Johannesson, D. G. Raffaelli and C. J. Hannaford Ellis (eds), Progress in Littorinid and Muricid Biology.
© 1990 Kluwer Academic Publishers.

# Distribution, abundance and shell morphology of *Littorina saxatilis* (Olivi) and *Littorina arcana* Hannaford Ellis at Robin Hood's Bay, North Yorkshire

C. Dytham, J. Grahame & P.J. Mill
*Department of Pure and Applied Biology, University of Leeds, Leeds LS2 9JT, England, UK*

*Key words:* colour polymorphism, rocky shores, gastropods

## Abstract

The distributions of the rough periwinkles *L. saxatilis* and *L. arcana* are considered in the northern half of Robin Hood's Bay, North Yorkshire, England. The relative proportions of these two species at different sites have been determined, using the criterion of the female reproductive system for their identification. The commonest shell colour morphs are grey (both species) and orange-banded (*L. saxatilis*). Hence information has been obtained not only on the proportion of banded animals in the total population, but on the proportion of banded *L. saxatilis*.

The northern end of the bay comprises a long, almost continuous, boulder field within which the proportion of *L. arcana* increases with greater exposure. Most of the boulder field comprises largely sandstone boulders but shale predominates at the most exposed site and, at the most sheltered site, which is separated from the rest of the boulder field by a short stretch of shingle, the boulders are exclusively shale. Banded morphs predominate in those regions where most of the boulders are sandstone but grey morphs predominate in the shale areas. A possible explanation for the distribution of these morphs is given in terms of gene flow. One site near the centre of the bay is unusual in that it contains *L. arcana* exclusively. However it comprises an artificial vertical substratum which is probably quite exposed.

## Introduction

Since it is only comparatively recently that *Littorina arcana* Hannaford Ellis has been separated from *Littorina saxatilis* (Olivi) (Hannaford Ellis, 1978), there is very little information on the relationships between the two species where they co-occur. The shells of both species are very variable and in most cases cannot be distinguished without reference to the body of the animal. Even then only mature females can be separated reliably, with males and immature females difficult or impossible to distinguish. Of the other two species in the *Littorina 'saxatilis'* complex, *L. nigrolineata* does not occur at Robin Hood's Bay and *L. neglecta* is found lower down the shore.

*L. saxatilis* and *L. arcana* are both direct developers (Hannaford Ellis, 1979, 1983); the young of *L. saxatilis* are brooded within the mantle cavity and emerge as fully formed 'crawlaways', while *L. arcana* lays an egg mass from which juveniles hatch. These modes of reproduction lead to rather slow dispersal of animals and hence of gene flow between populations, thereby promoting considerable variability between populations. Janson & Ward (1984) have shown that there can be gene frequency differences within a species over short distances; this has been supported by several subsequent studies (e.g. Ward *et al.*, 1986). Size frequency distributions and the shell morphology of populations of *L. saxatilis* have been shown to be affected by exposure (Janson & Sundberg, 1983), predation by crabs and birds (Pettitt, 1974; Hughes & Elner, 1979; Johannesson, 1986) and crevice availability (Raffaelli & Hughes, 1978).

As indicated above, the enormous variability of rough periwinkles has confounded efforts of taxonomy based solely on shell morphology (James, 1968; Heller, 1975). Positive identification of the two species in the field from shell characters alone is usually impossible. This study describes the relative abundance of the two species at different sites in the study area using the female reproductive system to identify them. However, at the sites described herein, shell characters can be used to identify some of the colour morphs to species without inspection of the soft parts. Thus field identification of unambiguous morphs ('orange-banded'), using a large number of animals, has been used to describe the colour variation in the population.

## Description of sites

Robin Hood's Bay is situated on the north-east coast of Britain, in the county of North Yorkshire. The bay is about five kilometres wide and faces north-east. It comprises a wide, wave-cut platform of shale (locally known as scars) stretching out to sea. It is backed by cliffs of shale and sandstone. Between the base of the cliff and the scars there is a region of sand or shingle beaches and large boulder fields on both the north and south sides of the Bay (Fig. 1).

For about 300 m north of the slipway (site 7) at Robin Hood's Bay village, the upper shore comprises sand and shingle. There is then a small boulder field (site 6) stretching for about 50 m along the shore. Northwards from this there is a shingle beach about 100 m long and then a larger,

continuous boulder field (sites 1 to 5) extending to beyond the northern limit of the bay.

At sites 1 and 6, the sea reaches the shale cliff face at high tide. In contrast, at sites 2 to 5 the tide never comes as high as the base of the cliff, which is a very unstable slope of small sandstone and shale rocks. Site 7 is a man-made seawall adjacent to the slipway and is reached by the sea at high tide.

## Materials and methods

Animals were collected from a single boulder at each site. In some cases several collections were made but always from the same part of the boulder. The technique involved removing all of the animals in a crevice to ensure no inadvertent bias towards particular morphs, although subsequently only those with a shell height of over 5 mm were used since smaller, immature animals were impossible to identify. In total 550 animals of appropriate size were obtained; these were boiled and the soft parts extracted for species identification (Hannaford Ellis, 1979). These data were used to obtain the relative proportions of *L. saxatilis* and *L. arcana* in the samples.

From amongst these animals, apex angle and seven linear variables of 30 banded *L. saxatilis* (see results) from each of sites 1–5 were measured using a digitiser. The linear variables were columella length (CL), shell width (SW), aperture height (AH), aperture width (AW), lip length (LL), width of penultimate whorl (WW1) and width of next to penultimate whorl (WW2), (Grahame & Mill, 1989). Animals from sites 6 and 7 were not used for shell measurements as there were not enough banded *L. saxatilis* available. The data were analysed using the Statistical Analytical System package (SAS: Statistics 1982).

In addition to the above mentioned collections, a total of 20 333 animals were scored for colour morph *in situ* from boulders around the 'site' boulders. The number of boulders used varied from site to site, depending on the local density of the animals, but was between 10 and 30. Again all

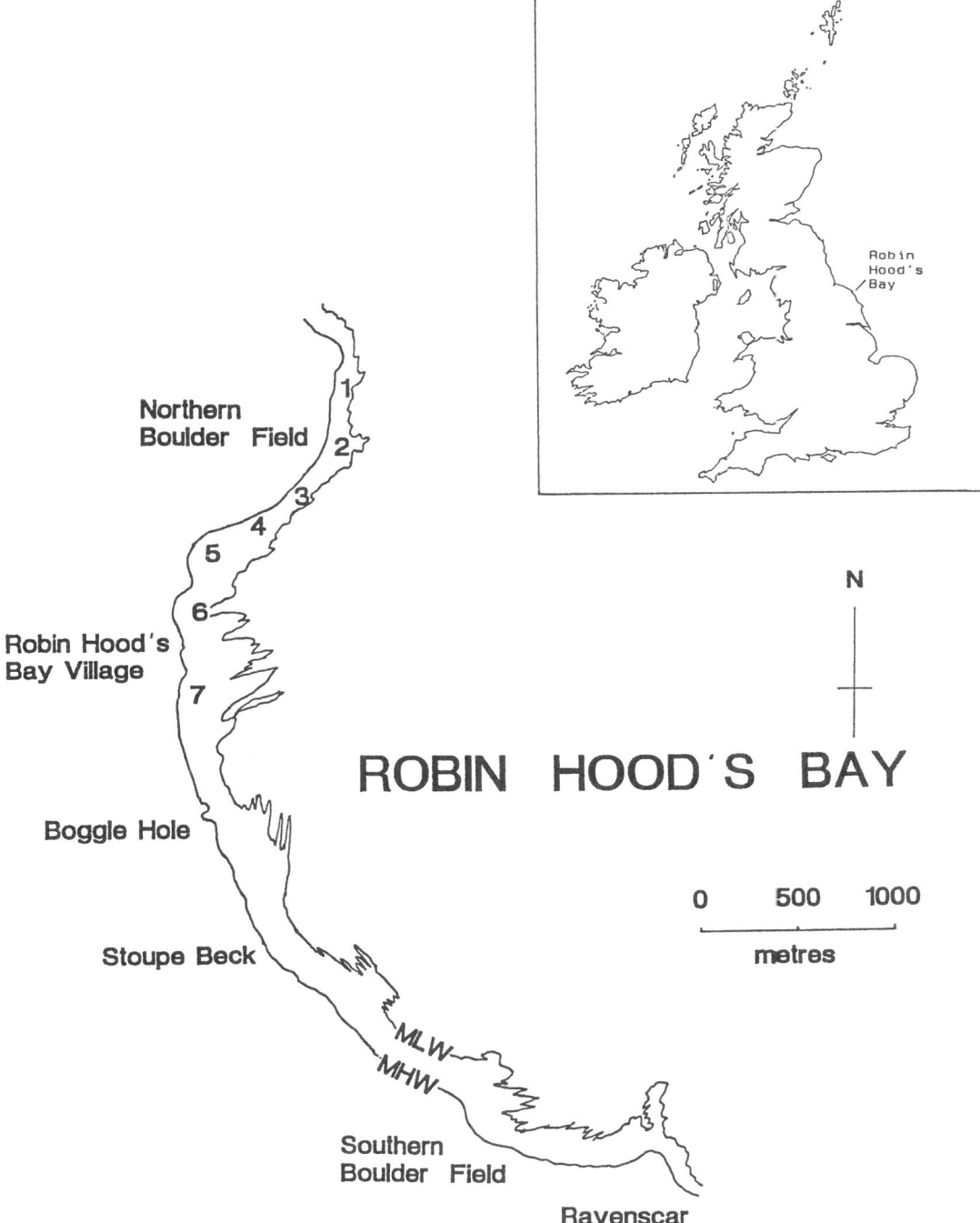

*Fig. 1.* Map of Robin Hood's Bay, indicating features of interest, together with (inset) map of the British Isles, showing the location of Robin Hood's Bay.

the animals from individual crevices were collected to avoid bias. Since the colour of the shell was being used as the criterion, all those animals in which the pattern could be readily determined, irrespective of size, were included. In some of the smaller animals the pattern was not clearly developed, and these were disregarded. The number discarded was in the region of 10% in each sample. From these data the number of banded animals at each site was estimated.

At each site between 105 and 120 boulders, including those from which the above collections were made, were scored as either sandstone or shale. The wave force was measured at sites 1 and 5. This was carried out by securing a dynamometer at each site, using the design of Denny (1983) with the slight modifications made by Grahame & Mill (1986). It was read once per day (two tides) over three days in June, 1988.

## Results

Both *Littorina saxatilis* and *L. arcana* were found at sites 1–6 but only the latter species was found at site 7. From Table 1 it can be seen that there are very few *L. saxatilis* at site 1. However, at site 2 about half the snails belong to *L. saxatilis* and the proportion steadily increases from here until almost all specimens belong to this species by the time site 6 is reached (Table 1, Fig. 2a). The mean

*Table 1.* Proportion of *L. saxatilis* in rough winkle samples, Robin Hood's Bay.

| Site No. | Sample size | *L. saxatilis* % population |
|---|---|---|
| 1 | 100 | 7 |
| 2 | 60 | 52 |
| 3 | 60 | 63 |
| 4 | 60 | 83 |
| 5 | 150 | 87 |
| 6 | 35 | 97 |
| 7 | 85 | 0 |
| Total | 550 | |

of the maximum wave forces recorded at site 1 was 5.4 N and that at site 5 was 1.8 N. This suggests a reduction in exposure from site 1 to site 5. The implications of this will be dealt with in the discussion.

In both species there are plain grey and plain yellow morphs, the former being the more common of these two in each species, while plain orange animals are always *L. saxatilis*. However, the plain yellow and orange animals form a very small proportion of the total population. Both species also have a morph with wide bands of colour on a grey background. In *L. saxatilis* the colour is orange, while in *L. arcana* it is cream.

Of the 20 333 animals scored for shell colour, 11 808 were orange-banded *L. saxatilis* (Table 2) and only 10 were cream-banded *L. arcana*. Con-

*Table 2.* Frequencies of banded morphs of *L. saxatilis* and sand stone boulders at Robin Hood's Bay sites.

| Site No. | Sample size | Banded *L. saxatilis* | Banded *L. saxatilis* % of total | Banded *L. saxatilis* % of *L. saxatilis* | % sandstone boulders |
|---|---|---|---|---|---|
| 1 | 2213 | 131 | 6 | 86 | 33 |
| 2 | 1972 | 811 | 41 | 79 | 96 |
| 3 | 3145 | 1638 | 52 | 83 | 92 |
| 4 | 5120 | 3732 | 73 | 88 | 85 |
| 5 | 6988 | 5458 | 78 | 90 | 91 |
| 6 | 490 | 38 | 8 | 8 | 0 |
| 7 | 405 | 0 | 0 | – | – |
| Totals | 22033 | 11808 | | | |

237

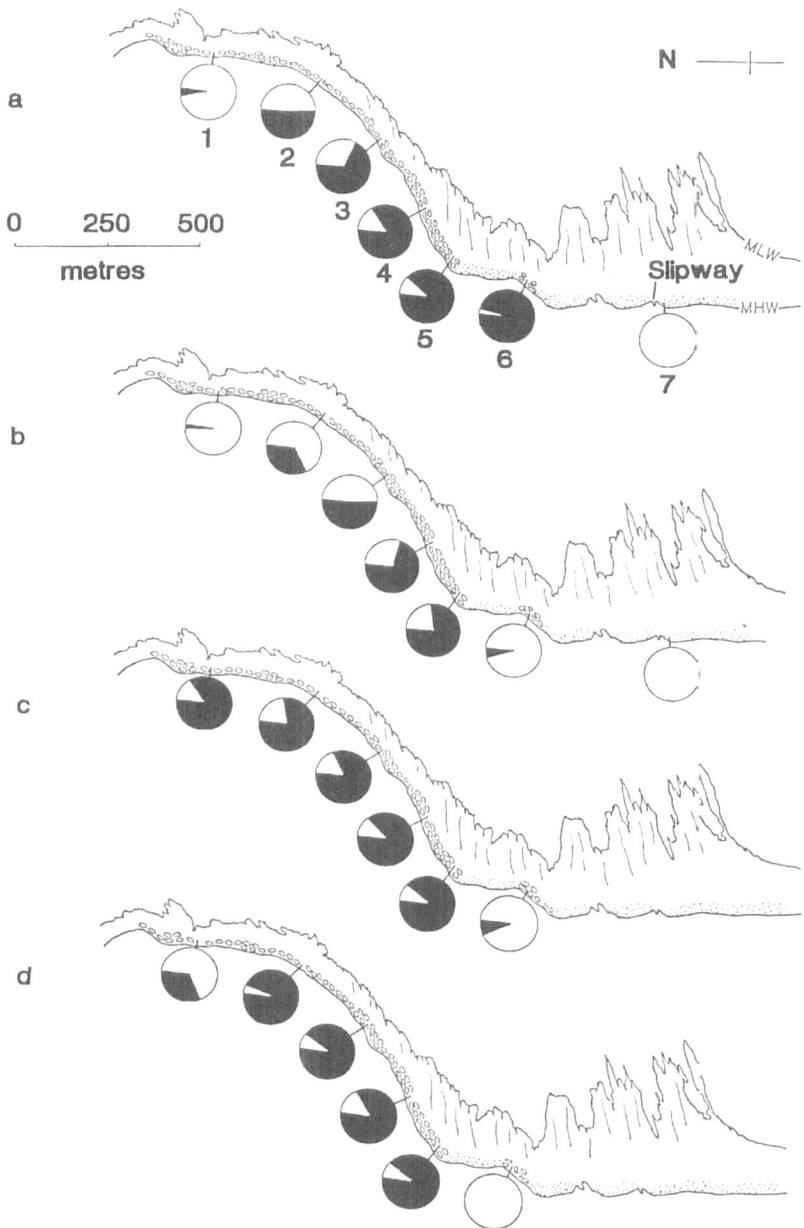

*Fig. 2.* a. Proportion of *L. saxatilis* (shaded) and *L. arcana*. b. Proportion of orange-banded *L. saxatilis* (shaded) and non-banded animals in the total population of *L. arcana* and *L. saxatilis*. c. Proportion of banded *L. saxatilis* (shaded) and other morphs of *L. saxatilis*. d. Proportion of boulders sandstone (shaded) and shale.

sidering the banded *L. saxatilis* as a proportion of the total number of rough periwinkles of both species, only a small percentage of the animals are banded at site 1 (where there are few *L. saxatilis*), while at site 2 this rises to nearly half of the population and then steadily increases to site 5 (Table 2, Fig. 2b), reflecting a parallel increase in

the proportion of *L. saxatilis* over these same sites (Table 1, Fig. 2a). However, at site 6 only 8% of the animals are banded even though *L. saxatilis* dominates the population here (97%). At site 7, where only *L. arcana* occurs, there are no banded animals. The percentage of banded *L. saxatilis* was fairly constant and high (between 79% and

238

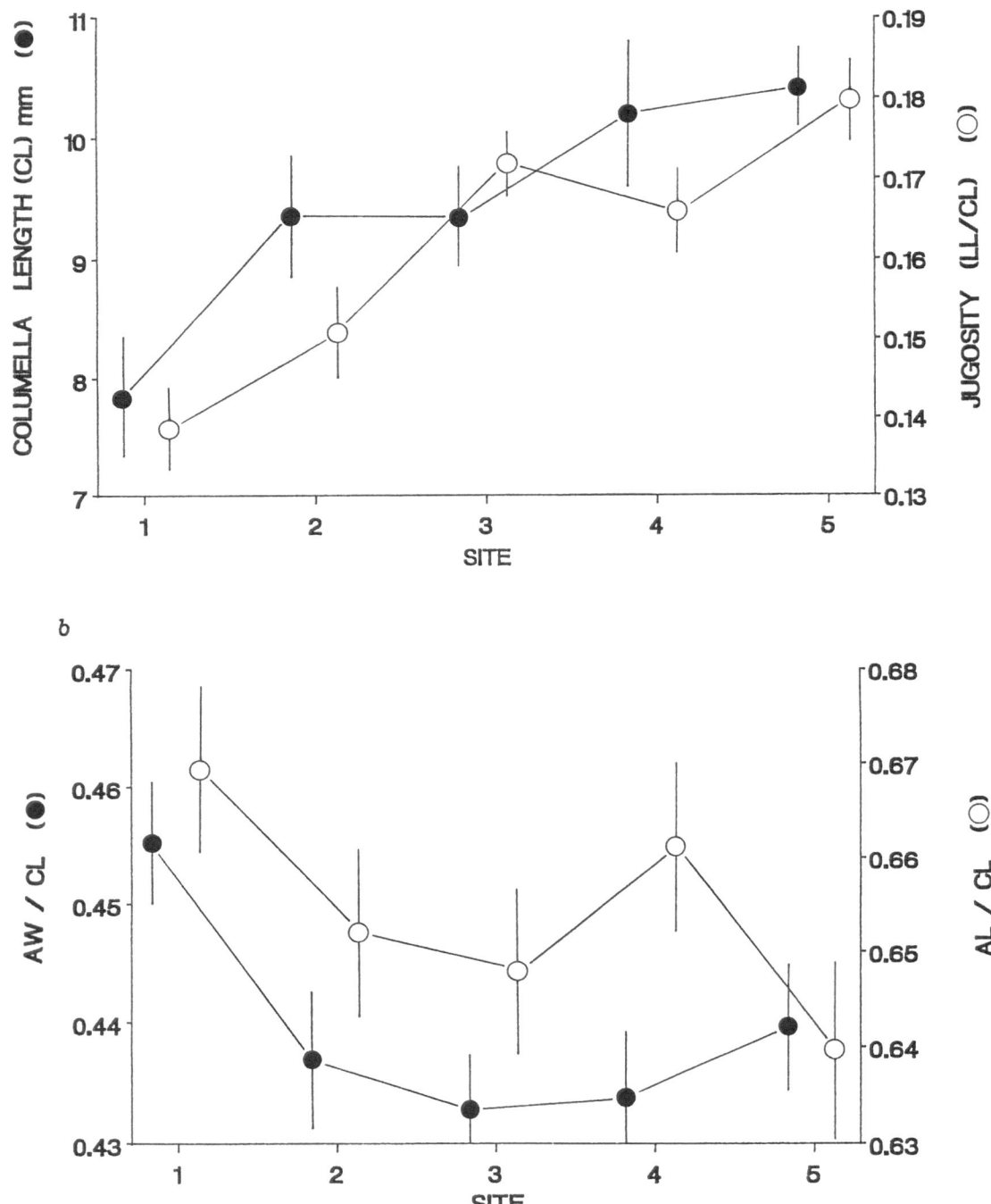

*Fig. 3.* a. Solid circles: size (columella length, CL); open circles: jugosity (ratio of lip length to columella length, LL/CL) – in orange-banded morphs of *L. saxatilis* at sites 1 to 5. Vertical bars show one standard error. b. Solid circles: relative aperture width (AW/CL); open circles: relative aperture length (AL/CL) – in orange-banded morphs of *L. saxatilis* at sites 1 to 5. Vertical bars show one standard error.

90%) at sites 1 to 5, but there were very few of this morph at site 6 (Table 2, Fig. 2c). At site 1, a third of the boulders are sandstone; at sites 2 to 5 they are almost all sandstone, while at site 6 they are all shale. There are no boulders at site 7 (Fig. 2d).

The shell measurements show that there is some change in the size and shape of banded *L. saxatilis* along the boulder field. The mean columella length (CL) increases as the exposure decreases from site 1 to site 5 (Fig. 3a). The relative size of the lip (LL/CL) also increases from site 1 to site 5, giving the animals from the more sheltered sites a jugose appearance. There is no trend in the relative size of the aperture (AL/CL and AW/CL) from sites 2 to 5. However, animals from the most exposed site (1) do have a slightly large relative aperture size than those from any other site (Fig. 3b).

## Discussion

Grahame & Mill (1986) noted that at Great Castle Head, Westdale Bay in Dyfed (Wales), *L. rudis* (a junior synonym of *L. saxatilis*) did not extend to the most exposed sites, which were occupied exclusively by *L. arcana*. It would appear that a similar situation occurs in the northern boulder field at Robin Hood's Bay, with *L. arcana* becoming relatively more abundant at more exposed sites.

Heller (1976) suggested that a larger aperture allows a larger foot and hence a greater adhesion to the rock surface to prevent dislodgement by waves. At more sheltered sites dislodgement is less of a problem but there may be increased predation by crabs, (Atkinson & Newbury, 1984; Pettitt, 1974), so a thicker shell and smaller aperture become common. Janson & Sundberg (1983) found that in Sweden (where *L. arcana* does not occur) *L. saxatilis* decreases in size and becomes more globose with increase in exposure. Grahame & Mill (1986) demonstrated that, on a rocky shore in Wales, both species showed an increase in foot area with exposure, although *L. arcana* always has the greater foot area. *L. arcana*, therefore, appears better suited to the most exposed sites and *L. saxatilis* to the most

sheltered. This is in accord with the increase in relative abundance of *L. arcana* in the present study along a transect from site 6 to site 1.

In this study the proportion of banded animals increases along the boulder field from sites 1 to 5 and this can be explained by the increase in proportion of *L. saxatilis* over these sites. However, at the sheltered, shale boulder field (site 6) the orange-banded morph of *L. saxatilis* again becomes rare even though almost all of the animals at this site are *L. saxatilis*. Thus it is possible that the incidence of this banded morph is related to the rock type, with grey shells of either *L. saxatilis* (site 6) or *L. arcana* (site 1) dominating populations on shale boulders. Grey shells may well have a cryptic advantage on the grey shale, but be more easily seen by visual predators on the sandstone (Pettitt, 1973; Atkinson & Warwick, 1983). The decrease in the proportion of grey morphs from site 5 to site 2 in the absence of any notable change in the proportion of sandstone boulders between these sites may be because the sandstone dominated boulder field is being constantly invaded by grey *L. arcana* from the shale dominated area to the north (site 1). There will presumably also be some migration of banded animals towards site 1 from this boulder field.

Site 6 is separated from the sandstone boulder field by a short shingle beach which may act as a partial barrier to the migration of animals from one to the other. Thus a predominantly grey population occurs at this site, protected from invasion by banded morphs from site 5 and having minimal effect upon the colour morph frequency in the site 5 population.

The banded morph of *L. arcana* is very rare. This could be explained by the fact that the bulk of the gene pool of this species is found in the shale dominated areas where a banded morph appears to be at a disadvantage. The similar morph of *L. saxatilis* fares very differently because the bulk of the gene pool of this species is in the sandstone dominated area where banding may be an advantage.

Site 7 appears to be unusual in that, although it is towards the centre of the bay it contains exclusively grey morphs of *L. arcana*. The sub-

240

strate at this site is an almost vertical sea wall, grey in colour; hence the grey morph may well be cryptic. The sea wall apparently presents a more exposed aspect than the boulders at, for instance, site 6. This may explain the absence of *L. saxatilis* from this site.

In a recent study Grahame & Mill (1989) have demonstrated that *L. saxatilis* on exposed promontories has a different shell shape depending on whether or not it is sympatric with *L. arcana*. The present study demonstrates both a decline in size and a change in shape of a single (orange-banded) morph of *L. saxatilis* along the boulder field with increasing exposure. It has been suggested by Sundberg (1988) that the major source of apparent shape variation is a side effect of size differences between samples. In the present study this problem has been met by using ratios, which are size-independent, to describe changes in shape. The changes we discuss here may indicate that stabilising selection is acting on the population by removing large animals and those with smaller apertures at more exposed sites. Alternatively, if the migration rate of animals between boulders is very slow then the changes could be due to gene frequency changes.

## Acknowledgements

One of the authors (CD) would like to thank the SERC for a research studentship, during the course of which this work was carried out.

## References

Atkinson, W. D. & S. F. Newbury, 1984. The adaptations of the rough winkle, *Littorina rudis*, to desiccation and to dislodgement by wind and waves. J. anim. Ecol. 53: 93–105.

Atkinson, W. D. & T. Warwick, 1983. The role of selection in the colour polymorphism of *Littorina rudis* Maton and *Littorina arcana* Hannaford Ellis (Prosobranchia: Littorinidae). Biol. J. linn. Soc. 20: 137–151.

Denny, M. W., 1983. A simple device for recording the maximum force exerted on intertidal organisms. Limnol. Oceanogr. 28: 1269–1273.

Grahame, J. & P. J. Mill, 1986. Relative size of the foot of two species of *Littorina* on a rocky shore in Wales. J. Zool. 208: 229–236.

Grahame, J. & P. J. Mill, 1989. Shell shape variation in *Littorina saxatilis* (Olivi) and *L. arcana* Ellis; a case of character displacement? J. Mar. biol. Ass. U.K. (in press).

Hannaford Ellis, C. J., 1978. *Littorina arcana* sp. nov: A new species of winkle (Gastropoda: Prosobranchia: Littorinidae). J. Conch. 29: 304.

Hannaford Ellis, C. J., 1979. Morphology of the oviparous rough winkle, *Littorina arcana* Hannaford Ellis, 1978, with notes on the taxonomy of the *L. saxatilis* species complex (Prosobranchia: Littorinidae). J. Conch. 30: 43–56.

Hannaford Ellis, C. J., 1983. Patterns of reproduction in four *Littorina* species. J. moll. Stud. 49: 98–106.

Heller, J., 1975. The taxonomy of some British *Littorina* species with notes on their reproduction (Mollusca: Prosobranchia). J. linn. Soc., Zool. 56: 131–151.

Heller, J., 1976. The effects of exposure and predation on the shell of two British winkles. J. Zool. 179: 201–213.

Hughes, R. N. & R. W. Elner, 1979. Tactics of a predator, *Carcinus maenas*, and the morphological responses of the prey, *Nucella lapillus*. J. anim. Ecol. 48: 65–78.

James, B. L., 1968. The distribution and keys of species in the family *Littorinidae* and of their digenian parasites in the region of Dale, Pembrokeshire. Field Studies 2: 615–650.

Janson, K. & P. Sundberg, 1983. Multivariate morphometric analysis of two varieties of *Littorina saxatilis* from the Swedish west coast. Mar. Biol. 74: 49–53.

Janson, K. & R. D. Ward, 1984. Microgeographic variation in allozyme and shell characters in *Littorina saxatilis* Olivi (Prosobranchia: Littorinidae). Biol. J. linn. Soc. 22: 289–307.

Johannesson, B., 1986. Shell morphology of *Littorina saxatilis* Olivi: the relative importance of physical factors and predation. J. exp. mar. Biol. Ecol. 102: 183–195.

Pettitt, C., 1973. An examination of the distribution of shell pattern in *Littorina saxatilis* (Olivi) with particular regard to the possibility of visual selection in this species. Malacologia 14: 339–343.

Pettitt, C., 1974. A review of the predators of *Littorina*, especially those of *L. saxatilis* (Olivi) [Gastropoda: Prosobranchia]. J. Conch. 28: 343–357.

Raffaelli, D. & R. N. Hughes, 1978. The effects of crevice size and availability on populations of *Littorina rudis* and *Littorina neritoides*. J. anim. Ecol. 47: 71–83.

SAS Institues Inc., 1982. SAS User Guide: Statistics. Cary North Carolina.

Sundberg, P., 1988. Microgeographic variation in shell characters of *Littorina saxatilis* Olivi – a question mainly of size? Biol. J. linn. Soc. 35: 169–184.

Ward, R. D., T. Warwick & A. J. Knight, 1986. Genetic analysis of ten polymorphic enzyme loci in *Littorina saxatilis* (Prosobranchia: Mollusca). Heredity 57: 233–241.

*Hydrobiologia* **193**: 241–260, 1990.
*K. Johannesson, D. G. Raffaelli and C. J. Hannaford Ellis (eds), Progress in Littorinid and Muricid Biology.*
© 1990 *Kluwer Academic Publishers.*

# Thermal tolerance, evaporative water loss, air–water oxygen consumption and zonation of intertidal prosobranchs: a new synthesis

Robert F. McMahon
*Section of Comparative Physiology, Department of Biology, Box 19498, The University of Texas at Arlington, Arlington, TX 76019, USA*

*Key words:* desiccation, emergence, intertidal zonation

## Abstract

Duration of emergence increases with tidal height on rocky shores therefore, emergence adaptations in intertidal species such as littorine and other prosobranch gastropods have been considered correlated with zonation patterns; temperature tolerance, desiccation resistance and aerial respiration rate all commonly assumed to increase progressively with increasing zonation level. Such direct correlations are rarely observed in nature. Maximal aerial gas exchange occurs in mid-shore, not high shore species. Temperature tolerance and desiccation resistance do not increase directly with shore height. Thus, hypotheses regarding physiological correlates of zonation require revaluation. A new hypothesis is presented that the high tide mark presents a single major physiological barrier on rocky shores. Above it, snails experience prolonged emergence and extensive desiccation; below it, predictable submergence and rehydration with each tidal cycle. Thus, desiccation stress is minimal below the high tide mark and maximal above it. Therefore, species restricted below high tide (the eulittoral zone) should display markedly different adaptive strategies to emergence than those above it (the eulittoral fringe). A review of the literature indicated that adaptations in eulittoral species are dominated by those allowing maintenance of activity and foraging in air including: evaporative cooling; low thermal tolerance; elevated aerial $O_2$ uptake rates; and high capacity for radiant heat absorption. Such adaptations exacerbate evaporative water loss. In contrast, species restricted to the eulittoral fringe display adaptive strategies that minimize desiccation and prolong survival of emergence including: foot withdrawal, preventing heat conduction from the substratum; aestivation in air; elevated thermal tolerance reducing necessity for evaporative cooling; position maintenance by cementation to the substratum and increased capacity for heat dissipation. In order to test of this hypothesis the upper thermal limits, tissue and substratum temperatures on emergence in direct sunlight and evaporative water loss and tissue temperatures on emergence in 40 °C were evaluated for specimens of six species of eulittoral and eulittoral fringe gastropods from a granite shore on Princess Royal Harbour near Albany, Western Australia. The results were consistant with adaptation to the proposed desiccation barrier at high tide. The eulittoral species, *Austrocochlea constricta, Austrocochlea concamerata, Nerita atramentosa* and *Lepsiella vinosa*, displayed adaptations dominated by maintenance of activity and foraging during emergence while the eulittoral fringe littorine species, *Bembicium vittatum* and *Nodilittorina unifasciata* displayed adaptations dominated by minization of activity and evaporative water loss during emergence. The evolution of adaptations allowing tolerance of prolonged desiccation have allowed littorine species to dominate high intertidal rocky shore gastropod faunas throughout the world's oceans.

## Introduction

Rocky shore prosobranchs display patterns of vertical distribution or zonation (Hughes, 1986; Newell, 1979; Underwood, 1979; Vernberg & Vernberg, 1972). As duration of emergence increases with tidal height, emergence adaptations of intertidal species have long been considered to be correlated directly with zonation level. Thus, temperature tolerance, desiccation resistance and ratios of aerial to aquatic gas exchange rates have all been commonly assumed to increase progressively with height of shore habitation (Newell, 1979; Underwood, 1979; Vernberg & Vernberg, 1972). In spite of the intuitive attractiveness of such hypotheses, direct correlations between physiological adaptations and zonation level are rarely observed in nature. Indeed, maximal capacity for aerial gas exchange generally occurs in mid-shore, not high shore meso- and neogastropods (McMahon, 1988), and temperature tolerance and desiccation resistance are not clearly correlated with shore height (Britton & McMahon, 1989; Broekhuysen, 1940; Brown, 1960; Cleland & McMahon, 1989; Evans, 1948; Frankel, 1966, 1968; McMahon & Britton, 1985; Stirling, 1982; Wolcott, 1973). Because the physiological adaptations of intertidal snails do not form distinct continua with increasing zonation height, hypotheses regarding their physiological correlates to zonation require revaluation.

A new hypothesis is presented suggesting that there are no continua of physiological adaptations with zonation height, rather, there is but a single major physiological barrier on rocky shores at the high tide mark above and below which distinctly different selective forces dominate. Certainly, the high tide mark centers on a zone of distinct transition in hard shore gastropod faunas. It is the upper limit for most eulittoral species (wetted each tidal cycle) and the lower limit for the majority of eulittoral fringe species (subjected to prolonged emergence between spring tides) (terminology from Newell, 1979) (Broekhuysen, 1940; Cleland & McMahon, 1989; Evans, 1948; Stephenson & Stephenson, 1972; Wolcott, 1973),

suggesting that it functions as a physiological barrier to species distributions. As modeled and reviewed by Johannesson (1989) for tidal shores, variation in time of emergence greatly increases above high tide, thus eulittoral fringe snails experience prolonged emergence and extensive desiccation; below it, variation in emergence is greatly reduced, thus eulittoral snails experience predictable submergence and rehydration with each tidal cycle, minimizing desiccation stress. As such, desiccation stress appears not to be a progressively increasing function of zonation. Rather, it is essentially minimal below the high tide mark and maximal above it with a relatively narrow transition zone between. Given a marked difference in desiccation pressure between these two habitats, eulittoral fringe species should display emergence strategies distinctly different from those of eulittoral species. Data on intertidal snails from the literature are utilized to support this new hypothesis regarding the relationship between physiological adaptation and zonation height in conjunction with results of an experiment testing this hypothesis with six species of rocky shore gastropods from southwestern Australia.

## The Hypothesis

Eulittoral gastropods, restricted below the high tide mark are wetted by high tides during each tidal cycle while eulittoral fringe species distributed above it must endure prolonged aerial exposures (>6–10 days between spring tides, Johannesson, 1989; McMahon, 1988). Therefore, it is hypothesized that eulittoral species rarely experience desiccation stress and generally should display adaptations dominated by maintenance of activity and foraging on emergence. Such adaptations could include: evaporative cooling; low thermal tolerance; elevated aerial $O_2$ uptake rates; large body size; darkly pigmented shells; and, perhaps, capacity for thermoregulation through radiant heat absorption and evaporative cooling. Such adaptations would maintain energy consumption and metabolic efficiency in air, but exacerbate evaporative water loss. However, high rates of water loss are of little consequence to eulittoral species since they rapidly rehydrate when wetted during each tidal cycle. In contrast, eulittoral fringe gastropods must endure prolonged and often unpredictable periods of emergence (Johannesson, 1989), thus, it is hypothesized that their adaptive strategies should be dominated by those that minimize water loss rates and prolong survival in air. Possible adaptations include: foot withdrawal, preventing heat conduction from the substratum; reduction of metabolic rate on air exposure; reduced size; lightly pigmented shells reflecting radiant heat; elevated thermal tolerance reducing the necessity for evaporative cooling; and position maintenance by cementation of the aperture edge to the substratum with mucus.

## Evidence from the literature

### Aerial gas exchange

There have been few comparative studies of the aerial and aquatic gas exchange rates of intertidal gastropods species from different shore levels (Houlihan, 1979; Houlihan & Innes, 1982; Innes & Houlihan, 1985; McMahon & Russell-Hunter, 1977; Micallef, 1967; Sandison, 1966; Toulmond, 1967a, b). While aerial respiratory rates in intertidal gastropods depend in part on degree of pallial water retention (Holulihan *et al.*, 1982) causing variability in aerial $O_2$ uptake rate determinations, a recent review of respiratory compensation in intertidal molluscs, summarizing all available data for intertidal meso- and neogastropods (McMahon, 1988), yielded the somewhat surprising result that the mean ratio of aerial to aquatic $O_2$ consumption rate in eulittoral meso- and neogastropod species was essentially 1 : 1 regardless of microzonation within the eulittoral. In contrast, that of eulittoral fringe species was depressed, being only 0.77 : 1. This result indicated that eulittoral species maintain similar metabolic rates in air and water and, by inference, similar levels of activity while the reduced ratio of

eulittoral fringe species (even in those well hydrated before $O_2$ uptake determinations) suggested suppression of metabolic rate and activity when emerged (McMahon, 1988).

While detailed comparative descriptions of the behaviour of intertidal snails during emergence are few, a specific pattern emerges on close examination. During relatively short emergence periods between tidal inundations, the majority of eulittoral meso- and neogastropods actively locomote and forage. Maintenance of high levels of activity result in both high aerial metabolic and evaporative water loss rates. However, as indicated by Underwood (1979), foraging in air can optimize energy consumption at a time when threat from aquatic predators is minimal. Associated with active foraging and locomotion is the distinct cost of elevated water loss rates, however some eulittoral species have evolved increased mantle cavity volumes allowing increased storage of extracorporal water to compensate for elevated aerial evaporative water loss rates (Vermeij, 1973). The water loss associated with maintenance of activity in air between tides is of little consequence to eulittoral species as they rapidly rehydrate with each incoming tide.

The much more prolonged periods of emergence experienced by eulittoral fringe species precludes maintenance of foraging activity as the resulting water loss could cause desiccation to lethal levels before rehydration occurred. Instead, eulittoral species generally become inactive, withdrawing into the shell and occluding or partially occluding the aperture with the operculum shorty after emersion. This reduction of activity in air results is a distinct decline in aerial $O_2$ uptake rate and greatly reduces exposure of moist surface epithelia to the atmosphere, minimizing water loss (McMahon, 1988). Withdrawal into the shell and aerial aestivation is, therefore associated with greatly reduced rates of desiccation in this group (Broekhuysen, 1940; Cleland & McMahon, 1989; McMahon, 1988; McMahon & Britton, 1985).

An example of emergence induced metabolic inhibition of metabolism in a eulittoral fringe species involves the littorines of the North American Atlantic Coast. Three species, *Littorina obtusata* (L.), *Littorina littorea* (L.) and *Littorina saxatilis* (Olivi) dominate North American Atlantic rocky shores. *L. obtusata* is associated with the macrophytic algae *Fucus* and *Ascophyllum* and, with them, is limited to the lower eulittoral. *L. littorea* is the most common shore species and found throughout the eulittoral, but not extending into the eulittoral fringe. *L. saxatilis* is the major species of the eulittoral fringe, but does not extend much below high tide (McMahon & Russell-Hunter, 1977). In air, both *L. obstusata* and *L. littorea* locomote and forage over the substratum. Such maintenance of activity is associated with an aerial $O_2$ consumption rate ($\dot{V}_{O_2}$) nearly equivalent to aquatic $\dot{V}_{O_2}$ at normal ambient temperatures (5–20 °C) (Innes & Houlihan, 1985; McMahon & Russell-Hunter, 1977). In contrast, soon after emersion (particularly at temperatures above 20 °C), specimens of *L. saxatilis*, while still well hydrated, cement the aperture edge to the substratum with mucus, withdraw into the shell and occlude the aperture with the operculum, behaviours associated with up to a 50% reduction in aerial relative to aquatic $\dot{V}_{O_2}$ (McMahon & Russell-Hunter, 1977).

*Temperature tolerance*

With increasing shore height, species spend progressively longer periods emerged in direct sunlight. Absorption of radiant solar energy causes emerged rocky substratum surfaces to warm well above air and water temperatures leading to the hypothesis that animals living at progressively higher shore levels should display increasingly greater thermal tolerance as an adaptation to elevated thermal stress associated with incremental emergence (reviewed by Hughes, 1986; Newell, 1979; Underwood, 1979). While superficially appealing, this hypothesis is empirically unsupportable. Both rocky substrata and gastropod tissues reach new elevated equilibrium temperatures within 30 min to one hour of tidal emergence into direct sunlight (Grainger, 1969; Southward, 1958). Therefore, intertidal gastro-

pods are exposed to approximately equal levels of acutely high temperatures on emergence regardless of shore position. Indeed, shore position is only correlated with the duration of such stress and not with the maximum temperature experienced (Cleland & McMahon, 1989; McMahon, 1988).

When carefully evaluated, the thermal tolerances of intertidal gastropod species do not form a discrete continuum with increasing shore height. Instead, species restricted to the eulittoral fringe generally have greatly elevated tolerances relative to those of eulittoral species. The mean heat coma temperatures (HCT) in response to a temperature increase of 1 °C 5 min$^{-1}$ of three eulittoral fringe gastropod species from a Hong Kong rocky shore were 3–6 °C greater than those of 5 eulittoral species whose HCT values all fell within 2 °C of each other. Further, HCT in eulittoral species was not correlated their relative positions within the eulittoral (Table 1, Cleland & McMahon, 1989). Similarly, on a rocky shore in False Bay, South Africa, the single eulittoral fringe species, *Nodilittorina africana* ('Kraus' Philippi), had an absolute HCT of 48.6 °C, 6–9 °C greater than that of five eulittoral species whose HCT values ranged from only 39.6 °C to 42.1 °C and were not correlated with their vertical distributions (Broekhuysen, 1940, Table 2). In addition, the mean HCT of 10 species of British rocky shore gastropods were unrelated to their vertical distributions, however, the eulittoral fringe species, *Littorina neritoides* (L.) and *L. saxatilis*, had HCT values (46.3 °C and 45.0 °C, respectively) considerably greater than those of the eight eulittoral species tested (Evans, 1948). Similar patterns of elevated thermal tolerance in eulittoral fringe species and lack of distinct correlation between thermal tolerance level and height of zonation within eulittoral or eulittoral fringe habitats have been reported for rocky intertidal snail faunas in Shirahama, Wakayama-Ken, Japan (Fraenkel, 1966) and Bimini, Bahamas (Fraenkel, 1968). Thus, on close examination, the thermal tolerance of intertidal gastropods does not appear to progressively increase with shore height. Instead, eulittoral fringe species have greatly elevated thermal

*Table 1.* Upper thermal limits measured as mean heat comma temperatures (HCT) while immersed with sample numbers and 95% confidence limits of the means for eight species of gastropods relative to height of zonation from a Hong Kong rocky shore (from Cleland & McMahon, 1989)[1].

| Species | Position on shore | °C mean heat coma temperature | Sample size | 95% confidence limits |
|---|---|---|---|---|
| *Nodilittorina pyramidalis* (Quoy & Gaimard) | Eulittoral Fringe | 46.5 | 34 | ± 0.26 |
| *Nodilittorina exigua* (Dunker) | Eulittoral Fringe | 44.8 | 36 | ± 0.49 |
| *Nerita chamaeleon* L. | Eulittoral Fringe | 43.3 | 17 | ± 0.35 |
| *Planaxis sulcatus* (Born) | Upper Eulittoral | 40.8 | 36 | ± 0.40 |
| *Batillaria sordida* (Gmelin) | Sublittoral fringe | 40.6 | 43 | ± 0.47 |
| *Monodonta labio* (L.) | Middle Eulittoral | 39.6 | 30 | ± 0.74 |
| *Morula musiva* (Kiener) | Lower Eulittoral | 39.6 | 60 | ± 0.18 |
| *Lunella coronata* (Gmelin) | Lower Eulittoral | 38.8 | 30 | ± 0.39 |

[1] Note lack of correlation in mean HCT in eulittoral species with only 2 °C separating the extreme HCT values of this group and a sublittoral species, *B. sordida*, having a greater HCT than three higher zoned eulittoral species, *M. labio*, *M. musiva* and *L. lunata*. Also note that the HCT values of eulittoral fringe species were 2.5–5.7 °C greater than that of the most temperature resistant eulittoral species, a greater separation of HCT than occurred across all the eulittoral species tested.

*Table 2.* Upper thermal limits measured as the range of temperatures at which death occurred while immersed in six species of gastropods relative to height of zonation on a South African rocky shore (from Broekhuysen, 1940).

| Species | Position on shore | Range of upper lethal limits in °C |
|---|---|---|
| *Nodilittorina africana* ('Kraus' Philippi) | Eulittoral Fringe | 47.4–48.6 |
| *Oxystele variegata* (Anton) | Eulittoral | 41.5–42.1 |
| *Thais dubia* (Krauss) | Eulittoral | 41.2–41.7 |
| *Oxystele tigrina* (Dillwyn) | Eulittoral | 38.8–40.5 |
| *Cominella cincta* (Roding) | Lower Eulittoral | 38.9–39.5 |
| *Oxystele sinensis* (Gmelin) | Lower Eulittoral | 38.0–39.6 |

tolerances relative to those of eulittoral species. As rapid tissue and substratum warming by insolation subjects gastropods to similar elevated temperatures regardless of shore position, elevated thermal tolerances in eulittoral fringe species cannot be a primary adaptation to acute temperature stress. Rather, it must be associated with tolerance of another factor such as degree of desiccation stress which, as discussed below, changes abruptly at the transition zone between eulittoral and eulittoral fringe habitats.

*Desiccation resistance*

The most obvious stress factor to change abruptly across the eulittoral/eulittoral fringe transition zone is that of desiccation stress. Within the eulittoral, snails are not subject to desiccation stress regardless of shore position because they are wetted and rapidly rehydrated with each high tide. In contrast, eulittoral fringe species experience extensive desiccation over prolonged periods of emergence (6–10 days) (McMahon, 1988; McMahon & Russell-Hunter, 1977). Recently, Byrne *et al.* (1988) demonstrated that in emerged specimens of the freshwater bivalve, *Corbicula fluminea* (Müller), temperature had a far greater effect on evaporative water loss rate than did relative humidity. Thus, the prolonged emergence in eulittoral fringe habitats at temperatures elevated by insolation should result in severe desiccation pressure.

There are few detailed comparative studies of the desiccation resistance and evaporative water loss rates of eulittoral and eulittoral fringe gastropod species on a single shore and none have been carried out at elevated temperatures associated with insolation. However, available data indicates that rate of evaporative water loss on emergence does not form a progressive continuum with increasing shore height. Rather, it is distinctly elevated in species restricted to the eulittoral fringe. Both Broekhuysen (1940) and Brown (1960) demonstrated that evaporative water loss rates in the eulittoral fringe littorine, *N. africana*, were 44–73% less than those of five eulittoral species after 100 hrs desiccation over $CaCl_2$. Nor was any correlation found between water loss rates and vertical distribution in the five eulittoral species. Similar lack of correlation between vertical distribution and water loss rate occurred in seven species of patellid limpets from South African rocky shores (Branch, 1975). Two species of western North American acmaeid limpets, *Acmaea digitalis* Eschscholtz and *A. persona* Eschscholtz, extending into the eulittoral fringe had desiccation rates 41–91% less than did two purely eulittoral species, *A. pelta* Eschscholtz and *A. testudinalis scutum* Eschscholtz (Wolcott, 1973). While initial water loss rates on emergence of a eulittoral fringe Hong Kong mangrove gastropod, *Cerithidea ornata* A. Adams, were elevated due to maintenance of activity, this species thereafter lost only 0.023% of total water $d^{-1}$ after becoming inactive (McMahon & Cleland, 1989), a rate of water loss many times less than that of six eulittoral species tested (McMahon & Britton, 1985). Very similar results have been reported for the south African rocky eulittoral fringe littorine, *N. africana* which sustained relatively high rates of water loss while active during the first two days of emergence, thereafter, water loss rate became minimal, being 3–10 times less than

that of six eulittoral species (Broekhuysen, 1940; Brown, 1960).

The exceptional capacity of eulittoral fringe species to endure prolonged emergence appears to be much more a result of their ability to limit evaporative water loss than to tolerate desiccation. Indeed, the absolute degree of water loss tolerated by eulittoral and eulittoral fringe species appears essentially similar. Thus, whether restricted to the eulittoral or eulittoral fringe, seven species of Hong Kong mangrove mesogastropods could tolerate a loss of at least 60% of total (corporal + extracorporal) water (McMahon & Cleland, 1989; McMahon & Britton, 1985). Similarly, seven South African intertidal species displayed an $LD_{50\%}$ of desiccation ranging from 16–38% of total water with the single eulittoral fringe species having a lower tolerance of absolute desiccation than two of six eulittoral species (Broekhuysen, 1940; Brown, 1960). Further, maximum tolerated water loss levels were essentially similar in six South African eulittoral and eulittoral fringe patellid speces regardless of shore position (Branch, 1975).

*Behaviour on emergence*

What then accounts for the major difference in the rate of water loss on emergence of eulittoral and eulittoral fringe gastropods? The answer lies in differences in their behavioural responses to emergence, particularly at elevated temperatures exacerbating desiccation. While studies are limited, evidence indicates that eulittoral species generally remain expanded with the foot attached to the substratum and often actively crawl and forage throughout normal emergence periods while eulittoral fringe species withdraw into the shell, cement the aperture to the substratum with mucus and become inactive within a relatively short time after air exposure (Britton & McMahon, 1989; McMahon, 1985, unpublished observations). The thermal tolerance values of eulittoral species generally fall below temperatures reached by their insolated substrata (Lewis, 1963; Markel, 1971; Vermeij, 1971; see above).

Since substratum and snail body temperatures rise rapidly above air temperatures due to insolation on emergence (Grainger, 1969; Southward, 1958), intertidal species are subjected equal levels of acute temperature stress regardless of shore position. As eulittoral species remain attached by the foot during emergence, heat is conducted from the substratum to their tissues. In contrast, withdrawal into the shell (with cementation of the aperture edge to the substratum), common in eulittoral fringe species, prevents conduction of heat from substratum to tissues, an adaptation also characteristic of pulmonate desert snails (Schmidt-Nielsen *et al.*, 1972). With the foot withdrawn, heat loads of emerged eulittoral fringe species would be much less than those of actively crawling eulittoral snails.

*Heat dissipation: evaporative cooling, shell morphology and pigmentation*

As the temperature of insolated substrata may rise well above a species' thermal tolerance limits, some intertidal gastropods have evolved a capacity for evaporative cooling of tissues below upper temperature limits. That tissue temperatures of tropical intertidal snails emerged in direct sunlight can be less than either ambient air or substratum temperatures indicate capacity for evaporative cooling (Vermeij, 1971). Eulittoral intertidal snails have been shown to survive one hour in air (Fraenkel, 1966) at temperatures 3–8 °C above

their acute upper thermal limits (Cleland & McMahon, 1989; Table 3) and, thus must thermoregulate by evaporative cooling. In contrast, the eulittoral fringe species, *Nodilittorina pyramidalis* (Quoy & Gaimard) could tolerate one hour exposure to an air temperature (Fraenkel, 1966) only 1.5 °C above its mean heat coma temperature (Cleland & McMahon, 1989) suggesting little or no capacity for evaporative cooling (Table 3).

Similarly, the Peruvian eulittoral fringe species, *Nodilittorina aspera* Philippi, maintained consistantly higher flesh temperatures when exposed to direct sunlight than did the eulittoral species, *Nodilittorina modesta* (Philippi) (Markel, 1971) suggestive of a reduced capacity for evaporative cooling in the eulittoral fringe species. Among acmaeid limpets, two eulittoral fringe species (*A. digitalis* and *A. scabra*) maintained tissue temperatures 4–11 °C above those of species restricted to the eulittoral (*A. pelta* and *A. testudinalis scutum*) when exposed to direct sunlight (Wolcott, 1973), again indicating a reduced capacity for evaporative cooling in eulittoral fringe species. Indeed, the two eulittoral fringe acmaeids sealed the edges of the aperture to the substratum with mucus, preventing convective air flow over moist epithelia, thus precluding evaporative cooling (Wolcott, 1973).

The reduced capacity of eulittoral fringe gastropods for evaporative cooling appears directly related to their behaviour. In air, eulittoral fringe mesogastropods withdraw into the shell, oc-

Table 3. Acute upper thermal limit measured as heat coma temperature (HCT) while immersed relative to maximum temperature survived for one hour in air for tropical Pacific intertidal mesogastropods (Note that differences between upper thermal limit and maximum temperature survived in air for one hour are much greater in the four eulittoral species than in *N. pyramidalis* (a eulittoral fringe species) suggestive of higher capacities of eulittoral species for evaporative cooling).

| Species | Position on shore | Heat coma temperature (°C) (Cleland & McMahon, 1989) | Maximum air temperature survived one hour (°C) (Fraenkel, 1966) |
|---|---|---|---|
| *Nodilittorina pyramidalis* (Quoy & Gaimard) | Eulittoral Fringe | 46.5 | 48.5 |
| *Planaxis sulcatus* (Born) | Eulittoral | 40.8 | 48.5 |
| *Littorina brevicula* (Philippi) | Eulittoral | 39.9 | 46.0 |
| *Lunella coronata* (Gmelin) | Eulittoral | 38.8 | 42.8 |
| *Monodonta labio* (L.) | Eulittoral | 39.6 | 42.8 |

cluding the aperture with the operculum and cementing the anterior aperture edge to the substratum with mucus (McMahon, 1985; McMahon & Britton, 1985). Occlusion of the aperture with the operculum functions to seal moist tissue surfaces from extensive air exposure reducing evaporative water loss and, thus capacity for evaporative cooling. In contrast, eulittoral species generally remain attached by the foot and often crawl on the substratum when emerged, exposing moist head-foot and mantle cavity epithelial surfaces directly to air currents, increasing capacity for evaporative cooling to dissipate excess heat gained by direct insolation and conduction from the substratum. Evaporative cooling allows maintenance of tissue temperature below deleterious levels on hot substrata, but results in high rates of water loss. While capacity for evaporative cooling allows eulittoral species to actively forage while emerged, it does not result in desiccation stress because individuals rapidly rehydrate with each successive tidal inundation. Evaporative cooling, thus allows eulittoral species to maximize energy consumption and assimilation during short-term tidal emergence. For this reason, the aerial metabolic rates of eulittoral meso- and neogastropods are essentially equivalent to aquatic rates (McMahon, 1988), reflecting maintenance of normal activity in either medium.

In contrast, eulittoral fringe species are emerged for 3–8 days between spring tides and could not tolerate the water loss associated with continual evaporative cooling and activity. Thus, behavioural adaptations in eulittoral fringe species centre on mechanisms minizing water loss. Withdrawal into the shell prevents conduction of heat from the substratum and occlusion of the aperture with the operculum greatly reduces evaporative water loss. Elevated thermal tolerance limits reduce the need for evaporative cooling. Indeed, some eulittoral fringe species appear incapable of evaporative cooling even as air and tissue temperatures approach lethal limits. Apparent aestivation during emergence greatly reduces metabolic demands reducing rate of utilization of limited energy stores and need to expose moist respiratory tissues to the external atmosphere.

Often associated with differences in physiological and behavioural responses of eulittoral and eulittoral fringe prosobranchs to emergence are distinct differences in shell morphology. In general, eulittoral fringe species tend to have much smaller maximum sizes relative to species restricted to the eulittoral, for example *Littorina littorea*, *L. saxatilis* and *L. neritoides* (L.) are found in the eulittoral, low eulittoral fringe and high eulittoral fringe, respectively, and represent a distinct gradient of declining maximum adult size with increasing level of zonation (Vermeij, 1973). Similar, reduction in adult size of eulittoral fringe species has been observed on rocky shores in both Hong Kong (Cleland & McMahon, 1989, unpublished observations of the authors) and the Albany area of Western Australia (see following section entitled 'A Test of the Hypothesis'). Another common observation on tropical shores is that the shells of eulittoral fringe species are often more lightly pigmented than those of eulittoral species. Reduced shell pigmentation allows eulittoral fringe species to reflect radiant solar heat while smaller size allows more rapid dissipation of heat by re-radiation, conduction and/or convection. Thus, both adaptions function to lower tissue equilibrium temperatures on exposure to direct sunlight relative to larger eulittoral snails whose more darkly pigmented shells and larger body sizes increase capacity for heat absorption and retention. Such lightly pigmented shells have similar adaptive value in desert land snails (Schmidt-Nielsen *et al.*, 1972).

In eulittoral fringe species, these morphological adaptations along with elevated thermal tolerance and ability to withdraw into the shell and aestivate in air greatly reduce or elminate necessity for evaporative cooling, thus minimizing water loss rates during prolonged emergence. In contrast, the larger body size and more darkly pigmented shells of eulittoral species enhance uptake and retention of radiant heat energy in direct sunlight. However, excess heat loads in eulittoral species are dissipated by evaporative cooling, allowing maintenance of body temperatures below deleterious levels and continual foraging during emergence. The water loss associated with maintenance of activity in air being compensated by periodic and predictable rehydration during submergence at successive high tides. As evaporative cooling prevents tissues from reaching lethal temperature levels, eulittoral species have distinctly lower upper thermal limits than do eulittoral fringe species.

Eulittoral and eulittoral fringe gastropod faunas, therefore display mutually exclusive adaptations to emergence associated with the contrasting levels of selection for maintenance of foraging activity and survival of desiccation when emerged presented by their respective habitats. Subject to periodic and predictable rehydration, eulittoral species have evolved adaptations dominated by maintenance of foraging activity, optimizing energy assimilation during relatively short bouts of tidal emergence, while the prolonged and often unpredictable emergence experienced by eulittoral fringe species has selected for adaptations dominated by minimization of water loss rates and, therefore desiccation stress in air.

## A test of the hypothesis

### Introduction

In the preceeding section, a hypothesis was presented stating that eulittoral gastropods should display adaptations associated with the maintenance of activity during emergence including increased capacity for evaporative cooling, reduced upper thermal limits, elevated capacity for radiant heat absorption and retention (*ie.*, darkly pigmented shells and increased adult size), and maintenance of foraging activity. In contrast, eulittoral fringe species should display adaptations associated with minimization of evaporative water loss in air, including withdrawal into the shell, occlusion of the aperture with the operculum, limited capacity for evaporative cooling, elevated upper thermal tolerance limits, and reduced capacity for radiant heat absorption and retention (*ie.*, lightly pigmented shells and reduced adult size). In order to test this hypo-

thesis, the tissue temperatures and behaviour on exposure to direct sunlight; upper thermal tolerance limits; and the behaviour, tissue temperatures and evaporative water losses in air at 40 °C were evaluated for four eulittoral and two eulittoral fringe gastropod species from a granite shore on Princess Royal Harbour, near Albany, Western Australia.

### Methods

The study site was on a gently sloping granite shore on the northern side of Geak Point, Vancouvers Peninsula in Princess Royal Harbour near Albany, Western Australia (35° 03′ 27″ S, 117° 03′ 27″ W). The site was 200 m north of the Camp Quaranup jetty. Camp Quaranup is a recreation camp at which all research was completed from 11–28 January 1988.

Princess Royal Harbour is a shallow lagoon approximately 7.6 km long and 4.3 km wide with a narrow 0.5 km opening to King George Sound. Its shores, including the collection site, were highly protected and experience wave surge only during storms. Lack of wave surge at the collection site allowed accurate evaluation of the zonation patterns of the six tested species as species' vertical distributions were not modified by wave action or continual splash. Tidal range at the site over the collection period was approximately one meter. Of the six species tested, four were strictly eulittoral in distribution including the trochids, *Austrocochlea constricta* (L.) and *Austrocochlea concamerata* (Wood), the nerite, *Nerita atramentosa* (Reeve) and the thaid, *Lepsiella vinosa* (L.). All four species were distributed throught the eulittoral from the low to high tide marks. *A. constricta*, *N. atramentosa* and larger adult *A. concamerata* were exposed to direct sunlight on rock faces by receding tides while specimens of *L. vinosa* were exposed to direct sunlight on sandy substratum or found under rocks when emerged. Juveniles and smaller specimens of *A. concamerata* also sheltered under rocks on emergence. Two littorines, *Bembicium vittatum* (Philippi) *fide* (Reid) and *Nodilittorina unifasciata*

(Gray) occurred in the eulittoral fringe. *B. vittatum* was restricted to the lower eulittoral fringe above the high water mark while *N. unifasciata* was distributed from the high tide mark to the upper extremities of the eulittoral fringe (for all six species see species identifications and descriptions in Wells, 1984 and Wells & Bryce, 1986).

The upper thermal tolerance limits of the six snail species were determined as mean heat coma tolerances (HCT) by a modification of the method (Cleland & McMahon, 1989; McMahon & Britton, 1985; McMahon & Russell-Hunter, 1981) used for other intertidal marine gastropods (Gowanloch, 1926; Broekhuysen, 1940; Evans, 1948; Southward, 1958; Stirling, 1982). Heat coma temperature in gastropods is the temperature at which normal nervous function is lost, manifested by cessation of locomotion, inability to remain attached to substrata, and a ventral-medial curling of the lateral edges of the foot (Cleland & McMahon, 1989; McMahon, 1976; McMahon & Britton, 1985; McMahon & Cleland, 1989; McMahon & Payne, 1980; McMahon & Russell-Hunter, 1981).

For determination of mean HCT, adult specimens of a single species were collected at low tide and returned immediately to the laboratory. Determination of HCT began within 30 min of collection. The shell lengths (SL) of 30 specimens representative of the size range of the adult population were measured to the nearest 0.1 mm with dial calipers and the sample subdivided into 6 subgroups of five individuals each such that the SL of all individuals in a subgroup fell within 2 mm of each other. Thus, each subgroup had a distinctly different mean SL and the mean SL of the six subgroups spanned that of the natural adult population.

Each subgroup was placed separately into 500 ml screw-top glass jars on which caps were screwed while the jars were submerged in seawater from the collection site to prevent inclusion of air bubbles. The jars were then placed in a water bath to a depth at which all but jar caps were immersed. During determinations, bath water was vigorously circulated by aeration to insure temperature uniformity. The bath heater switch was manually controlled to raise bath temperature and, therefore jar water temperature, approximately 1 °C 5 min$^{-1}$. This rate of temperature increase makes lag between jar water and snail tissue temperatures negligible in snails of sizes similar to those tested (Broekhuysen, 1940). Jar water temperature was monitored with a thermistor probe and YSI model 43-TD tele-thermometer through a small hole in the jar cap.

Each HCT determination began at room ambient temperature (19–22 °C) roughly equivalent to ambient water temperature at the collecting site (20.0–24.6 °C). Snails were allowed to attach to jar walls and actively locomote before increasing jar water temperature. Individuals remaining inactive during the experiment where not included in HCT computations. The number of individuals displaying symptoms of heat coma in each jar was recorded for every 1 °C increase in jar water temperature. After all snails in all jars displayed heat coma, jars were removed from the water bath, the water they contained poured off, and the snails allowed to cool to room temperature. Snails were, thereafter observed at 1 and 12 hr post-heat treatment and the number expanded and actively locomoting recorded.

Ambient air, water, substratum and snail tissue temperatures were recorded on the shore on 14, 15 and 21 January 1988 when low tides in mid- to late afternoon exposed snails to maximum levels of insolation. Temperatures were recorded with a small, rapidly responding, flexible YSI thermister probe (No. 402) with a tip diameter of 3.18 mm. Air temperature was recorded in the shade approximately 3 cm above the substratum and water temperature at a depth of 3–5 cm at the waters edge. The tissue and substratum temperatures were recorded for 20 individuals of each species emerged in direct sunlight. Substratum temperature was recorded by application of the momentarily shaded thermister tip to the rock surface within 0.5 cm of an emerged specimen. That specimen was lifted from the substratum, noting if it was attached by an expanded foot or withdrawn into the shell, and the tip of the thermister forced rapidly past the operculum into its mantle cavity to record tissue temperature while the snail was momentarily shaded. The SL of each individual was measured to the nearest 0.1 mm with dial calipers.

In order to determine evaporative water loss and tissue temperatures at an elevated constant temperature, individuals of each species were emerged at 40 °C in a drying oven. Fifty individuals of each species were collected from the shore, individually numbered with india ink, wet weighed to the nearest mg, and divided randomly into five subgroups of 10 individuals each, with the exception of *L. vinosa*, whose relative rarity allowed collection of only enough specimens for subgroups of five individuals each (total *n* = 25). The SL of each individual was measured to the nearest 0.1 mm with dial calipers. Each subgroup was placed in a glass petri dish and all dishes placed simultaneously in the oven at 40 °C. Every 30 min thereafter, the tissue temperature of one subgroup were measured with a thermister as described above. That particular subgroup was then removed from the oven and each individual again wet weighed. Individuals were then placed in a jar with a small amount of sea water at room temperature and allowed to recover. Any individual actively locomoting or attached by the foot to the jar walls after one hour was considered recovered. Subsequently, all individuals were dried to constant weight at 95 °C (> 12 hr). Subtraction of wet weight after emer-

gence in 40 °C from initial wet weight yielded an estimate of evaporative water loss. Subtraction of total dry weight from initial total wet weight gave the total (corporal + extracorporal) water weight of each specimen, allowing water loss to be expressed as per cent of total water. This analysis provided estimates of tissue temperature and per cent of total water lost for individual subgroups of each species after 30, 60, 90, 120 and 150 min exposure in air at 40 °C.

## Results

There was no correlation between subgroup size and HCT for any tested species allowing thermal tolerance to be expressed as mean HCT values. The mean HCT values of the four eulittoral species were lower than those of the two eulittoral fringe species (Table 4, Fig. 1A–F). Mean HCT values for the four eulittoral species were 35.6, 37.7, 38.9 and 39.2 °C for *A. concamerata*, *A. constricta*, *N. atramentosa* and *L. vinosa*, respectively. Mean HCT for the eulittoral fringe littorines was 41.0 °C for *B. vittatum* and 41.3 °C for *N. unifasciata*. All individuals of all species recovered within one hour after exposure to the maximum temperature tolerated by a particular species (Table 4).

Least squares linear regression analysis indicated no significant correlations between tissue

*Table 4.* Mean heat comma temperature (HCT) for six species of intertidal snails from a granite shore in Princess Royal Harbour near Albany Western Australia, *n* = sample size, s.d. = standard deviation, 95% Conf. Limits = 95% confidence limits of the mean and Range = minimum to maximum values.

| Species | Zonation[1] | Mean HCT (°C) | *n* | s.d. | 95% conf. limits | Range |
|---|---|---|---|---|---|---|
| *Austrocochlea constricta* | E | 37.7 | 30 | ± 0.89 | ± 0.33 | 36–39 |
| *Austrocochlea concameratra* | E | 35.6 | 30 | ± 0.88 | ± 0.37 | 34–38 |
| *Nerita atramentosa* | E | 38.9 | 30 | ± 1.33 | ± 0.50 | 36–42 |
| *Lepsiella vinosa* | E | 39.2 | 30 | ± 0.98 | ± 0.39 | 37–41 |
| *Bembicium vittatum* | EF | 41.0 | 26 | ± 1.08 | ± 0.44 | 39–43 |
| *Nodilittorina unifasciata* | EF | 41.3 | 30 | ± 1.23 | ± 0.46 | 38–44 |

[1] E = occur throughout, but restricted to the eulittoral zone (area between tide marks), EF = restricted to the eulittoral fringe (area above the high tide mark) with *B. vittatum* restricted to the lower eulittoral fringe and *N. unifasciata* extending to the upper extreme of the eulittoral fringe.

252

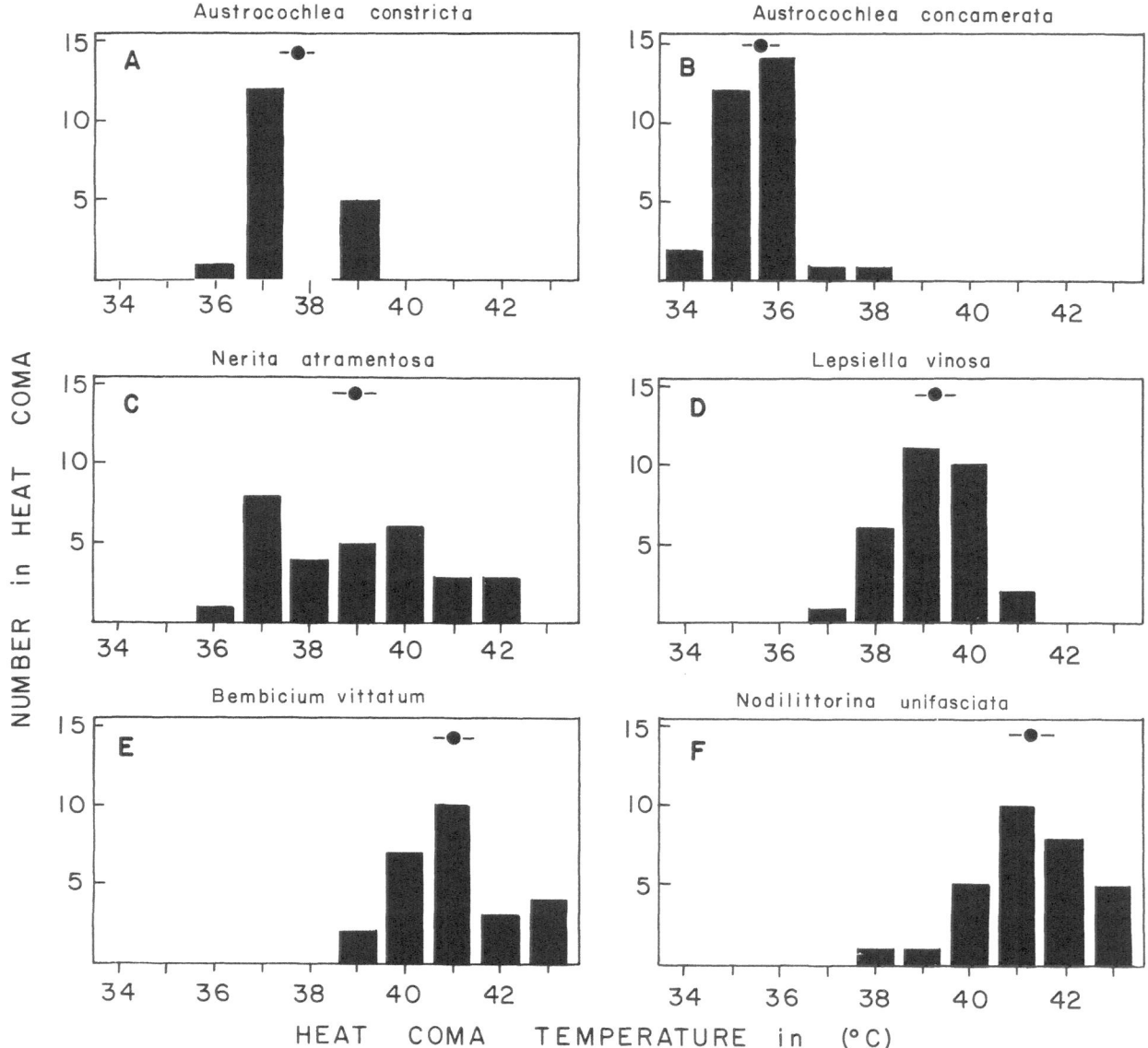

*Fig. 1.* Upper thermal limits measured as mean heat coma temperatures (HCT) in six species of eulittoral or eulittoral fringe gastropods from a granite shore on Princess Royal Harbour near Albany, Western Australia. Horizontal axis in all figures is the temperatures at which experimental individuals displayed symptoms of heat coma when exposed to temperature increasing at a rate of 1 °C 5 min$^{-1}$. The vertical axis represents the number of individuals in the sample displaying heat coma symptoms at any one temperature. The solid circle at the top of each figure represents a species' mean HCT value in °C and the horizontal bars about the mean, 95% confidence limits of the mean. A. *Austrocochlea constricta* (eulittoral) B. *Austrocochlea concamerata* (eulittoral) C. *Nerita atramentosa* (eulittoral) D. *Lepsiella vinosa* (eulittoral) E. *Bembicium vittatum* (Eulittoral fringe) F. *Nodilittorina unifasciata* (Eulittoral fringe).

temperature and SL for specimens exposed in direct sunlight on the shore ($P > 0.05$), thus mean tissue temperatures were computed for each species. The relationship between tissue tempera-

tures and substratum temperatures were distinctly different in the eulittoral and eulittoral fringe species when exposed to direct sunlight (Table 5, Fig. 2). For all six species, paired t-tests revealed

*Table 5.* Ambient air, water, mean substratum ( ± standard deviation) and tissue temperature[1] (°C) ( ± standard deviation) for six species of intertidal snails from a granite shore on Princess Royal Harbour near Albany, Western Australia, n = number in sample, s.d. = standard deviation, 95% conf. limits = 95% confidence limits of the mean, range = minimum to maximum values.

| Species | Zonation[2] | Air temp. | Water temp. | Mean substratum temp. ( ± s.d.) | Mean tissue temp. | n | s.d. | 95% conf. limits | Range |
|---|---|---|---|---|---|---|---|---|---|
| *Austrocochlea constricta* | E | 18.0 | 22.3 | 22.2( ± 1.74) | 26.3 | 20 | ± 2.25 | ± 0.70 | 22.8–28.0 |
| *Austrocochlea concamerata* | E | 22.0 | 24.1 | 26.6( ± 1.39) | 31.8 | 20 | ± 0.97 | ± 0.23 | 29.8–33.5 |
| *Nerita atramentosa* | E | 21.1 | 20.0 | 23.5( ± 1.30) | 26.6 | 20 | ± 1.77 | ± 0.83 | 23.0–30.0 |
| *Lepsiella vinosa* | E | 24.5 | 24.1 | 27.4( ± 1.62) | 31.0 | 20 | ± 1.65 | ± 0.77 | 28.5–33.6 |
| *Bembicium vittatum* | EF | 23.1 | 24.1 | 35.2( ± 1.57) | 34.4 | 20 | ± 0.66 | ± 0.31 | 33.3–35.5 |
| *Nodilittorina unifasciata* | EF | 28.2 | 24.5 | 36.1( ± 1.02) | 33.1 | 20 | ± 1.17 | ± 0.55 | 31.5–34.9 |

[1] Paired t-test analysis indicated a significant difference ($P < 0.0001$) between mean substratum and tissue temperatures in all six species.

[2] E = distributed throughout, but restricted to the eulittoral zone (area between tide marks), EF = restricted to the eulittoral fringe (area above high tide mark) with *B. vittatum* restricted to the lower eulittoral fringe and *N. unifasciata* extending to the upper extreme of the eulittoral fringe.

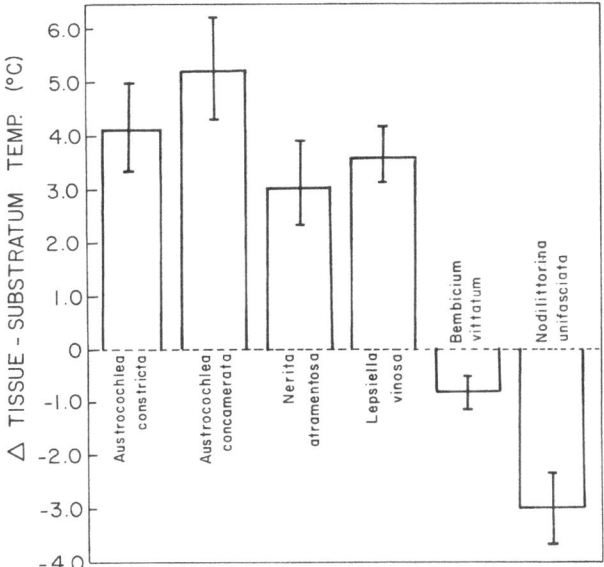

*Fig. 2.* Difference between tissue temperature and substratum temperature when exposed to direct sunlight by receding tides in specimens of six species of eulittoral or eulittoral fringe gastropods on a granite shore in Princess Royal Harbour near Albany, Western Australia. Horizontal axis is the mean value of tissue temperature minus substratum temperature for samples of 20 specimens of each species as presented by the vertical histograms. Species tested are indicated at the base of each histogram. Vertical bars at the top of each histogram represent 95% confidence limits of the means. Paired t-tests indicated that mean substratum and tissue temperatures were significantly different in all six

that mean substratum and tissue temperatures were significantly different ($t$ range = 5.4–13.8, $n = 40$, $P < 0.0001$). The eulittoral fringe species (*A. constricta*, *A. concamerata*, *N. atramentosa* and *L. vinosa*) all had mean tissue temperatures 3–6 °C above mean substratum temperatures. In contrast, the eulittoral fringe species, *B. vittatum* and *N. unifasciata* had tissue temperatures 0.8 °C and 3.0 °C below that of the substratum, respectively (Table 5, Fig. 2). All emerged specimens of the four eulittoral species were found attached to the substratum with an expanded foot while all specimens of the eulittoral fringe species had the foot withdrawn into the shell occluding the aperture with the operculum and the anterior aperture edge cemented to the substratum with mucus.

Least squares linear regression analyses indicated no significant correlation between SL and tissue temperature or per cent total water lost ($P > 0.05$) in any of the six species emerged at 40 °C, thus tissue temperature and water loss values were computed as means for each species

species ($P < 0.0001$). The vertical distributions of *Austrocochlea constricta*, *Austrocochlea concamerata*, *Nerita atramentosa* and *Lepsiella vinosa* were limited to the eulittoral zone while those of *Bembicium vittatum* and *Nodilittorina unifasciata* were restricted to the eulittoral fringe.

subgroup. When emerged at 40 °C, the eulittoral species, *A. constricta*, *A. concamerata* and *N. atramentosa*, kept the foot expanded and attached to the substratum. All three species generally maintained tissue temperatures 1–4 °C below ambient air temperature (40 °C) throughout the 2.5 hr exposure period (Fig. 3A–C). In all three species, there was continual evaporative water loss recorded throughout the exposure period with approximately 20–30%

of total water lost after 2.5 hrs emergence (Fig. 3A–C). A single exception to this generality was *N. atramentosa* in which specimens appeared not to lose water between 60 and 90 min emergence and whose mean body temperature rose to 39.9 °C at 90 min exposure, suggestive of lack of evaporative cooling during this period (Fig. 3C). Other than the above exception, both *A. constricta* and *N. atramentosa* generally evaporatively cooled tissues below HCT values (37.7 and

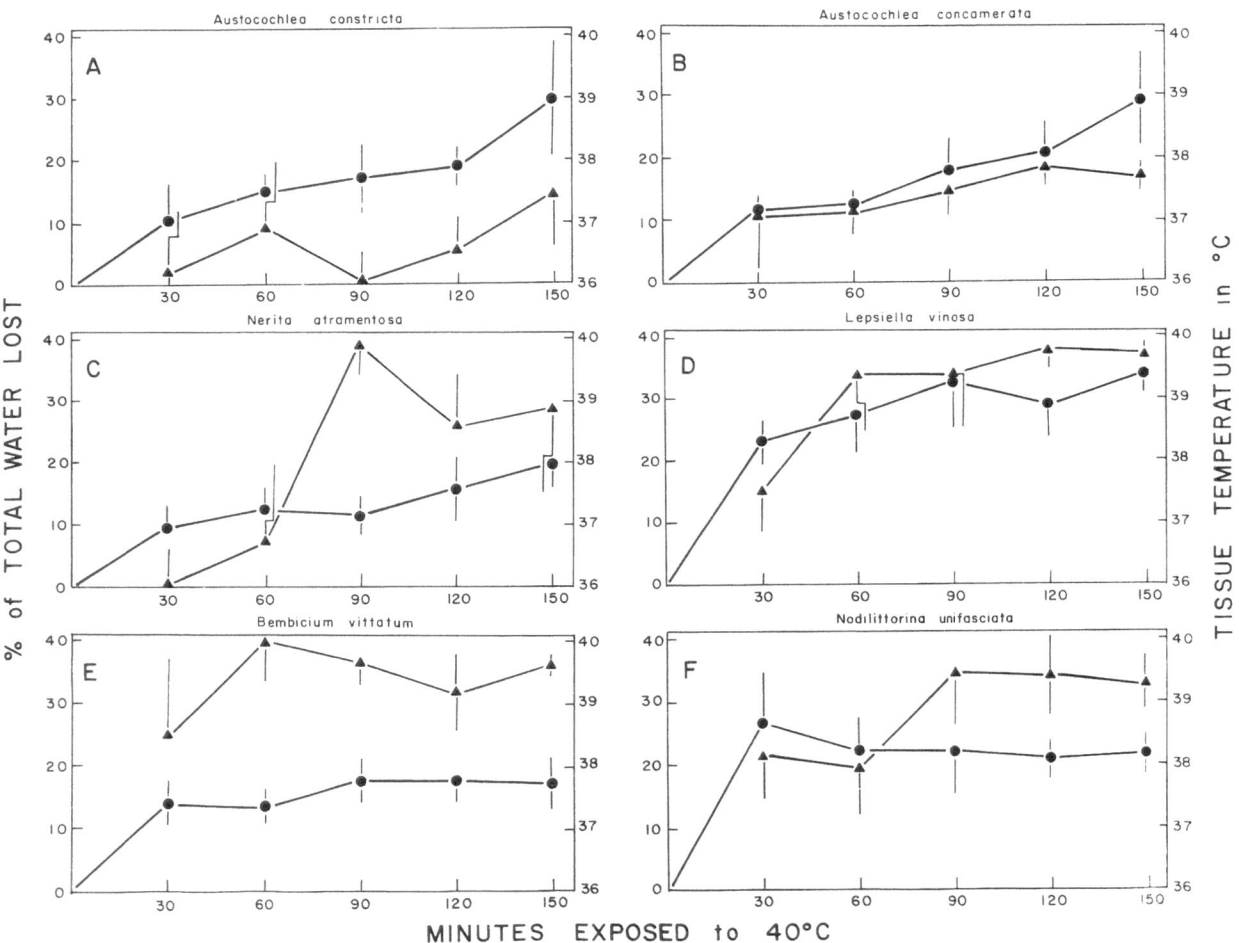

*Fig. 3.* Mean tissue temperatures and evaporative water loss on emergence at a constant temperature of 40 °C of subsamples of six species of eulittoral or eulittoral fringe gastropods from a granite shore on Princess Royal Harbour near Albany Western Australia. The horizontal axis is duration of emergence in minutes at 40 °C experienced by separate subsamples of 10 individuals (5 individuals for *L. vinosa*). The left vertical axis is mean per cent of total (corporal and extracorporal) water lost by a subsample during an emergence period (solid circles) while the right vertical axis is the mean tissue temperature of each subsample (solid triangles). Vertical bars about each mean represent 95% confidence limits of the mean. A. *Austrocochlea constricta* (eulittoral) B. *Austrocochlea concamerata* (eulittoral) C. *Nerita atramentosa* (eulittoral) D. *Lepsiella vinosa* (eulittoral) E. *Bembicium vittatum* (eulittoral fringe) F. *Nodilittorina unifasciata* (eulittoral fringe).

38.9 °C, respectively) throughout the exposure period (Fig. 3A & C) and completely recovered on cooling after 2.5 hr at 40 °C. In contrast, *A. concamerata* maintained tissue temperatures 1.5–2 °C above their mean HCT of 35.6 °C (Fig. 3B) and while able to completely recover after 2 hr at 40 °C, none recovered after 2.5 hr at 40 °C.

The two eulittoral fringe species, *B. vittatum* and *N. unifasciata*, and the eulittoral species, *L. vinosa*, did not remain attached to the substratum throughout the 40 °C air exposure period. The evaporative water losses of these three species and tissue temperatures were closely correlated with their behavioural responses. In *B. vittatum* there was little further water loss after 30 min exposure when specimens withdrew deeply into the shell. After 30 min exposure, mean tissue temperatures were 1.5 °C below ambient (however tissue temperatures were highly variable at this time, Fig. 3E), indicative of continued evaporative cooling in at least some specimens. Thereafter, specimens withdrew deeply into the shell sealing the aperture with the operculum, following which water loss became negligible over the remaining exposure period and tissue temperatures closely approximated 40 °C (Fig. 3E). A similar pattern was evident for *N. unifasciata* in which no detectable evaporative water loss occurred after 30 min exposure when the majority of specimens withdrew deeply into the shell. In *N. unifasciata*, initial water loss was associated with cooling of tissues approximately 2 °C below ambient (40 °C) for the first 60 min of exposure, thereafter water loss ceased and tissue temperatures approached 40 °C (Fig. 3F). Evaporative water loss in *L. vinosa* occurred only during the initial stages of emergence at 40 °C, 27% of total water being lost after 60 min of exposure. This initial water loss occurred during a period when individuals were gaping the operculum from the aperture and tissue temperatures were maintained 1–2 °C below ambient, indicative of evaporative cooling. After 60 min, individuals withdrew deeply into the shell, sealing the aperture with the operculum. Thereafter, there was little further evaporative water loss and tissue temperatures

approached 40 °C (Fig. 3D). Tissue temperatures remained near or below mean HCT in all three species throughout the 2.5 hr aerial exposure to 40 °C (Table 4, Fig. 3D–F) and specimens of all three species completely recovered within one hour of rehydration and cooling to room temperature after 2.5 hr emergence at 40 °C.

## Discussion

The results clearly indicate that there are major differences in the physiological responses of emerged eulittoral and eulittoral fringe rocky shore gastropods. As predicted by the hypothesis, eulittoral species with the exception of *L. vinosa* had characteristically lower upper thermal limits than had eulittoral fringe species. Three of the four eulittoral species (*A. constricta*, *A. concamerata* and *N. atramentosa*) remained attached with the foot to the substratum even when ambient air temperatures were above HCT values and evaporatively cooled to keep tissue temperatures below lethal limits. Evaporative cooling was associated with continual water loss, but allowed survival of at least 2.5 hrs in air at 40 °C, with the exception of *A. concamerata* whose relatively low thermal tolerance may have been associated with its capacity to withstand only two hours of exposure. In contrast, the two eulittoral fringe species withdrew the foot deeply into the shell and sealed the aperture with the operculum after the initial 30 min of emergence. Sealing the aperture with the operculum resulted in cessation of evaporative water loss and near complete equilibration of tissue and ambient air temperatures (40 °C, Fig. 3E & F). Apparently the elevated upper thermal limits of these two species (HCT > 41 °C) allowed survival of 2.5 hr emergence at 40 °C without reliance on evaporative cooling as occurred in three of the eulittoral species.

However, the eulittoral species, *L. vinosa*, displayed behaviours intermediate between the other three eulittoral species and the eulittoral fringe species. Like the eulittoral fringe species, specimens of *L. vinosa* withdrew the foot into the shell on exposure to 40 °C in air, a behaviour perhaps related to its tendency to seek sandy substrata when emerged. But, unlike the eulittoral fringe species, it gaped the operculum from the aperture for the first 60 min of emergence, a behaviour associated with considerable water loss and evaporative cooling of tissues well below ambient temperature (40 °C) (Fig. 3D). Thereafter, this species behaved more like eulittoral fringe species by further withdrawing the foot and sealing the aperture with the operculum, reducing further water loss and allowing tissue temperatures to approach 40 °C (Fig. 3D). This latter behaviour may be associated with the rather high mean HCT of this species (39.2 °C) which approximated the exposure temperature of 40 °C and therefore, precluded necessity for evaporative cooling.

Most interesting were data indicating that, when emerged on the shore in direct sunlight, the four eulittoral species had tissue temperatures 3–7 °C above substratum temperatures (Fig. 2, Table 5). These data suggest that these species have high capacities for absorbence and retention of radiant solar energy (even greater than the rocky sustratum). All four eulittoral species (*A. constricta*, *A. concamerata*, *N. atramentosa* and *L. vinosa*) had dark brown or black pigmented shells likely to absorb radiant heat. Capacity for absorption of radiant heat, along with increased capacity for heat retention due to larger body size relative to the eulittoral fringe species, may account for tissue temperatures of eulittoral species being greater than those of the substratum when emerged in direct sunlight. In contrast, the shells of the two eulittoral fringe species were more lightly pigmented, those of *B. vittatum* being light yellow or orange and those of *N. unifasciata*, light gray to chalky white, making them reflective of radiant heat. Reduced absorption of radiant heat, lack of direct tissue contact with the substratum and smaller relative size minimize eat absorption and maximize heat dissipation and, therefore account for the low equilibrium tissue temperatures recorded for individuals of these two eulittoral fringe species relative to the substratum in direct sunlight.

That eulittoral species maintained tissue temperatures higher than the substratum suggests capacity for absorption and retention of radiant heat energy. That these and other eulittoral species (McMahon, 1988; McMahon & Britton, 1985; Vermeij, 1971) actively forage when emerged may make ability to sustain body temperatures above ambient by absorption and retention of radiant heat energy highly adaptive as it would allow maintenance of metabolic rates at optimal levels for foraging and energy assimilation. Such elevation of foraging activity in air could partially compensate for depressions in metabolic rate associated with temperature reductions experienced during tidal submergence in cooler seawater (Table 5). Capacity for both evaporative cooling and radiant heat absorption in eulittoral species suggests capacity for limited thermoregulation at metabolically efficient body temperatures during emergence in direct sunlight. Certainly, that body temperatures were considerably greater than either ambient substratum or air temperatures in these Australian and other tropical eulittoral species (Lewis, 1963) clearly indicates that thermoregulation in eulittoral gastropods warrants further investigation.

## Conclusions

The literature review and data for Australian species presented herein clearly indicate that the physiological and behavioural adaptations of intertidal prosobranch snails do not form a continuum with increasing level of zonation on rocky shores as previously assumed by many investigators. Indeed, in terms of desiccation, there appears to be a distinct physiological barrier between the eulittoral zone and the eulittoral fringe at the high tide mark. As eulittoral gastropods are submerged periodically and predictably by high tides allowing periodic and predictable rehydration, they are not subject to desiccation stress regardless of height of habitation within the eulittoral, accounting for the relatively wide distributions of most eulittoral species between tide marks (for distributions see Broekhuysen, 1940; Chambers, 1980; Cleland & McMahon, 1989;

Evans, 1948; Takenouchi, 1985; for a review see Stephenson & Stephenson, 1972). As indicated by the data presented for Australian species, without desiccation stress, the major physiological and behavioural adaptations of eulittoral meso- and neogastropods are dominated by those associated with maintenance of optimal foraging activity when emerged accounting for the relatively high aerial oxygen consumption of eulittoral species (McMahon, 1988). Chief among these is capacity for evaporative cooling as demonstrated in this study and suggested by others (Fraenkel, 1966; Lewis, 1963; Markel, 1971; Southward, 1958; Vermeij, 1971). Capacity for evaporative cooling allows eulittoral species to avoid short-term temperature stress induced by insolation and, thus, remain active in air while cooling tissues below their characteristically reduced upper lethal temperature limits. Further, eulittoral species often have more darkly pigmented shells and larger adult sizes than eulittoral fringe species (particularly in the tropics) increasing their capacity for radiant heat absorption and retention. Radiant heat absorption allows maintenance of tissue temperatures above air and substratum levels (Lewis, 1963; this study) and in conjunction with capacity for evaporative cooling may allow limited thermoregulation when emerged in direct sunlight at levels metabolicly optimal for foraging, maximizing ingestion before body temperatures are reduced below optimal levels by tidal submergence in low temperature sea water. While maintenance of foraging activity and thermoregulation in air are associated with high levels of water loss, predictable resubmergence and rehydration with each tidal cycle prevent excessive desiccation in eulittoral species. Such high levels of water loss may account for the massive mortalities reported for eulittoral gastropods exposed in air for extended periods by exceptionally low water and/or to exceptionally high temperatures (Wolcott, 1973; for other examples see Underwood, 1979).

In direct contrast to eulittoral species, eulittoral fringe species experience prolonged and often unpredictable periods of emergence on exposed shores between spring tides (Johannesson, 1989).

Therefore, their physiological and behavioural responses center on mechanisms that minimize evaporative water loss allowing survival of prolonged air exposure. Chief among these are aestivation associated with deep withdrawal into the shell, sealing the aperture with the operculum to minimize evaporative water loss and upper thermal tolerance limits that are generally greater than the ambient substratum or tissue temperatures experienced even when exposed to direct sunlight precluding necessity for evaporative cooling common in eulittoral species with lower temperature limits. The ability of eulittoral fringe snails to cement the aperture edge to the substratum when emerged (McMahon & Britton, 1985; Vermeij, 1971) allows maintenance of position while withdrawing the foot to prevent conduction of heat from the substratum to tissues. The more lightly pigmented shells characteristic of many eulittoral fringe species reduce radiant heat absorption and their relatively small adult size allows greater heat dissipation via radiation, conduction and/or convection, maintaining tissue temperatures below deleterious levels, eliminating necessity for evaporative cooling when emerged in direct sunlight. Therefore, while the actual levels of desiccation tolerated by eulittoral fringe species are essentially equivalent to those of eulittoral species, their greatly reduced rates of evaporative water loss in air considerably extend the duration of emergence tolerated, allowing survival of much more prolonged exposures.

Aestivation of eulittoral fringe species in air greatly reduces foraging time and energy assimilation and would be a poor strategy in the eulittoral were submergence with each tidal cycle precludes desiccation stress, but aestivation does allow survival of the prolonged emergence periods associated with life on the high shore. Conversely, eulittoral species, which sustain high water losses due to maintenance of foraging activity and evaporative cooling while in air, could not survive the prolonged periods of emergence encounted in the eulittoral fringe habitat. Thus, the distributions of intertidal gastropod species do not appear to be divided into a series of narrow vertical bands up the shore, each occupied by species with progressively greater capacities for temperature and desiccation tolerance as previously proposed. Rather zonation patterns are primarily dipartite, the eulittoral and eulittoral fringe habitats separated by a distinct desiccation barrier at the high tide mark selecting for adaptations associated with maintenance of activity while emerged in eulittoral species and desiccation resistance in eulittoral fringe species. That gastropod species are broadly sympatric within both of these habitats, but display little or no species overlap between them and that species restricted within these two habitats have evolved mutually exclusive physiological adaptations to emergence is strong evidence for the existence of a physiological barrier between them and the primarily dipartite nature of intertidal zonation.

Throughout the world's oceans, members of the family Littorinidae constitute the vast majority of eulittoral fringe mesogastropod species (Stephenson & Stephenson, 1972). As indicated by the results presented in this study for *B. vittatum* and *N. unifasciata*, and reviewed here from previously published literature, high shore littorines have evolved a suite of characteristics that make them ideally adapted for life in the eulittoral fringe including: small adult sizes; reflective shells; elevated upper lethal limits; capacity to aestivate on emergence; and ability to maintain position while withdrawn into the shell by cementation of the aperture edge to the substratum with mucus preventing heat conduction from substratum to tissues. This suite of characteristics common to high intertidal littorines is rare in other intertidal species being recorded in only a few members of the family Potamididae from the tropical Pacific (McMahon, 1985; McMahon & Britton, 1985; McMahon & Cleland, 1989). This suite of adaptations allows high shore littorine species to dominate the eulittoral fringe gastropod faunas of the world's oceans.

## Acknowledgements

Appreciation is extended to Dr. Fred E. Wells, Curator of Mollusks at the Western Australian Museum, for his support of this research at the 'International Workshop of the Marine Biology of the Albany Area, Western Australia'. Travel to the Workshop and the Second European Meeting on Littorinid Biology was partially supported by funds from the Department of Biology and the College of Science of the University of Texas at Arlington. Dr. Wells and Dr. David Reid of the British Museum (Natural History) confirmed identifications of experimental species. Dr. David Raffaelli of the University of Aberdeen and Dr. Kerstin Johannesson of the Tjarno Marine Biological Laboratory, Sweden, provided valuable editorial comments. Susanna Lamers assisted with computation and statistical analysis of the data.

## References

Branch, G. M., 1975. Ecology of *Patella* species from the Cape Peninsula, South Africa. IV. Desiccation. Mar. Biol. 32: 179–188.

Britton, J. C. & R. F. McMahon, 1989. The relationship between vertical distribution, evaporative water loss rate, behaviour, and some morphometric parameters in four species of rocky intertidal gastropods from Hong Kong. In B. Morton (ed.), The Marine Flora and Fauna of Hong Kong and Southern China. II, vol. 3. Hong Kong University Press, Hong Kong. In press.

Broekhuysen, G. J., 1940. A preliminary investigation of the importance of desiccation, temperature and salinity as factors controlling the vertical distribution of certain intertidal marine gastropods in False Bay, South Africa. Trans. r. Soc. S. Afr. 28: 255–291.

Brown, A. C., 1960. Desiccation as a factor influencing the vertical distribution of some South African gastropoda from intertidal rocky shores. Port. Acta biol. 7B: 11–23.

Byrne, R. A., R. F. McMahon & T. H. Dietz, 1988. Temperature and relative humidity effects on aerial exposure tolerance in the freshwater bivalve *Corbicula fluminea*. Biol. Bull. 175: 153–260.

Chambers, M. R., 1980. Zonation, abundance and biomass of gastropods from two Hong Kong rocky shores. In B. Morton (ed.), The Malacofauna of Hong Kong and Southern China. Hong Kong University Press, Hong Kong: 139–148.

Cleland, J. D. & R. F. McMahon, 1989. Upper thermal limit of nine intertidal gastropod species from a Hong Kong rocky shore in relation to vertical distribution and desiccation associated with evaporative cooling. In B. Morton (ed.), The Marine Flora and Fauna of Hong Kong and Southern China. II. vol. 3. Hong Kong University Press, Hong Kong. In press.

Evans, R. G., 1948. The lethal temperatures of some common British littoral molluscs. J. anim. Ecol. 17: 165–173.

Fraenkel, G., 1966. The heat resistance of intertidal snails at Shirahama, Wakayama-Ken, Japan. Publs Seto mar. biol. Lab. 14: 185–195.

Frankel, G., 1968. The heat resistance of intertidal snails at Bimini, Bahamas; Ocean Springs, Mississippi; and Woods Hole, Massachusetts. Physiol. Zool. 41: 1–13.

Gowanloch, J. N., 1926. Contributions to the study of marine gastropods II. The intertidal life of Buccinum undatum, a study in non-adaptation. Contributions to Canadian Biology and Fisheries: Being Studies from the Biological Stations of Canada 3: 169–177.

Grainger, J. N. R., 1969. Factors affecting the body temperature of Patella. Verh. dt. zool. Ges. 1968: 479–487.

Houlihan, D. F., 1979. Respiration in air and water of three mangrove snails. J. exp. mar. Biol. Ecol. 41: 143–161.

Houlihan, D. F. & A. J. Innes, 1982. Respiration in air and water of four Mediterranean trochids. J. exp. mar. Biol. Ecol. 57: 35–54.

Houlihan, D. F., A. J. Innes & D. G. Dey, 1982. The influence of mantle cavity fluid on the aerial oxygen consumption of some intertidal gastropods. J. exp. mar. Biol. Ecol. 49: 57–68.

Hughes, R. N., 1986. A functional biology of marine gastropods. Johns Hopkins University Press, Baltimore, Maryland, 245 pp.

Innes, A. J. & D. F. Houlihan, 1985. Aquatic and aerial oxygen consumption of cool temperate gastropods: A comparison with some Mediterranean species. Comp. Biochem. Physiol. 82A: 105–109.

Johannesson, K., 1989. The bare zone of Swedish rocky shores: why is it there? Oikos 54: 77–86.

Lewis, J. B., 1963. Environmental and tissue temperatures of some tropical intertidal marine animals. Biol. Bull. 124: 277–284.

Markel, R. P., 1971. Temperature relations in two species of tropical west American littorines. Ecology 52: 1126–1130.

McMahon, R. F., 1976. Effluent-induced interpopulation variation in the thermal tolerance of Physa virgata Gould. Comp. Biochem. Physiol. 55A: 23–28.

McMahon, R. F., 1985. The relationships between morphometric parameters and the vertical distribution patterns of seven species of turbinate gastropods on mangrove trees in Hong Kong. In B. Morton & D. Dudgeon (eds), The Malacofauna of Hong Kong and Southern China. II, Vol. 1. Hong Kong University Press, Hong Kong: 199–215.

McMahon, R. F., 1988. Respiratory response to periodic emergence in intertidal molluscs. Am. Zool. 28: 97–114.

McMahon, R. F. & J. C. Britton, 1985. The relationship between vertical distribution, thermal tolerance, evaporative water loss rate, and behaviour on emergence in six species of mangrove gastropods from Hong Kong. In B. Morton & D. Dudgeon (eds), The Malacofauna of Hong Kong and Southern China. II, vol. 2. Hong Kong University Press, Hong Kong: 563–582.

McMahon, R. F. & J. D. Cleland, 1989. Thermal tolerance, evaporative water loss and behaviour during prolonged emergence in the high zoned mangrove gastropod Cerithidea ornata: Evidence for atmospheric water uptake. In B. Morton (ed.), The Marine Flora and Fauna of Hong Kong. II, vol. 3. Hong Kong University Press, Hong Kong. In press.

McMahon, R. F. & B. S. Payne, 1980. Variation of thermal tolerance limits in populations of Physa virgata Gould (Mollusca: Pulmonata). Am. Midl. Nat. 103: 218–230.

McMahon, R. F. & W. D. Russell-Hunter, 1977. Temperature relations of aerial and aquatic respiration in six littoral snails in relation to their vertical zonation. Biol. Bull. 152: 182–198.

McMahon, R. F. & W. D. Russell-Hunter, 1981. The effects of physical variables and acclimation on survival and oxygen consumption in the high littoral salt-marsh snail, Melampus bidentatus Say. Biol. Bull. 161: 246–269.

Micallef, H., 1967. Aerial and aquatic respiration of certain trochids. Experimentia 23: 52.

Newell, R. C., 1979. Biology of intertidal animals. Marine Ecological Surveys Ltd., Faversham, Kent, 781 pp.

Sandison, E. E., 1966. The oxygen consumption of some intertidal gastropods in relation to zonation. J. Zool., Lond. 149: 163–173.

Schmidt-Nielsen, K., C. R. Taylor & A. Shkolnik, 1972. Desert snails: Problems of survival. In G. M. O. Maloiy (ed.), Comparative Physiology of Desert Animals. Academic Press, Lond.: 1–13.

Southward, A. J., 1958. Note on the temperature tolerances of some intertidal animals in relation to environmental temperatures and geographical distribution. J. mar. biol. Ass. U.K. 37: 49–66.

Stephenson, T. A. & A. Stephenson, 1972. Life between tide marks on rocky shores. W. H. Freeman & Company, San Francisco, Calif., 425 pp.

Stirling, H. P., 1982. The upper temperature tolerance of prosobranch gastropods of rocky shores at Hong Kong and Dar Es Salaam, Tanzania. J. exp. mar. Biol. Ecol. 63: 133–144.

Takenouchi, K., 1985. A boulder shore gastropod fauna in Hong Kong. In B. Morton and D. Dudgeon (eds), The Malacofauna of Hong Kong and Southern China. II, vol. 2. Hong Kong University Press, Hong Kong: 413–419.

Toulmond, A., 1967a. Consommation d'oxygène, dans l'air et dans l'eau, chez quatre gastéropodes du genre Littorina. J. Physiol., Paris 59: 303–304.

Toulmond, A., 1967b. Étude de la consommation d'oxygène en fonction du poids, dans l'air et dans l'eau, chez quatre espèces du genre Littorina (Gasteropoda: Prosobranchiaita). C. r. Acad. Sci., Paris, Ser. D 264: 636–638.

Underwood, A. J., 1979. The ecology of intertidal gastropods. Adv. mar. Biol. 16: 111–210.

Vermeij, G. J., 1971. Temperature relationships of some tropical Pacific intertidal gastropods. Mar. Biol. 10: 308–314.

Vermeij, G. J., 1973. Morphological patterns in high-intertidal gastropods: Adaptive strategies and their limitations. Mar. Biol. 20: 319–346.

Vernberg, W. B. & F. J. Vernberg, 1972. Environmental physiology of marine animals. Springer-Verlag, New York; Heidelberg; Berlin, 346 pp.

Wells, F. E., 1984. A guide to the common molluscs of southwestern Australian estuaries. Western Australian Museum, Perth (W.A.), 112 pp.

Wells, F. E. & C. W. Bryce, 1986. Seashells of Western Australia. Western Australian Museum, Perth (W.A.), 207 pp.

Wolcott, T. G., 1973. Physiological ecology and intertidal zonation in limpets (*Acmaea*): A critical look at 'Limiting Factors'. Biol. Bull. 145: 389–422.

*Hydrobiologia* **193**: 261–270, 1990.
*K. Johannesson, D. G. Raffaelli and C. J. Hannaford Ellis (eds), Progress in Littorinid and Muricid Biology.*
© 1990 *Kluwer Academic Publishers.*

# Mate discrimination in *Littorina littorea* (L.) and *L. saxatilis* (Olivi) (Mollusca:Prosobranchia)

Marianne Saur
*Tjärnö Marine Biological Laboratory, Pl. 2781, S-452 00 Strömstad, Sweden*

*Key words:* mating behaviour, copulation duration, interspecific copulation, intrasexual copulation

## Abstract

The ability of males of *Littorina littorea* and *L. saxatilis* to discriminate between mates of different sex, species and size was examined. In partner choice experiments males of *L. littorea* had the possibility to initiate a copulation with either a female or a male. The males did not show a preference for either sex. There was therefore no evidence that they could determine the sex of a conspecific prior to copulation. The duration of intrasexual copulation was considerably shorter than for intersexual copulation, both in the field and in laboratory experiments. For the two species, intersexual copulations were far more frequent than intrasexual ones. This can partly be explained by the difference in copulation time.

Few interspecific copulating pairs were found on the shore. This may reflect a low interspecific encounter rate rather than a mechanism of species recognition. On all of these occasions, however, the active male was of *L. saxatilis*. It is argued that selection against precopulatory species and sex recognition is a more likely explanation than an hypothesis that states that the required mutations for precopulatory mate identification has not yet occurred.

*L. littorea* males copulated longer with large than with small females. Copulation time was short with parasitized females, which are sterile or of low fecundity. The allocation of mating effort by males is discussed.

## Introduction

There are numerous examples in the literature of how visual, olfactory auditory and behaviour signals are used for sexual advertising and identification of mates prior to copulation. Copulations with individuals of other species usually result in a waste of resources such as energy, time and gametes, and fitness should be reduced for undiscriminating individuals compared to that of more discriminating individuals. Selection for mechanisms that enable individuals to accurately identify an appropriate mate can, in general, be expected. The benefit of species recognition and the intensity of selection for it, should be dependent not only on the costs of interspecific copulation and the cost of identification, but also on the frequency of interspecific encounters.

Copulation attempts between males have been described for fish, birds and mammals (Schutz, 1966). Among social mammals mounting is often used as dominance display, but it is very doubtful whether it could be assigned any adaptive value in non-social species (a noteworthy exception is however homosexual rape to diminish mating opportunities of competing males in acanthocephalan worms, Abele & Gilchrist, 1977). Although attempts at intrasexual copulations are not common in the literature, this behaviour may simply have been overlooked or seen as a rather infrequent and insignificant event. Indeed, there is no reason to expect recognition mechanisms to be perfect. For species with promiscuous mating systems in which male reproductive success is primarily dependent on how many matings a male can achieve, a trade-off between accuracy of mate identification and speed of sequestering a mate, can be anticipated. Dungflies of the genus *Scatophaga* provide a good example of this mating system, and, although the mating behaviour of the male has successfully been analyzed in terms of optimal mating strategies (Parker, 1978), it is not surprising that misidentification of sex and species is frequent.

Within several species of *Littorina* intrasexual copulations and interspecific copulations are a consistent feature of the mating behaviour. Raffaelli (1977) reported that on the shore, males of *Littorina saxatilis* (Olivi) and *Littorina nigrolineata* Gray copulated with intraspecific males or individuals of the other species to the same extent as with females of their own species. Male-male copulations occur commonly also in *L. keenae* (Rosewater) ( = *L. planaxis* Philippi) (Gibson, 1965) and *Littoraria scabra* (L.) (Struhsaker, 1966). This is in sharp contrast to the behaviour of the two sympatric species *Littoraria pintado* (Wood) and *Nodilittorina hawaiiensis* Rosewater & Kadolsky ( = *L. picta* Philippi) where no interspecific or intrasexual copulations were found in 1000 copulating pairs of each species (Struhsaker, 1966). In these studies copulation was defined as the penis of one individual being inserted into the mantle cavity of another. Whether sperm was actually transferred is not known.

On the Swedish west coast four species of *Littorina* live in intergrading habitats, which results in a varying degree of distributional overlap. The present study considers *L. saxatilis*, which occurs mainly above mean water level in exposed and sheltered habitats and *L. littorea* (L.) which inhabits sheltered shores, from about mean water level to a few metres depth. *Littorina saxatilis* reproduces throughout the year, is ovoviviparous and the offspring are brooded and released as 'crawlaways'. *Littorina littorea* has a more restricted reproductive period from March to the beginning of July. During this time matings take place and the females release great numbers of pelagic egg capsules, from which veliger larvae hatch.

The purpose of this study is to establish if males of both species mate randomly rather than assortatively with respect to sex, species and size of partner.

## Material and methods

Mating patterns of the two species *L. littorea* and *L. saxatilis* was studied between March and July 1984, on Saltö on the Swedish west coast. The study areas was a boulder shore facing south west and a rock-pool on the cliffs close to the boulder shore.

To estimate frequencies of intersexually (male-female) and intrasexually (male-male) copulating pairs, random samples (sample sizes: *L. littorea* = 129; *L. saxatilis* = 124) were collected on the boulder shore. A pair was considered copulating if one individual (referred to as 'active') had its penis inserted into the mantle cavity of the partner ('passive'). This could be seen when the pair was carefully picked up and gently separated. Pairs found copulating were brought back to the laboratory where shell lengths were measured and the sex of the passive partner was established by dissection. Sex ratio in the general population was estimated by collecting all snails from an area of approximately 1.5 m$^2$.

Mating behaviour on the shore was observed in a small rock-pool (1.5 × 0.6 m) inhabited by both species. All *L. littorea* (30 males and 36 females) and the largest individuals of *L. saxatilis* (18 males and 31 females) were brought to the laboratory and sexed (by the presence or absence of a penis), individually marked with enamel paint and their shell lengths measured. The snails were returned to the pool two days after collection, and observations made on copula durations. As the penis could seldom be observed, copula duration was approximated to the time the pair spent in copula position. Observations of copulations in the laboratory, where the actual copulation time could be observed and compared with the time spent in copula position, suggests that the approximation will give a slight but consistent over-estimate of copula duration of about 20 seconds. In order to activate the snails within the pool, and thus increase the encounter rate and pair formation, snails of *L. littorea* were removed from the substrate and then quickly replaced at random.

In addition, laboratory experiments with *L. littorea* were set up to assess duration of male-female compared to male-male copulations. Snails were sexed and either one male and a female, or two males, were placed in a small glass bowl (80 × 45 mm) half filled with sea water, chilled with running sea water at 10–15 °C. Copulation time was taken from the moment the active male's penis entered the mantle cavity of the other individual until the tip of the penis left the partner's mantle cavity. Copulations were frequent, and the first copulation of a pair usually occurred within fifteen minutes.

Mating preference was recorded in an experiment using the equipment described above, but with two males and one female of *L. littorea* within each bowl. Each male could thus begin to copulate with either a male or a female. The snails were placed equidistantly in the bowl with animals withdrawn into their shells and the nature of first copulation in each bowl recorded. Snails used in this experiment were sexed and individually marked before the experiment and dissected afterwards to confirm sexual identity. All the snails used in these experiments were collected from the boulder shore, and were used several times. Males and females were kept in separate aquaria, and provided with fucoid algae as food.

Copula duration was assessed in bowls in June 1985. Each male (size range: 19.6–21.7 mm) was allowed to copulate once on two consecutive days with a large (22.9–26.0 mm) and a small female

(15.0–18.5 mm). Half of the males copulated first with a large female and the other first with a small one. Females were mated only once a day. These winkles were collected from the harbour at Tjärnö where the population has a higher incidence of digenean parasites.

Animals were dissected after copulation and the degree of parasitism of females assessed, using the following scoring system; 0 = not parasitized; 1 = parasitized, but still with pink ovary-tissue (digenean sporocysts observed in the digestive gland and sometimes also in the reproductive tract); 2 = heavily parasitized, without any pink ovary tissue (digenean sporocysts observed in the digestive gland, and usually also in the reproductive tract). Heavily infected females were sterile, as were heavily parasitized males which had lost the penis. Two males were parasitized but still possessed a penis and were given a score of 1. Two males had pene that appeared abnormal, in that they were not withdrawn and held in the normal 'resting position', but were slightly protruding backwards when the males crawled. These males were asigned scores according to their degree of parasitizm.

## Results

### Description of the mating act

Mating behaviour of *L. littorea* was observed in detail in the laboratory, and was similar to the behaviour observed in the wild. When a male encounters another snail, he crawls onto the other individual's shell, feeling his way with the tentacles. Males can mount from any side of the partner, but always crawl counter-clockwise on the shell to the right side of the partner. When in copulation position the active male's right shell margin contacts the substrate, and the male can, if necessary, push to make the slit between the partner's shell margin and the substrate wider. The elongated penis is then inserted into the mantle cavity of the passive individual. When the penis has been withdrawn, the active male sometimes at once inserts it again, but more often

264

crawls once round the partners shell before copulating again. Both these behaviours are referred to as 'repeated copulation'.

No difference was detected between copulations with females and those with males, except in duration. The 'passive' partner usually was inactive, and seldom crawled or made feeding movements. On two occasions, however, I observed females positively biting males that made attempts to insert the penis for repeated copulations. In one case the male intended a third, repeated copulation and in the other the male tried to start a fourth copulation after three in a row with the same female. The males were rejected in less than 20 seconds on both occasions. The female aimed accurately with her radula at the penis, and after a few probings the male gave in and crawled off with minor tissue damage, having had a penial gland rasped away.

*Sex of partners in copulating pairs*

Since the frequencies of intra- and intersexual copulations are likely to be influenced by the sex ratio of adults on the shore, the sex ratio was estimated and tested for deviation from 1:1. The sex ratio did not significantly deviate from 1:1 in either of the species (*L. littorea* 122 males, 107 females, $\chi^2 = 0.369$, df = 1, $P > 0.05$; *L. saxatilis* 60 males, 99 females, $\chi^2 = 3.07$, df = 1, $P > 0.05$), although the sex ratio of *L. saxatilis* approached one with a significant female bias.

Of a total of 129 copulating pairs of *L. littorea* collected on the shore 9 turned out to be male-male copulations. In *L. saxatilis* 33 out of totally 124 copulations were intrasexual ones. For both species there is a significant difference in the frequencies of intra- and intersexual copulations (*L. littorea* $\chi^2 = 56.5$, df = 1, $P < 0.001$; *L. saxatilis* $\chi^2 = 13.4$, df = 1, $P < 0.001$), demonstrating an ability of males to distinguish between the sexes. The frequency of intrasexual copulations differs between the two species ($\chi^2 = 16.2$, df = 1, $P < 0.001$) suggesting differences in sexual behaviour.

During laboratory experiments males of *L. littorea* showed no ability to determine the sex of mates prior to copulation as 25 male-female and 30 male-male copulations were recorded as first occurring copulation ($\chi^2 = 0.454$, df = 1, $P > 0.05$).

*Copula duration*

Intersexual copulations of *L. littorea* lasted longer than intrasexual ones (Table 1 and Fig. 1). Durations obtained in the field and the laboratory were pooled since there was no significant effect of location, and log-transformation of data was made to reduce differencies in sample variances. Repeated copulations of male-female pairs had shorter durations than first copulations (Table 2), but there was no tendency of this in male-male copulations (Table 3). Five durations of inter-

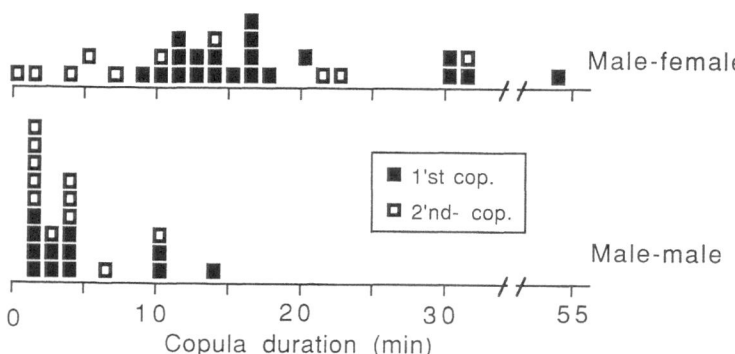

*Fig. 1.* Copula durations of first and repeated inter- and intrasexual matings of *Littorina littorea*. Copula duration approximated from the time the pair was observed to form copula position. Laboratory and field observations are pooled.

*Table 1.* Effect of sex of partner and locality (field or laboratory) on copula durations of *Littorina littorea*.

Two-way ANOVA:

| Source of variation | df | MS | F ratio | P |
|---|---|---|---|---|
| Sex of partner | 1 | 11.26 | 19.52 | < 0.0001 |
| Locality | 1 | 0.14 | 0.25 | 0.62 |
| Interaction | 1 | 0.11 | 0.11 | 0.67 |
| Error | 49 | 0.58 | | |

*Table 2.* Effect of sequence of copulation (first copulation compared to repeated copulations) on copula duration of *Littorina littorea*, male-female copulation.

One-way ANOVA:

| Source of variation | df | MS | F ratio | P |
|---|---|---|---|---|
| Between groups | 1 | 2.73 | 4.63 | 0.040 |
| Within groups | 30 | 0.14 | 0.59 | |
| Total | 31 | | | |

*Table 3.* Effect of sequence of copulation (first copulation compared to repeated copulations) on copula duration of *Littorina littorea*, male-male copulation.

One-way ANOVA:

| Source of variation | df | MS | F ratio | P |
|---|---|---|---|---|
| Between groups | 1 | 0.03 | 0.08 | 0.781 |
| Within groups | 19 | 0.42 | | |
| Total | 20 | | | |

specific copulations between an active male of *L. saxatilis* and a passive partner of *L. littorea* were recorded in the field. The durations of interspecific female-male copulations were 1, 1 and 17 minutes, while interspecific male-male copulations lasted for 3 and 5 minutes.

The mating durations with large compared to small females were tested but there was no significant difference (Fig. 2). However, there was a tendency for matings with large females to last longer (Mann-Whitney U-test, $U = 86.5$, $n = 16$; U value of 83 or less indicates a significant difference in a one-tailed test). If females with no remaining functional ovary tissue, i.e. females scored as severely parasitized, are excluded there is in fact a significant difference ($U = 36.5$, $n = 13$, $P < 0.025$, one-tailed test).

*Size relations*

There was no sexual dimorphism in shell lengths of *L. littorea*, either in copulating pairs or within the random sample of adults from the shore (Table 4). The female of a copulating pair of *L. saxatilis* was on average larger than the male and passive males tended to be larger than active males (Table 5).

There was no relationship between the size of mates of intersexual copulating pairs of *L. littorea* on the shore (Fig. 3). In *L. saxatilis*, however,

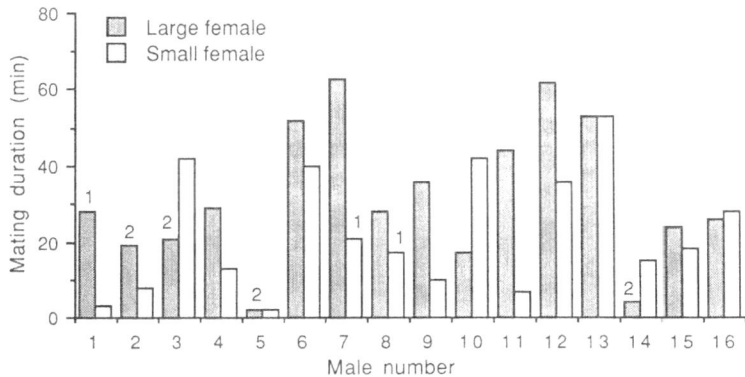

*Fig. 2.* Durations of first and repeated copulations of matings of *Littorina littorea*. The male mating a large and a small female on consecutive days. Parasitized females are indicated as parasitized (1) and severely parasitized (2) on top of the bars. (Males 13, 14 and 15 were parasitized and males 15 and 16 were considered 'abnormal'; see text.)

*Table 4.* Mean shell length ($\bar{x}$) of males and females in a random sample of *Littorina littorea* in the field, and of males and females which formed copulating pairs at the same occasion. Differences between sexes are tested with a Student's t-test.

|  | Sex | n | $\bar{x}$ (mm) | s | d | P |
|---|---|---|---|---|---|---|
| Random sample | Males | 122 | 18.8 | 2.14 | 0.23 | > 0.10 |
| | Females | 107 | 18.8 | 2.33 | | |
| Copulating pairs | Males | 49 | 19.1 | 1.80 | 1.08 | > 0.10 |
| | Females | 49 | 19.6 | 2.37 | | |

*Table 5.* Mean shell length ($\bar{x}$) of males and females in intersexually copulating pairs, and of active and passive males in intrasexually copulating pairs of *Littorina saxatilis*.

|  | Sex | n | $\bar{x}$ (mm) | s | d | P |
|---|---|---|---|---|---|---|
| Intersexual pairs | Male | 40 | 10.2 | 1.39 | 2.71 | < 0.01 |
| | Female | 40 | 11.0 | 1.48 | | |
| Intrasexual pairs | Active male | 14 | 10.5 | 1.07 | 1.73 | < 0.10 |
| | Passive male | 14 | 11.2 | 1.87 | | |

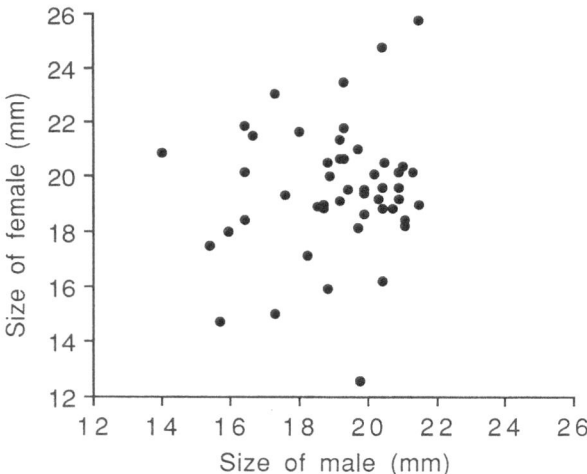

*Fig. 3.* Sizes of mates in intersexual pairs of *Littorina littorea*. (Spearman's correlation coefficient, $r_s = 0.079$, df = 47, $P > 0.20$).

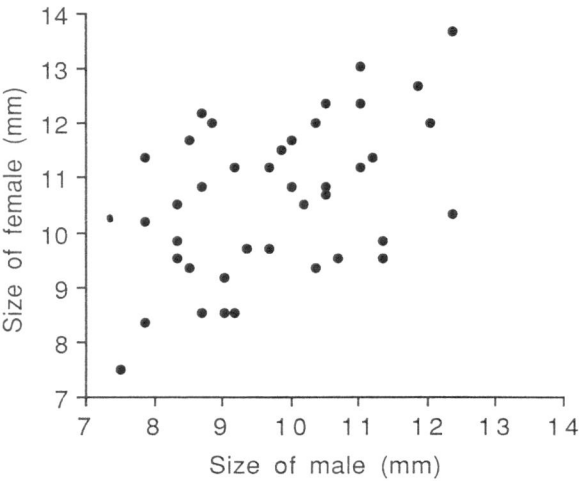

*Fig. 4.* Sizes of mates in intersexual pairs of *Littorina saxatilis*. (Spearman's correlation coefficient, $r_s = 0.43$, df = 38, $P < 0.01$).

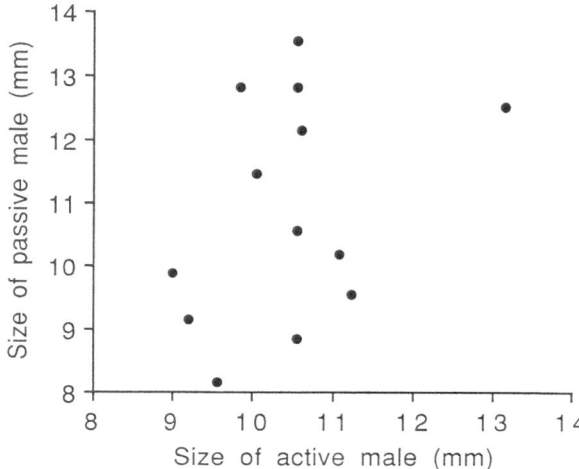

*Fig. 5.* Sizes of mates in intrasexual pairs of *Littorina saxatilis*. (Spearman's correlation coefficient, $r_s = 0.44$, df = 12, $P > 0.05$).

large males were more often found copulating with large females, and small males with small females, resulting in a significant correlation between the shell lengths of mates (Fig. 4). This trend was also apparent for male-male copulating pairs of *L. saxatilis* but this was not significant (Fig. 5).

## Discussion

Chemical cues, acting as sexual attractants, have been demonstrated in several gastropod taxa (Croll, 1983). In many prosobranchs with communal spawning it appears that males are attracted by a substance released by females prior to copulation and spawning (Cate, 1968; Edwards, 1968; Martel et al., 1986). Males were also reported to follow mucus trails recently laid by females, while females did not (Edwards, 1968; Cate, 1968). Both females and males of L. keenae and Littoraria irrotata (Say) follow mucus trails during the mating season irrespective of the sex of the trail layer, and trails of snails of other genera are also followed to a lesser extent (Peters, 1962).

It has been suggested that female Littorina release pheromones in their mucus trails (Dinter & Manos, 1972; Struhsaker, 1966). Although this may be the case for some species (Struhsaker, 1966), it seems less likely for others (Raffaelli, 1977; Peters, 1962; Struhsaker, 1966). In my partner choice experiments, where males of L. littorea had the possibility of copulating with either a female or a male, initially the males showed no sex preference, suggesting that males do not distinguish between sexes before copulation. However, following any mucus trails could obviously increase the encounter rate of potential mates, but this behaviour may also enhance foraging (Connor, 1986).

During copulation, males of L. littorea did seem to distinguish between the sexes, since intersexual copulations lasted about four times longer than intrasexual ones. The short copulations are probably attempts which failed, and it is less likely that sperm is actually ejaculated on these occasions. Despite interrupting an intrasexual copulation after a short time, a male often repeated the attempt, indicating that males do not obtain definite information about a mate's sex at the first attempt. This is supported by the observation that short copulations with females sometimes preceded copulations of normal intersexual duration. Success in containing the bursa copulatrix with the tip of the penis may be the only way in which a male can tell that the partner is a female, and this would not always be obtained at the first attempt.

Clearly, as intersexual copulations last longer than intrasexual ones, the number of male-female copulating pairs should be greater than the number of male-male copulating pairs at any moment on the shore. In the present study, intersexual copulations were most numerous both in L. littorea and L. saxatilis. In L. littorea, however, the copula duration difference is not large enough to account for the much greater frequency of intersexual copulations that was observed. This study agrees with others in that precopulatory sex identification seems unlikely and that other factors, apart from the duration difference, should contribute to the high frequency of intersexual copulations on the shore. One possible explanation is that the relative encounter rate of females at instances of intense sexual activity would be increased by intrasexual copulations, as they result in fewer males being available as mating partners, while intersexual copulations diminish the number of available females and males equally. It is not known whether the level of sexual activity was sufficiently high to contribute to the overall percentage of intersexual copulations recorded on the shore. Another possibility is that males may have a greater tendency to stay with a female than with a male partner after copulation, and that successive copulations between the same mates take place, which would increase the frequency on intersexual copulating pairs on the shore. Similar behaviour has been observed in L. keenae resulting in high frequencies of intersexually copulating pairs on the shore. Intersexual pairs may persist for 10 hours, copulating only for short intervals during this time, while intrasexual copulations last just for a few minutes and then the pair split up (Gibson, 1965).

## Species discrimination

Compared to the data on interspecific copulations of L. saxatilis and L. nigrolineata (Raffaelli, 1977) I found few interspecific pairs. The low frequency of interspecific copulations does not, however, prove an ability of males to distinguish between species, as the interspecific encounter rate may be correspondingly small, since the two species tend to exploit different microhabitats in Sweden. Unsuccessful attempts of male L. littorea to mate with female L. saxatilis in the laboratory suggest,

268

however, that the great size difference between individuals of the two species contributes to incompatibility.

*Littoraria pintado* and *Nodilittorina hawaiiensis* deviate from the mating pattern reported for other littorinids, as no intrasexual or interspecific copulations were found within samples from an area where the two species were sympatric (Struhsaker, 1966). These two species thus seem capable of both species- and sex pre-copulatory identification, both of which may be achieved by sexual signals. Thus there seem to be some species within the littorinids that have evolved a pre-copulatory identification mechanism, while others have not. A possible explanation for this is that the required mutations in the non-discriminating species have not yet occurred. An alternative explanation is that these mutations have occurred, but they have not been favoured by selection in some species. This might be the case if a female that releases sexual attractants is not favoured by selection, or if a male responding to these cues by investing more sperm or time in such females leaves less descendents than males that do not respond.

The pay-off for different behaviours will be dependent not only of the behaviour of conspecifics but also of behaviours and encounter rates of individuals of the opposite sex of sympatric species. If males of sympatric species respond to female smells, 'smelly' females could avoid interspecific matings. Such matings were rare in this investigation and the short duration of those that did occur suggests that no sperm was transferred. The risk of wasted eggs seem small under such circumstances, and the benefit of sexual attractants insignificant in this respect. If conspecific males respond to female smell, 'smelly' females might gain a greater number of matings compared to females that emitted less attractants. If extra copulations are costly, 'smelly' females would not be favoured. Many different costs have been associated with matings for the female (Daly, 1978). Although *Littorina* females are passive during copulation, there may be costs such as reduced time for feeding and a greater risk of losing the grip on the substratum. In this respect my observation of female rejection

in *L. littorea* is interesting, indicating that many repeated copulations may not be advantageous for the female. The fact that the rejection was so easily performed and so rarely seen despite hours of close observation, suggests however that the costs involved at most matings are small.

Females can also derive benefits from multiple matings (Knowlton & Greenwell, 1984), and one of these is of course supply of sperm. Females of *Littorina* store sperm in a *receptaculum seminis* that has an epithelium which seems to orientate, nourish and digest sperm (Fretter & Graham, 1962). Sperm storage and the rate at which males of *L. saxatilis* and *L. littorea* copulate with females probably result in an excess of sperm to the female. After being isolated from males, females of *L. saxatilis* continue to reproduce for more than a year (Johannesson, pers. commun.). Ejaculates of *Littorina* contain energy-rich compounds (Buckland-Nicks & Chia, 1977), and females may be able to utilize ejaculates and excess sperm for nutrition. Such a direct benefit could outweigh the costs associated with copulation.

In promiscuous species, male reproductive success is expected to be limited by the number of matings a male can achieve (Parker, 1978). Intrasexual copulations are likely to reduce the time available for intersexual copulations and feeding, and increase energy costs. Thus, an early identification of the partner's sex could be profitable for a male. It is, however, questionable whether males that were able to trace females by smell would be favoured in competition with males that did not, for the following reason. Females are likely to vary in the amount of chemical stimulus they release. If a number of males responded accordingly by allocating more mating effort, for example sperm, to 'smelly' females, they would suffer more from sperm competition than males that did not respond to female smells. And if the cost of a mating attempt with another male is comparatively small, and interspecific encounters are rare, males that use female smells as cues would not be favoured by selection.

Thus, even though pre-copulatory sex identification may be advantageous for males, it is possible that it has not evolved because it would be disadvantageous for females, and sperm competition may also select against it. For species with high interspecific encounter rates and probabilities of fertilization, the situation is quite different and mutations for sexual attractants may be favoured; this might be the case in *Littoraria pintado* and *Nodilittorina hawaiiensis*.

Finally, there may be physical constraints on copulation. The morphology of the penis differs between the two species, as does the *bursa copulatrix* of the females (Reid, this volume). This is likely to cause a degree of incompatibility between the different species and, more importantly, may prevent fertilization.

*Mate size discrimination*

Sperm production has not been regarded as costly, and hence not a limiting resource for males. But considering promiscuous species such as *Littorina littorea*, in which sperm competition should be intense, Dewsbury's (1981) view seem more appropriate. He states that the number of ejaculates of spermatophores a male can produce should actually be limited, and therefore some discrimination with respect to the males' pattern of allocating spermatophores should be expected. Male preference for large, more fecund females has been demonstrated both in species in which males make contributions of nutrients (Gwynne, 1981; Rutowsky, 1982) and in species in which they do not (Verell, 1982). In many cases, preferred females are courted or accepted, while the less fecund females are ignored or rejected. In other species, however, the preference is more graded. For example, virgin females are courted more intensely or for a longer time (Wiklund & Forsberg, 1985).

In my experiments male *L. littorea* showed a clear tendency to invest more time in matings with large females than in matings with small females. Fecundity is proportional to the cube of female shell length in *L. saxatilis* (Janson, 1985) and *L. littorea* (Hughes & Answer, 1982). Using the latter author's data concerning egg capsule pro-

duction during a mating season as a rough estimate, the fecundity of a 'large' female used in my experiments should be about four times greater than the fecundity of a 'small' female. Copula duration has been shown to reflect the amount of sperm investment (Svärd & Wiklund, 1986), and this aspect needs further exploration.

The short copula durations with parasitized females may indicate that completion of mating is in some way inhibited or that males are able to discriminate between uninfected and parasitized females. The winkles are not infected with these parasites until they reach sexual maturity, and thereafter the probability of any individual being infected increases with age and hence size (Hughes & Answer, 1982). From the males point of view, large females should constitute both extra-rich resource patches and extra-risky ones. If male reproductive success is dependent on prudent allocation of limiting resources, males discriminating between females of different fecundity and allocating their resources accordingly, should be favoured. Since parasitic castration has been observed in high and varying frequencies, and affects fecundity severely (Hughes & Answer, 1982), the possibility that males can discriminate between parasitized and non-parasitized females should be investigated.

*Assortative mating*

The tendency of passive males of intrasexual copulating pairs of *L. saxatilis* to be larger than the active male could be due to males investing more time in copulation attempts with larger than smaller individuals. Such size-differential behaviour could be adaptive, since within *L. saxatilis* the average size of females is larger than that of males (pers. obs.), and a large individual consequently is more likely to be a female.

The positive correlation found between shell lengths of mates within intersexual copulating pairs of *L. saxatilis* is at variance with the results of Raffaelli (1977). He found no such relationship within *L. saxatilis* and the larger, closely related species of *L. nigrolineata*. The positive correlation found in the present study has several possible explanations. Firstly, there could be a non-random distribution of size classes on the shore. Secondly, there might be some degree of mechanical incompatibility of snails of great size differences. This appears unlikely, as Raffaelli (1977) found no correlation despite a greater variance of sizes of both females and males on his shore compared to Saltö. A third possibility is that the correlation might reflect some mate discriminating behaviour of males or females.

In summary, the two species of *Littorina* do show a type of assortative mating, with respect to sex and species, although random encounter seems to be the rule. It is still puzzling that mechanisms of precopulatory sex- and species identification do not seem to have evolved and I suggest that selection pressures, perhaps contradictory for the two sexes, may be responsible.

## Acknowledgements

I thank Kerstin Johannesson for constructive critiscism and David Raffaelli for valuable discussion of the results. I am grateful to the support from colleagues and staff at Tjärnö Marine Biological Laboratory.

## References

Abele, L. G. & S. Gilchrist, 1977. Homosexual rape in acanthocephalan worms. Science 197: 81–83.

Buckland-Nicks, J. A. & F. S. Chia, 1977. On the nurse cell and the spermatozeugma in *Littorina sitkana*. Cell and Tissue Research 179: 347–356.

Cate, J. M., 1968. Mating behavior in *Mitria idae* Melvill, 1893. Veliger 10: 247–252.

Connor, V. M., 1986. The use of mucous trails by intertidal limpets to enhance food resources. Biol. Bull. 171: 548–564.

Croll, R. P., 1983. Gastropod chemoreception. Biol. Rev. 58: 293–319.

Daly, M., 1978. The cost of mating. Am. Nat. 112: 771–774.

Dewsbury, D. A., 1982. Ejaculate cost and male choice. Am. Nat. 119: 601–610.

Dinter, I. & P. J. Manos, 1972. Evidence for a pheromone in the marine periwinkle *Littorina littorea* Linnaeus. Veliger 15: 45–47.

Edwards, D. C., 1968. Reproduction in *Oliviella biplicata*. Veliger 10: 297–304.

Fretter, V. & A. Graham, 1962. British prosobranch molluscs. Ray Society, London, 755 pp.

Gibson, D. G., 1965. Mating behavior in *Littorina planaxis* Philippi (Gastropoda: Prosobranchiata). Veliger 7: 134–139.

Gwynne, D. T., 1981. Sexual difference theory: mormon crickets show role reversal in mate choice. Science 213: 779–780.

Hughes, R. N. & P. Answer, 1982. Growth, spawning and trematode infection of *Littorina littorea* (L.) from an exposed shore in north Wales. J. moll. Stud. 48: 321–330.

Janson, K., 1985. Variation in the occurence of abnormal embryos in females of the intertidal gastropod *Littorina saxatilis* Olivi. J. moll. Stud. 51: 64–68.

Knowlton, N. & S. R. Greenwell, 1984. Male sperm competition avoidance mechanisms: The influence of female interests. In R. L. Smith (ed.), Sperm competition and the evolution of animal mating systems. Academic Press, Orlando (Florida): 62–83.

Martel, A., D. H. Larrivée, K. R. Klein & J. H. Himmelman, 1986. Reproductive cycle and seasonal feeding activity of the neogastropod *Buccinum undatum*. Mar. Biol. 92: 211–221.

Parker, G. A., 1978. Searching for mates. In J. R. Krebs & N. B. Davies (eds), Behavioural ecology an evolutionary approach. Blackwell Scientific Publications, Oxford: 214–244.

Peters, R. S., 1962. Function of the cephalic tentacles in *Littorina* Philippi (Gastropoda: Prosobranchiata). Veliger 7: 143–148.

Raffaelli, D. G., 1977. Observations of the copulatory behavior of *Littorina rudis* Maton and *Littorina nigrolineata* Gray (Gastropoda: Prosobranchia). Veliger 20: 75–77.

Rutowsky, R. L., 1982. Epigamic selection by males as evidenced by courtship partner preferences in the checkered white butterfly (*Pieris protodice*). Anim. Behav. 30: 108–112.

Schutz, F., 1966. Homosexualität bei Tieren. Studium Generale 19: 273–285.

Struhsaker, J. W., 1966. Breeding, spawning periodicity and early development in the Hawaiian *Littorina*: *L. pintado* (Wood), *L. picta* Philippi and *L. scabra* (Linné). Proc. malac. Soc. Lond. 37: 137–166.

Svärd, L. & C. Wiklund, 1986. Different delivery strategies in first versus subsequent mating in the swallowtail butterfly *Papilio machaon* L. Behav. Ecol. Sociobiol. 18: 325–330.

Verell, P. A., 1982. Male newts prefer large females as mates. Anim. Behav. 30: 1254–1255.

Wiklund, C. & J. Forsberg, 1985. Courtship and male discrimination between virgin and mated females in the orange tip butterfly *Anthocharis cardamines*. Anim. Behav. 34: 328–332.

*Hydrobiologia* **193**: 271–273, 1990.
*K. Johannesson, D. G. Raffaelli and C. J. Hannaford Ellis (eds), Progress in Littorinid and Muricid Biology.*
© 1990 *Kluwer Academic Publishers.*

# Epilogue

Dave Raffaelli
*Culterty Field Station, University of Aberdeen, Newburgh, Aberdeen AB4 0AA, Scotland, UK*

## Introduction

The purpose of the Tjärnö meeting was to bring together as many researchers as possible currently working on Littorinids and other intertidal snails. In reviewing the papers it became clear that Littorinid research has taken a number of directions, and that substantial progress has been made in two main areas: systematics and variation, and the ecological role of Littorinids in the wider community. This paper is an attempt to synthesise those papers presented at Tjärnö which are relevant to these main themes.

The Littorinidae have long attracted scientific interest. They are one of the most obvious components of rocky shores, salt marshes and mangrove areas and few naturalists can have failed to notice their great variety in shell colour, shape and sculpture. Over the last 20 years there has been much effort directed towards understanding the causes of this variation and to assessing the usefulness of shell and other morphological characters in Littorinid systematics. Much of this research was carried out in the 1970's and showed that shell characteristics could be quite variable within a species and could be modified by a range of habitat dependent selection pressures, such as exposure and predation. This in turn prompted a then quite radical revision of the systematics of several species. Since then other characteristics, particularly the morphology of the reproductive tract and allozyme frequencies, have helped to define the species more precisely and have greatly improved our understanding of the evolutionary relationships within the group.

## Systematics and variation

Research on the ecology and systematics of European Littorinids up to the beginning of the 1980's was the subject of an earlier review (Raffaelli, 1982), in which the close association between ecological research and taxonomic developments was noted. Often, however, the systematics lagged behind the ecological work, and taxonomic revisions frequently invalidated prior ecological research. The description of species groups, such as that including *Littorina saxatilis* Olivi, *L. arcana* Hannaford Ellis and *L. neglecta* Bean, within which the species have very similar shell characters, was not welcomed wholeheartedly by all field biologists. Indeed, it took a considerable time for some species to be accepted and during that period research on their population and community ecology was perceived to be risky. This situation was not improved by suggestions that some variants should be given specific status. All of these have since been shown to be ecomorphs of other species (see below).

Since the end of the 1970's these taxonomic problems have been largely resolved. For instance, it is now clear that within the *L. saxatilis*-complex there exists at least the following species: *L. arcana, L. saxatilis* and *L. nigrolineata* Gray. This is clearly seen from studies of biochemical genetics (Ward, this volume) and the comparitive morphology of the reproductive tract and other characters of a wider range of species (Reid, this volume). Both these approaches confirm that the fourth entity within the *L. saxatilis*-complex, *L. neglecta*, is very closely related to

*L. saxatilis* itself. *Littorina neglecta* has often been distinguished from *L. saxatilis* in the field on shell and habitat characteristics. However, detailed investigations of the morphometrics and biochemical genetics of *L. saxatilis* and *L. neglecta* show that there are in fact few quantifiable differences between the two taxa and that the latter could merely be a variant of the former (B. Johannesson & K. Johannesson, this volume; K. Johannesson & B. Johannesson, this volume). They rightly ask the question 'Is *L. neglecta* a good species?'. Boulding (this volume) has suggested that despite high immigration rates genetic differentiation can occur between Littorinids at the extremes of an environmental cline, and this model may apply to the case of *L. saxatilis* and *L. neglecta*.

The biochemical genetics approach has been particularly successful for rationalising the European taxa. Thus *L. rudis* Maton (*sensu* J.E. Smith, 1981) and *L. tenebrosa* Montagu are synonomous with *L. saxatilis*, and *L. aestuarina* with *L. obtusata*, whilst the Venitian lagoonal population of *L. saxatilis* is seen as an isolated population of the North Atlantic species. (see Ward, this volume for a review). Notwithstanding these advances, the field ecologist still has the problem of identifying the different entities in the field. For several European species it is only possible to identify females with any degree of certainty. The identity of males can only be assigned a probability and juveniles must be guessed at. A good key or field guide to European species would help considerably, and we were pleased to hear at the Tjärnö meeting that at least two were in preparation.

It is clear that there is still an emphasis in Littorinid research on assessing variation within and between species and attempting to understand its underlying causes. The Tjärnö meeting heard several papers dealing with shell shape variation in closely related sympatric species (e.g. Dytham *et al.*, this volume; Grahame *et al.*, this volume; B. Johannesson & K. Johannesson, this volume; Boulding, this volume) and we saw the power of multivariate techniques such as canonical variate analysis, principal components analy-

sis and multidimensional scaling for teasing out patterns from complex morphometric data sets. Multivariate approaches have also been used to assess the taxonomic distances between and within populations of a wide range of species, using both morphometric and allozymic data (Reid, this volume; Sundberg *et al.*, this volume). Not only have these studies demonstrated so lucidly evolutionary phenomena, such as founder effects and genetic bottlenecks, but they have also given us a very clear picture of the phylogeny of the group. In this respect the work of David Reid cannot be praised too highly. In a very short space of time he has made a series of remarkable contributions to the study of the evolutionary systematics of the Littorinidae. By using the techniques of cladistics to describe the taxonomic affinities of a wide range of genera and sub-genera from the Atlantic and the Pacific, Reid (this volume) has been able to suggest possible mechanisms to explain the present day biogeographical distributions of species. Interestingly, his work indicates that western European shores have an unusually high diversity of *Littorina* species.

## Littorinids in the community

The main thrust of ecological research on European Littorinids in the 1970's concerned comparisons of the population dynamics of species with contrasting reproductive modes and presumed different life history strategies. Those species occuring on British shores seemed particularly attractive because comparisons could be made between those that produced 'crawlaways', those that laid benthic egg masses and those with planktonic larvae. However, it was not possible to interpret many aspects of reproductive biology strictly in the context of life history theory. Other selection pressures were thought to be more important in determining reproductive traits (Raffaelli, 1982) and, as Reid (this volume) has stressed, the phylogeny of species should also be considered when comparing such basic features as reproductive mode.

Since then attention seems to have switched

towards interactions between Littorinid grazers and their algal resources. Plant-herbivore interactions are difficult to study in any ecosystem, and this is particulary true for the intertidal zone where both macro- and micro-algae have to be considered. Norton *et al.* (this volume) reviewed the chemical and physical properties of algae in relation to the feeding capabilities of herbivores. Several Littorinids can discriminate between different algal types, often from a considerable distance, but the form of the radula may be important in determining what kind of algae can be ingested. Of particular interest is the extent to which *Littorina* take the endolithic algae that are common in the upper shore (Voltolino & Sacchi, this volume).

*Littorina* are abundant on many shores and could play an important ecological role in controlling the composition and structure of algal communities. Whilst this appears to be true for some North American shores which lack limpets, this may not be the case on most European shores. Norton *et al.* (this volume) argue that there is a need for more work on both sides of the North Atlantic on shores with contrasting grazer regimes and we should be cautious of extrapolating research findings from one region to another. Grazing by Littorinids, such as *L. mariae*, may also reduce the fouling of macro-algal fronds by smaller epiphytic species. Williams (this volume) suggested that if the performance of macro-algae is substantially affected by fouling organisms, then epiphytic grazers could be significant in the structuring of seaweed communities.

## Conclusions

In a previous review of research on European *Littorina* it was clear that there were several areas where greater effort was required. One of these areas was the taxonomic muddle of the North Atlantic Littorinidae. This has now largely been resolved and a number of interesting evolutionary and phylogenetic relationships have been highlighted in the process. In this respect, the maintenance of laboratory breeding populations of intertidal snails is vital for examining the degree of reproductive isolation between species and for assessing the relative importance of genetic and environmental influences on morphological traits (Palmer, this volume, Warwick *et al.*, this volume). In my earlier review it was evident that many of the larger programmes on *Littorina* initiated in the 1970's were coming to an end, and hopes were expressed that the 1980's would be as productive as the previous decade. It was clear from the Tjärnö meeting that the 1980's certainly have been productive, and that significant progress has been made in Littorinid research.

## References

Raffaelli, D. G., 1982. Recent ecological research on some European species of *Littorina*. J. moll. stud. 48: 342–354.
Smith, J. E., 1981. The natural history and taxonomy of shell variation in the periwinkles *Littorina saxatilis* and *Littorina rudis*. J. mar. biol. Ass. U.K. 61: 215–241.

# Index of genera and species

# General index